Nanotechnology for Sustainable Manufacturing

Nanotechnology for Sustainable Manufacturing

EDITED BY

DAVID RICKERBY

CRC Press
Taylor & Francis Group
Boca Raton London New York

CRC Press is an imprint of the
Taylor & Francis Group, an **informa** business

CRC Press
Taylor & Francis Group
6000 Broken Sound Parkway NW, Suite 300
Boca Raton, FL 33487-2742

Printed on acid-free paper
Version Date: 20140407

International Standard Book Number-13: 978-1-4822-1482-6 (Hardback)

Library of Congress Cataloging-in-Publication Data

Nanotechnology for sustainable manufacturing / editor, David Rickerby.
pages cm
Includes bibliographical references and index.
ISBN 978-1-4822-1482-6
1. Nanotechnology--Industrial applications. 2. Green products. 3. Sustainable engineering. I. Rickerby, David G., editor of compilation.

T174.7.N37375 2014
620'.5--dc23 2014002785

Visit the Taylor & Francis Web site at
http://www.taylorandfrancis.com

and the CRC Press Web site at
http://www.crcpress.com

To Mark and Andrew

Contents

Preface

Nanotechnology could potentially play an important role in increasing the sustainability of a wide range of industrial sectors. This book contains chapters discussing some of the recent progress in this area written by leading experts in relevant research fields from Europe, North America, Asia, and Australia. It draws on my personal experience from organizing two scientific workshops on the environmental impact of nanotechnology on behalf of the European Commission in collaboration with the Institute of Nanotechnology, coauthoring a chapter in the United Nations Environment Programme (UNEP) 2007 GEO Year Book on nanotechnology and the environment, and involvement in the 2009 Organisation for Economic Co-operation and Development (OECD) Conference on Potential Environmental Benefits of Nanotechnology.

The scope of the book comprises the areas of energy and materials efficiency related to resource savings and conservation of raw materials, which are drivers for the application of nanotechnology in the industrial context. It provides an innovative perspective with regard to the synergies between different subjects connected to the nanotechnology theme by bridging the gap between academic research and industry. The topics covered give an overview of recent advances in the respective fields, including electronics, agrifood, aerospace, pulp and paper manufacturing, batteries, catalysts, solar energy, fuel cells, drinking water and construction materials. While this list is by no means exhaustive, it does provide insights into the diverse industries that are being and are likely to be impacted by current developments in nanotechnology and nanomaterials. In addition, Chapters 11 and 12 deal with the way in which life cycle analysis can be used to perform sustainability assessment for nanotechnology-based products and processes.

My task as editor has been assisted by the dedicated team of reviewers, whose work has greatly contributed to the quality of the final volume. I should, therefore, like to take this opportunity to thank R.H. Balasubramanya, S. Gössling-Reisemann, E. Golin, J. Huang, K. Kendall, R. Maguire, C. McCullough, K. McGuigan, M. Meiler, A.N. Skouloudis, and M. Steinfeldt for their efforts.

David G. Rickerby
Ispra, Italy

Editor

David G. Rickerby, PhD, is a senior scientific officer in the Institute for Environment and Sustainability at the European Commission Joint Research Centre, Ispra, Italy. After earning a doctoral degree from the University of Cambridge, he carried out postdoctoral research at the Pennsylvania State University. He was a visiting professor at the University of Quebec for a number of years and has taught graduate courses at the University of Trento and the University of Venice. His present research interests involve evaluation of the potential risks and benefits of nanotechnologies, including development of risk assessment tools and methodologies. He was one of the group of international experts who coauthored a chapter on nanotechnology and the environment for the UNEP GEO Year Book and is a member of the OECD Working Party on Manufactured Nanomaterials, Steering Group 9 on the Environmentally Sustainable Use of Nanotechnology.

Contributors

John Anthony Byrne
Nanotechnology and Integrated BioEngineering Centre
University of Ulster
Newtownabbey, United Kingdom

Vicente Cortés Corberán
Institute of Catalysis and Petroleumchemistry
Consejo Superior de Investigaciones Científicas
Madrid, Spain

Michael B. Cortie
Institute for Nanoscale Technology
University of Technology Sydney
New South Wales, Australia

Shangfeng Du
School of Chemical Engineering
University of Birmingham
Birmingham, United Kingdom

Sophia Fantechi
European Commission
Directorate-General for Research and Innovation
Key Enabling Technologies
D.3—Advanced Materials and Nanotechnologies
Brussels, Belgium

Pilar Fernandez-Ibañez
Plataforma Solar de Almería (CIEMAT)
Tabernas, Spain

Bertrand Fillon
CEA/LITEN
Grenoble, France

Gregory Heness
Institute for Nanoscale Technology
University of Technology Sydney
Sydney, Australia

Prateek Jain
Central Institute for Research on Cotton Technology
Mumbai, India

Frans W.H. Kampers
Wageningen UR
Wageningen, The Netherlands

Seongsin Margaret Kim
Electrical and Computer Engineering Department
The University of Alabama
Tuscaloosa, Alabama

Patrick Kung
Electrical and Computer Engineering Department
The University of Alabama
Tuscaloosa, Alabama

Larry Larson
Ingram School of Engineering
Texas State University-San Marcos
San Marcos, Texas

Zhihong Liu
Ingram School of Engineering
Texas State University-San Marcos
San Marcos, Texas

Eduardo Martínez-Tamayo
Institute of Material Sciences of the University of Valencia
València, Spain

Natalia V. Mezentseva
Boreskov Institute of Catalysis
Russian Academy of Sciences
Novosibirsk, Russia

Padraig G. Moloney
Rice University
Houston, Texas

Mark Morrison
The Institute of Nanotechnology
Glasgow, Scotland

Deb Newberry
NSF NanoLink Center
Dakota County Technical College
Rosemount, Minnesota

Bruno G. Pollet
School of Chemical Engineering
University of Birmingham
Birmingham, United Kingdom

David G. Rickerby
European Commission
Joint Research Centre
Institute for Environment and Sustainability
Ispra, Italy

Vicente Rives
GIR-QUESCAT, Department of Inorganic Chemistry
University of Salamanca
Salamanca, Spain

Vladislav A. Sadykov
Boreskov Institute of Catalysis
Russian Academy of Sciences
Novosibirsk, Russia

Prasad Satyamurthy
Central Institute for Research on Cotton Technology
Mumbai, India

Thomas P. Seager
School of Sustainable Engineering and the Built Environment
Arizona State University
Tempe, Arizona

Geoffrey B. Smith
Institute for Nanoscale Technology
University of Technology Sydney
Sydney, Australia

Michael Steinfeldt
Universität Bremen
Bremen, Germany

Nicholas Stokes
Institute for Nanoscale Technology
University of Technology Sydney
Sydney, Australia

Walt Trybula
The Trybula Foundation, Inc.
Austin, Texas

and

Ingram School of Engineering
Texas State University-San Marcos
San Marcos, Texas

Nadanathangam Vigneshwaran
Central Institute for Research on Cotton Technology
Mumbai, India

Iain Weir
Optimat Ltd.
Glasgow, United Kingdom

Ben A. Wender
School of Sustainable Engineering and the Built Environment
Arizona State University
Tempe, Arizona

Leonard L. Yowell
NASA Johnson Space Center
Houston, Texas

Quigkai Yu
Ingram School of Engineering
Texas State University-San Marcos
San Marcos, Texas

1 Introduction

David G. Rickerby and Mark Morrison

CONTENTS

1.1 NANOSCIENCE AND NANOTECHNOLOGY

Research on nanotechnology and its applications has increased rapidly in recent years, and growing numbers of nanotechnology-based products are already being commercialized (Maynard 2007). Products currently marketed generally tend to be of the incremental or evolutionary type in which either nanoparticles, carbon nanotubes, or nanocomposites are incorporated in the product, or a nanotechnology-enabled process such as nanofabrication or nanopatterning is used in manufacturing. As a result of competition to improve existing products by developing higher performance materials, at the same time reducing costs and waste, the industrial applications of nanotechnology should increase rapidly in number. However, the nanomaterials used in products represent only a fraction of the total cost. The economic impact of nanotechnology results mainly from the added value that these materials impart when used in manufacturing products rather than the intrinsic value of the materials themselves.

The term nanotechnology is used to refer to engineering at the atomic or molecular level. It covers an array of technologies that enable the manipulation of matter at the nanoscale to create innovative materials, structures, and devices.

At this length scale, the chemical and physical properties of solid matter are modified remarkably, allowing the fabrication of stronger, lighter, and more efficient structures. Nanomaterials demonstrate exceptional properties due to their increased surface area to volume ratio, which causes them to have greater chemical reactivity, and due to quantum effects that result in novel optical, electrical, and magnetic properties. Engineered nanomaterials are designed to exploit the different properties deriving from small dimensions.

There is at present, however, no universal consensus on a precise definition of what is nanotechnology. The National Science and Technology Council (NSTC 2011) defines it as "the understanding and control of matter at dimensions between approximately 1 and 100 nanometers, where unique phenomena enable novel applications." A British government commissioned report (The Royal Society and The Royal Academy of Engineering 2004) makes the distinction between nanoscience: "the study of phenomena and manipulation of materials at atomic, molecular and macromolecular scales, where properties differ significantly from those at a larger scale" and nanotechnologies, which involve "the design, characterization, production and application of structures, devices and systems by controlling shape and size at nanometer scale."

Neither is there an internationally agreed definition of what constitutes a nanomaterial. The Organisation for Economic Co-operation and Development (OECD) Working Party on Manufactured Nanomaterials (OECD 2008) has provided a definition: "nanomaterials intentionally produced to have specific properties or specific composition, a size range typically between 1 nm and 100 nm and material which is either a nano-object (i.e., that is confined in one, two, or three dimensions at the nanoscale) or is nanostructured (i.e., having an internal or surface structure at the nanoscale)." The European Commission (2011) has adopted a recommended definition of nanomaterial as "a natural, incidental or manufactured material containing particles, in an unbound state or as an aggregate or as an agglomerate and where, for 50% or more of the particles in the number size distribution, one or more external dimensions is in the size range 1 nm–100 nm." Nanomaterials may thus be one dimensional (surface films or coatings), two dimensional (nanotubes), or three dimensional (nanoparticles). Common nanomaterials include metals or metal oxides, semiconductor quantum dots, carbon nanotubes, and fullerenes.

1.2 IMPLICATIONS FOR SUSTAINABILITY

Reducing the impact of industry on the environment has been widely recognized as an important priority for achieving sustainability (Roome 1998; Sarkis 2001; Elkins 2002; Labuschagne et al. 2005). As a consequence, products and production methods are being modified, supply chains are evolving, and increased attention is paid to the disposal of waste and recycling. Nanotechnology can play a key role in these developments because it has the potential to confer substantial societal, economic, and environmental benefits through more efficient energy generation and storage systems, reduction of emissions, resource saving, and substitution of hazardous substances (Rickerby and Morrison 2007). It can thus be expected to have a significant influence on a wide range of manufacturing sectors in the future. The aim of the present volume is to provide an overview of selected industrial applications of nanotechnology with special reference to their implications for sustainable development.

Nanotechnology can contribute to industrial innovation and competitiveness by reducing demand for raw materials, energy, and water in manufacturing. Both top-down processes (such as lithography, etching, electrospinning, and milling) and bottom-up methods (such as vapor phase deposition, liquid phase techniques, and self-assembly) may be used (Şengül et al. 2008). Industrial areas in which innovative applications of nanotechnology are contributing to increasing sustainability include electronics, aerospace, automobiles, energy, water, agrifood, papermaking and packaging, construction, and petrochemicals, that is, sectors that require high direct or indirect consumption of resources. Underpinning these developments are applications of nanomaterials for fabricating more efficient catalysts, batteries, electronic devices, photovoltaic panels, fuel cells and high-strength lightweight engineering materials, and the use of nanoscale processing techniques to conserve energy and raw materials.

1.2.1 Resource Efficiency

The mounting pressure on natural resources as a result of industrialization and population growth has created awareness of the potential of nanotechnologies to contribute to eco-innovation by increasing resource efficiency in manufacturing processes (Lang-Koetz et al. 2010). Resource-efficient products must be designed to be more material and energy efficient over their entire life cycle to reduce their environmental impact and ensure sustainability. Nanotechnology manufacturing techniques require a reduced input of raw materials to achieve the same or improved performance and can thus increase efficiency. However, energy and material consumption during extraction, processing, fabrication, and use must be considered to determine if nanotechnologies are truly sustainable or simply shift the environmental burden upstream or downstream.

Although nanotechnological processes can undoubtedly reduce resource consumption in manufacturing, the actual products themselves may present problems for recycling because of the complexity of separating out and recovering the finely divided materials they contain. Improving recycling rates, particularly of less abundant elements such as rare earths, is desirable not only to promote more efficient use of raw materials but also to reduce the environmental damage caused by energy-intensive mining, ore extraction, and refining (Graedel et al. 2011). However, to increase the percentage of materials recycled requires intelligent design of products to facilitate the disassembly of components and recovery of materials.

A further important way in which nanotechnology could potentially contribute to more sustainable production is by helping to minimize waste (Kassim 2005). Besides the serious environmental hazard it can create, waste also results in a significant economic cost due to the loss of expensive raw materials. Rather than relying on traditional *end-of-pipe* solutions, it is preferable to avoid the generation of waste in the first place by source reduction. This could be achieved by using more efficient manufacturing methods, such as bottom-up nanotechnological processes, which require smaller material inputs and produce less pollution than conventional top-down methods.

1.2.2 Substitution of Hazardous Chemicals

It has been suggested that nanotechnology could provide solutions for replacement or reduction of the use of dangerous chemical substances (Fiedeler 2008). The candidate substances include toxic, carcinogenic, or persistent chemicals such as heavy metals, pesticides, and chlorinated hydrocarbons. Chemical substitution, in the narrowest sense, is understood to refer simply to the replacement of a substance with an alternative one, but substitution of a hazardous substance can also be achieved by modification of a process or by adoption of an entirely new technology to eliminate its use. It is imperative, however, to consider whether the benefits of substitution outweigh any risks due to the environmental release and toxicity of nanomaterials (Ellenbecker and Tsai 2011).

Despite the obvious potential there have been until now few examples of nanotechnology or nanomaterials being used to substitute hazardous chemicals. Achieving equivalent functionality and performance is generally a complex problem in practice and this has tended to slow down progress in the field. One prospective application is the use of nanomaterials to replace highly toxic chemicals in antifouling coatings (Rosenhahn et al. 2008; Wouters et al. 2010). Instead of using a biocide to destroy the microorganisms that cause fouling, antiadhesive, nanocomposite coatings are applied to the surface to prevent the microorganisms from attaching to it. Nanoscale surface roughness is a key parameter influencing the hydrophobic and antifouling properties of such coatings (Scardino et al. 2009; Wouters et al. 2010).

1.3 INDUSTRIAL APPLICATIONS

The present volume covers some of the main applications of nanotechnologies in various industrial sectors with emphasis on their contribution to sustainability. The use of cleaner manufacturing processes and more efficient or alternative energy generation and storage systems, incorporating nanotechnologies and nanomaterials, can assist in reducing the environmental impacts of industry and transport by enabling increased efficiency in the consumption of raw materials, energy and water, combined with decreased emissions and waste. According to the type of product, the effects may be direct or indirect and may be associated with the production phase, the use phase, or both.

1.3.1 Catalysts

Catalysts perform an indispensable function in numerous industrial, energy generation, and transport applications. The true economic value of catalysts greatly exceeds their original cost due to their widespread application in the chemical and petrochemical industries. Catalytic activity increases as the particle size is decreased due to the higher proportion of coordinatively unsaturated surface atoms. It is dependent, however, not only on the particle size and composition but also on the local environment surrounding active sites. Advances in surface science techniques have led to improved understanding of catalytic reactions and the complex relationships between the local atomic structure and catalyst performance (Bell 2003; Besenbacher et al. 2007). This has assisted in the formulation of nanostructured catalysts with enhanced activity and selectivity using novel synthesis methods (Cortés Corberán 2009).

The main trends in current research and development are synthesis and characterization of highly structured micro- and mesoporous materials such as zeolites, preparation of metal and metal oxide particles with controlled sizes, facets, and compositions, together with growing emphasis on theoretical modeling of catalyst structure and reaction kinetics (Davis 2011). Nanostructured catalysts could potentially make an immense contribution to cleaner, environmentally friendlier production of energy and chemicals. Priority areas for future work are the development of catalysts for conversion of coal and natural gas to liquid fuels and chemicals such as ethylene and propylene (Hu et al. 2011). The use of these natural resources in this way, instead of wastefully using them to produce energy, would constitute a major step forward toward a more sustainable economy.

1.3.2 Energy Generation and Storage

Development of renewable energy systems is necessary to allow conversion from nuclear and fossil fuel energy to more sustainable solar and hydrogen power generation. Potential improvements in the performance of renewable energy technologies such as fuel cells, wind turbines, and photovoltaics can be achieved by using nanomaterials or nanotechnological components (Guo 2012). This should enable these alternative energy sources to become more economically viable by reducing costs while increasing efficiency.

Nanoporous carbon–supported catalysts, nanocomposite membranes, and nanostructured membrane electrode assemblies will be used in the next generation of fuel cells (Pak et al. 2010). Carbon nanotube catalyst supports are being developed with the aim of improving the stability, durability, and catalytic activity (Lee et al. 2006), while nanotechnology is crucial to optimizing membrane performance (Thiam et al. 2011). Nanoporous silicon–based membranes have several advantages, such as better proton conductivity and higher power density, compared to conventional polymer membranes (Moghaddam et al. 2010). The enhanced diffusion properties of nanocrystalline materials can, in addition, be used to improve the storage capacity and sorption kinetics of metal hydrides (Dornheim et al. 2007; Sakintuna et al. 2007), which are leading candidates for solid-state hydrogen storage in mobile fuel cell applications.

Nanomaterials are also contributing to improving the efficiency of the high-performance rare earth magnets that are components in wind turbine generators (Hadjipanayis et al. 2006). Anisotropic magnetic nanoparticles aligned along their easy magnetization axes can be used to construct permanent magnets with higher maximum energy products by combining two different magnetic materials, one with a high coercive field and the other with a high saturation magnetization, to form a nanocomposite (Balamurugan et al. 2012). Improved magnetic properties can be achieved with these composite materials while reducing consumption of scarce materials, such as neodymium, used to manufacture magnets.

Although solar power appears extremely promising as a low-carbon energy source, its future expansion will be dependent on the ability to deliver high efficiency at reduced cost per kilowatt-hour, which requires less-expensive alternatives to be found to crystalline silicon (Wadia et al. 2009). Photovoltaic efficiency can be improved and costs reduced by using ultrathin nanostructured absorbers (Dittrich et al. 2011). Thin-film solar cells fabricated with cadmium tellurium, copper indium diselenide, and copper indium gallium selenide have efficiencies approaching those obtained with crystalline silicon (Green et al. 2011). However, their potential for widespread application is limited by the relative scarcity of tellurium and indium. There is also concern with regard to the toxicity of these materials (Fthenakis 2009), which make their sustainability questionable unless effective recycling and recovery processes are in place to avoid release to the environment.

Organic solar cells have attracted interest due to the improvement in efficiency that can be achieved by the use of novel materials and innovative device structures (Hoppe and Sariciftci 2004; Benanti and Venkataraman 2006). They have the advantages of low cost and simple fabrication techniques such as vacuum evaporation/sublimation and solution cast or printing technologies (Günes et al. 2007). The most efficient organic solar cells use heterojunction construction techniques. However, improvements in the present ~3% power conversion efficiencies of the best organic solar cells will be necessary for them to become competitive with inorganic devices, for which efficiencies are currently in excess of 20%. Hybrid solar cells are being developed that incorporate organic and inorganic materials to combine the low production costs of organic photovoltaics with the higher efficiencies of inorganic devices (Wright and Uddin 2012). Current work on fabrication of polymer solar cells on cellulose nanocrystal substrates offers promise for the development of sustainable photovoltaic technologies based on easily recyclable and abundant, renewable materials (Zhou et al. 2013).

The principal disadvantage of many renewable energy generation technologies is that the output is intermittent and cannot be increased or decreased with demand as in the case of conventional power plants, so that some way is needed to store the electricity generated. The energy densities of existing energy storage systems are insufficient for this purpose, but nanostructured materials are under development for high-performance batteries and supercapacitors to provide the required storage capability (Serrano et al. 2009).

Lithium-ion batteries with nanostructured electrodes provide shorter diffusion paths for lithium ions, resulting in increased charge/discharge rates and higher energy densities (Armand and Tarascon 2008). Nanocomposite cathodes consisting of transition metal nanoparticles embedded in a Li_2O matrix can be produced in situ from a transition metal oxide by the conversion reaction. The use of nanomaterials results in transfer of a larger number of electrons per transition metal atom compared to the conventional insertion reaction (Amatucci and Pereira 2007). The lithium ions are furthermore accommodated with lower strain in the electrode lattice, thereby extending the life of the battery and increasing its safety. The possibilities for development of more efficient anode materials based on nanostructured TiO_2 and SnO_2 (Deng et al. 2009; Serventi et al. 2012) or carbon nanotubes (Landi et al. 2010) are also being explored.

1.3.3 Transport

Opportunities for the application of nanomaterials exist in both the automotive (Presting and König 2003) and aerospace (Agee et al. 2009) sectors. These include high-strength construction materials, engine components, sensor and avionics systems, and functional coatings. Of particular relevance to increasing the sustainability of transport systems are lightweight, carbon nanotube-reinforced composites (Bogue 2011), nano catalysts (Fino 2007), tribological coatings (Dahotre and Nayak 2005), and fuel additives (Wakefield et al. 2008). Nanotechnological innovation designed to improve fuel cell performance could help overcome the present technical and economic limitations in automotive applications (von Helmholt and Eberle 2007).

Weight reduction by replacement of structural materials with lighter composites is an area of increasing interest. Polycarbonate composites offer weight savings of up to 50% compared with conventional glass but have tended to suffer from poor ultraviolet (UV) and scratch resistance in the past. These problems are being solved, respectively, through the inclusion of nanoparticles of silica within the polymer matrix and application of thin-film coatings containing inorganic oxides, developments that have led industry experts to predict that 20% of the automotive glazing market will be polycarbonate by 2020 (Soutar et al. 2008; Sitja 2010).

Applications of nanotechnology in tires are driven by needs for better performance (abrasion resistance and grip) and have reduced fuel consumption (by decreasing rolling resistance). Carbon black and silica have been included in tires for many decades to improve mechanical strength, braking distance, and rolling resistance. In the last decade, further developments of silica have improved both rolling resistance and braking distances in the wet, through nanoscale interactions with the rubber matrix (Vasiliadis 2011). More recent innovations include nanoscale rubber particles that improve the binding of silica to the rubber matrix and are being incorporated in new eco-tires. Other advances include a nanostructured coating that is applied to the surface of tires to reduce heat generation and nanoclays incorporated into tires to improve stiffness, wear, and handling properties.

Fuel catalysts such as nanoparticulate additives containing CeO improve the combustion of diesel fuels while decreasing the emission of particulates (Wakefield et al. 2008). Nanotechnology can also have a further impact in reducing exhaust gas emissions through improvements in the activity of catalytic converters and decreasing the use of noble metals such as Pt or in the future, perhaps even allowing replacement of these with more readily available materials (Stafford 2007).

Power and electrical systems are also benefiting from advances in nanotechnology, including improved batteries, supercapacitors, and fuel cells (Fuel Cell Today 2012; Reddy et al. 2012; Vasiliadis 2012) to provide energy storage (also for recovered waste energy from braking), power to drive, and operate equipment in hybrid and electric vehicles.

1.3.4 Drinking Water and Wastewater Treatment

There are remarkable opportunities for nanotechnology in drinking water purification (Brame et al. 2011) and wastewater treatment (El Saliby et al. 2008). The performance of membranes, oxidants, and absorbents for water treatment can be greatly improved by nanoscale engineering (Bottero et al. 2006). Precise control of membrane nanostructure allows selectivity to be increased and costs reduced, while the higher redox reaction rates obtained with nanoparticle catalysts and photocatalysts enable pollutants to be degraded more rapidly and completely.

Prospective applications of nanostructured membranes include drinking water filtration, desalination, ultrapure water production, and industrial wastewater treatment. Embedding reactive nanoparticles in polymeric and ceramic membranes can help to improve their efficiency by reducing fouling and increasing the flux (Kim and Van der Bruggen 2010). Nanocomposite membranes offer relatively small increases in efficiency but are already commercially available; bioinspired membranes have high potential but are still far from the market, whereas catalytically enhanced membranes will require further development before commercialization (Pendergast and Hoek 2011).

Photocatalysis is a convenient method for purification of drinking water and wastewater treatment (Bahnemann 2004). It can inactivate aquatic microorganisms more quickly and efficiently than UV irradiation alone (Sichel et al. 2007) and is effective even for chlorine-resistant bacteria. In addition, it is able to degrade organic chemicals typically present as contaminants in the groundwater and surface water used to supply drinking water (McMurray et al. 2006). Wider application of photocatalytic water treatment will be largely determined by its cost effectiveness compared to established processes already on the market. Because the operating costs are mainly due to the power consumed by the UV light source, a significant amount of research is carried out in the area of solar photocatalysis.

1.3.5 Electronics

Reducing the dimensions of electronic circuits provides savings in raw materials, weight, and power consumption. The current approach to miniaturization in the semiconductor industry by simply decreasing device size cannot, however, be extended indefinitely and novel device technologies and architectures need consequently to be explored (Avouris et al. 2007). Many of these new concepts are based on carbon electronics using nanotubes, graphene layers, nanowires, and nanoribbons. Bottom-up approaches to device fabrication, entailing molecular-level control of structure and composition, may also be adopted to overcome the limitations of top-down methods (Lu and Lieber 2007). Single-molecule devices might ultimately be constructed by using organic synthesis techniques (Moth-Poulsen and Bjørnholm 2009).

From the sustainability perspective, the most significant applications of nanoelectronics are in the fields of photovoltaics and lighting. Extensive use of solid-state lighting could drastically reduce electricity consumption for this purpose (Haitz and Tsao 2011). Light-emitting diodes are more energy efficient than halogen or incandescent lamps and have a smaller environmental impact because of the reduction in resource consumption due to a longer lifetime. They contain no mercury and less phosphorous than fluorescent lamps. Technical problems that still need to be solved include achieving high-intensity emission over the entire visible spectrum and maintaining efficiency during high-power operation (Crawford 2009). The use of innovative phosphors and low-dimensional nanostructures could result in significant improvements in performance (Schubert et al. 2006).

Recent work has demonstrated the potential of carbon nanotubes for use as transistors in integrated circuits. Up to 10,000 can be self-assembled on a single silicon substrate by conventional semiconductor technology (Park et al. 2012). The potential implications of this research for sustainability are increased computational power with reduced materials and energy consumption.

1.3.6 Agrifood and Forest Products

Nanotechnology can contribute to solutions to the problems of agricultural sustainability and food security (Datta 2008). Some examples of the ways in which it could assist in the introduction of more efficient and environmentally friendly farming methods have been reviewed (Chen and Yada 2011). Controlled delivery and release of fertilizers, pesticides, and herbicides by nanoscale carriers would prevent overuse and reduce pollution due to run off. Deployment of nanosensor networks could enable real-time monitoring of crops and soil and weather conditions to allow more efficient use of resources for improved crop yields. Major improvements in the resistance of crops to drought and disease could be possible by application of the knowledge obtained through current progress in genome research.

In the food industry, nanoparticles can be used to enhance properties, such as solubility and absorption, of bioactive compounds in functional foods, whereas nanoencapsulation may be used to protect flavors, additives, and nutrients (Sozer and Kokini 2009; Sekhon 2010). Nanosensors have applications for monitoring and controlling food quality and safety, in the processing chain or embedded in packaging (Duncan 2011; Neetirajan and Jayas 2011), for the purpose of detecting spoilage caused by excess moisture or oxygen and the presence of microorganisms. Multilayer polymer nanocomposite barriers can be used in biodegradable packaging materials and may additionally incorporate silver or titanium dioxide nanoparticles to provide additional antibacterial protection (Duncan 2011). Packaging materials can be made from natural biodegradable polymers including chitosan, cellulose, collagen, and zein in which nanoclays may be incorporated as gas barriers (Robinson and Salejova 2010).

Nanocellulose (also known as microfibrillated cellulose) is derived from a variety of plant (such as wood, grasses, and fibrous vegetables) and microbial sources through mechanical and/or chemical treatment of pulp (cellulose fibers). Nanocellulose fibrils have diameters between 5 and 60 nm and are up to several micrometers in length. Nanocrystalline cellulose is produced by acid hydrolysis and has shorter, more rigid fibers. The usefulness of both forms is due to the strong intermolecular bonding between fibrils that can improve structural strength and rigidity in several applications including plastic composites and paper (Klemm et al. 2011). Research into resource use has provided some evidence that energy and material requirements are lower when manufacturing products from nanocellulose (Siró and Plackett 2010; Walter 2012). Many vegetable waste streams, such as sugar beet fiber, are a potential source of nanocellulose.

1.3.7 Construction

The construction industry was one of the first to recognize that nanotechnology could potentially bring huge benefits (Zhu et al. 2004). Various nanoparticulate materials, such as TiO_2, SiO_2, and $CaCO_3$, are extensively used in coatings, paints, adhesives, sealants, and composites. Other applications include nanostructured reinforced steel, self-cleaning and antigraffiti coatings, and inexpensive flexible plastic solar panels. There are numerous eco-innovative nanotechnologies in the construction sector in areas such as insulation and load-carrying materials, architectural coatings, climate control and lighting systems, air purification, and renewable energy systems (Andersen and Geiker 2009). These confer considerable environmental advantages with regard to durability, recyclability, energy and resource efficiency, indoor and outdoor air quality, and so on.

Because concrete is the most common construction material, its sustainable use presents specific challenges with respect to resource consumption and environmental pollution (Van Vliet et al. 2012). These include carbon dioxide emission due to the combustion that provides heat to the kiln during production and the decomposition of the limestone, which is a major constituent of cement. Nanoparticulate additives can be used to enhance the mechanical and thermal properties of cement and make it more sustainable by increasing its durability and extending its service life (Sanchez and Sobolev 2010).

Aerogels are typically made from silica and have the highest insulation rating known. They thus offer significant energy savings for heating of buildings (Baetens et al. 2011). However, they are costly to manufacture and suffer from poor resistance to mechanical and environmental stresses. Innovative methods based on the use of organic materials, including nanocellulose, are being developed to solve these problems and can contribute to sustainability through manufacturing processes requiring less energy and fewer chemicals that result in more durable materials (Klemm et al. 2011).

1.4 SUSTAINABILITY ASSESSMENT: LIFE-CYCLE ANALYSIS

To establish whether nanotechnology could lessen the environmental impact of industry, it is necessary to determine whether the potential benefits outweigh the risks associated with the potential release of nanomaterials into the environment. Although it is generally accepted that life-cycle analysis could be a useful tool in assessing the impact of nanomaterials (Seager and Linkov 2008), comparatively little work has been carried out in this area until now (Bauer et al. 2007). Such analysis must include the total energy and material inputs during the whole life cycle and outputs in the form of emissions and waste. However, essential information is lacking regarding the possible release routes for nanomaterials during production, use, and final disposal or recycling (Shatkin 2008).

The use of cycle thinking provides a systematic method of analyzing data on material and energy flows at every stage of industrial production including raw materials extraction, refinement, manufacturing and use, waste disposal, and recycling (Dhingra et al. 2010). Among the aspects that have to be considered in assessing the sustainability of a process are raw materials and energy consumption, material efficiency as a result of increased recyclability, and substitution of hazardous materials. Any additional energy requirements in manufacturing nanomaterials must therefore be included when carrying out life-cycle analysis (Theis et al. 2011). In developing sustainability criteria it is important to include the possible influence of incidental or unintentional effects. It needs to be examined, for example, whether a new product might indirectly lead to greater resource use, because it either enables or promotes a new activity or causes an increase in existing activity.

Integration of life-cycle concepts at an early stage in the development of nanomaterials and nanotechnology products is essential to ensure their long-term sustainability. The use of a tiered approach has been recommended for assessing the impact of nanotechnologies (von Gleich et al. 2008), taking into account technological, ecological, health, safety, and environmental aspects. A survey of the existing information on the use of nanomaterials in industrial products (Meyer et al. 2009) provides insights into the applications and types of nanomaterials that need to be considered. Additional factors that should be included in a sustainability assessment for nanomaterials are toxicity, nonrenewable energy use, and greenhouse gas emissions.

ACKNOWLEDGMENTS

This work received funding from the European Union Seventh Framework Programme (FP7/2007-2013) under grant agreement number 247989.

REFERENCES

Agee FJ, Lozano K, Gutierrez JM et al. (2009). Nanotechnology research for aerospace applications. In: Szu HH, Agee FJ (eds) *Independent Component Analyses, Wavelets, Neural Networks, Biosystems, and Nanoengineering VII*. Proceedings of SPIE Vol. 7343, SPIE Press, Bellingham, WA.

Amatucci GG, Pereira N (2007). Fluoride based electrode materials for advanced energy storage devices. *J Fluorine Chem* 128: 243–262.

Andersen MM, Geiker MR (2009). Nanotechnologies for climate friendly construction—Key issues and challenges. *Nanotechnol Construct* 3: 199–207.

Armand M, Tarascon J-M (2008). Building better batteries. *Nature* 451: 652–657.

Avouris P, Chen Z, Perebeinos V (2007). Carbon-based electronics. *Nat Nanotechnol* 2: 605–615.

Baetens R, Jelle BP, Gustavsen A (2011). Aerogel insulation for building applications: A state-of-the-art review. *Energy Build* 43: 761–769.

Bahnemann D (2004). Photocatalytic water treatment: Solar energy applications. *Sol Energy* 77: 445–459.

Balamurugan B, Sellmyer DJ, Hadjipanayis GC et al. (2012). Prospects for nanoparticle-based permanent magnets. *Scr Mater* 67: 542–547.

Bauer C, Buchgeister J, Hischier R et al. (2007). Towards a framework for life cycle thinking in the assessment of nanotechnology. *J Photochem Photobiol A* 189: 239–246.

Bell AT (2003). The impact of nanoscience on heterogeneous catalysis. *Science* 299: 1688–1691.

Benanti TL, Venkataraman D (2006). Organic solar cells: An overview focusing on active layer morphology. *Photosynth Res* 87: 73–81.

Besenbacher F, Lauritsen JV, Wendt S (2007). STM studies of model catalysts. *Nano Today* 2(4): 30–39.

Bogue R (2011). Nanocomposites: A review of technology and applications. *Assem Autom* 31: 106–112.
Bottero J-Y, Rose J, Wiesner MR (2006). Nanotechnologies: Tools for sustainability in a new wave of water treatment processes. *Integr Environ Assess Manag* 2: 391–395.
Brame J, Li Q, Alvarez PJJ (2011). Nanotechnology-enabled water treatment and reuse: Emerging opportunities and challenges for developing countries. *Trends Food Sci Technol* 22: 618–624.
Chen H, Yada R (2011). Nanotechnologies in agriculture: New tools for sustainable development. *Trends Food Sci Technol* 22: 585–594.
Cortés Corberán V (2009). Nanostructured oxide catalysts for oxidative activation of alkanes. *Top Catal* 52: 962–969.
Crawford MH (2009). LEDs for solid-state lighting: Performance challenges and recent advances. *IEEE J Sel Top Quantum Electron* 15: 1028–1040.
Dahotre NB, Nayak S (2005). Nanocoatings for engine application. *Surf Coat Technol* 194: 58–67.
Datta PS (2008). Nano-agrobiotechnology: A step towards food security. *Curr Sci* 94: 22–23.
Davis R (2011). *International Assessment of Research and Development in Catalysis by Nanostructured Materials*. Imperial College Press, London.
Deng D, Kim MG, Lee JY et al. (2009). Green energy storage materials: Nanostructured TiO_2 and Sn-based anodes for lithium-ion batteries. *Energy Environ Sci* 2: 818–837.
Dhingra R, Naidu S, Upreti G et al. (2010). Sustainable nanotechnology: Through green methods and life-cycle thinking. *Sustainability* 2: 3323–3338.
Dittrich T, Belaidi A, Ennaoui A (2011). Concepts of inorganic solid-state nanostructured solar cells. *Sol Energy Mater Sol Cells* 95: 1527–1536.
Dornheim M, Doppiu S, Barkhordarian G et al. (2007). Hydrogen storage in magnesium-based hydrides and hydride compounds. *Scr Mater* 56: 841–846.
Duncan TV (2011). Applications of nanotechnology in food packaging and food safety: Barrier materials, antimicrobials and sensors. *J Colloid Interface Sci* 363: 1–24.
Elkins P (2002). *Economic Growth and Environmental Sustainability: The Prospects for Green Growth*. Routledge, London.
Ellenbecker M, Tsai S (2011). Engineered nanoparticles: Safer substitutes for toxic materials, or a new hazard? *J Cleaner Prod* 19: 483–487.
El Saliby IJ, Shon HK, Kandasamy J et al. (2008). Nanotechnology for wastewater treatment: In brief. In: Vigneswaran S (ed), *Water and Wastewater Treatment Technologies*, Vol. 3, Encyclopedia of Life Support Systems (EOLSS), Oxford, UK.
European Commission (2011). Commission recommendation of 18 October 2011 on the definition of nanomaterial. *Off J Eur Union L* 275: 38–40. Available at http://eur-lex.europa.eu/LexUriServ/LexUriServ.do?uri=CELEX:32011H0696:EN:NOT. Accessed 21 October 2012.
Fiedeler U (2008). Using nanotechnology for the substitution of hazardous chemical substances. *J Ind Ecol* 12: 307–315.
Fino B (2007). Diesel emission control: Catalytic filters for particulate removal. *Sci Technol Adv Mater* 8: 93–100.
Fthenakis VM (2009). Sustainability of photovoltaics: The case for thin-film solar cells. *Renew Sust Energy Rev* 13: 2746–2750.
Fuel Cell Today (2012). Fuel cell electric vehicles: The road ahead. Available at http://www.fuelcelltoday.com/media/1711108/fuel_cell_electric_vehicles_-_the_road_ahead_v3.pdf. Accessed 23 November 2012.
Graedel TE, Allwood J, Birat J-P et al. (2011). What do we know about metal recycling rates? *J Ind Ecol* 15: 355–366.
Green MA, Emery K, Hishikawa Y et al. (2011). Solar cell efficiency tables (version 37). *Prog Photovolt Res Appl* 19: 84–92.

Günes S, Neugebauer H, Sariciftci NS (2007). Conjugated polymer-based organic solar cells. *Chem Rev* 107: 1324–1338.

Guo KW (2012). Green nanotechnology of trends in future energy: A review. *Int J Energy Res* 36: 1–17.

Hadjipanayis G, Liu J, Gabay A et al. (2006). Current status of rare-earth permanent magnet research in USA. *J Iron Steel Res Int* 13 (Suppl 1): 12–22.

Haitz R, Tsao JY (2011). Solid-state lighting: 'The case' 10 years after and future prospects. *Phys Status Solidi* 208: 17–29.

Hoppe H, Sariciftci NS (2004). Organic solar cells: An overview. *J Mater Res* 19: 1924–1945.

Hu EL, Davis M, Davis R et al. (2011). Applications: Catalysis by nanostructured materials. In: Roco M, Hersam MC, Mirkin CA (eds) *Nanotechnology Research Directions for Societal Needs in 2020*. Springer, Dordrecht.

Kassim TA (2005). Waste minimization and molecular nanotechnology: Towards total environmental sustainability. In: Hutzinger O (ed) *Handbook of Environmental Chemistry*. Springer, Berlin.

Kim J, Van der Bruggen B (2010). The use of nanoparticles in polymeric and ceramic membrane structures: Review of manufacturing procedures and performance improvement for water treatment. *Environ Pollut* 158: 2335–2349.

Klemm D, Kramer F, Moritz S et al. (2011). Nanocelluloses: A new family of nature-based materials. *Angew Chem Int Ed* 50: 5438–5466.

Labuschagne C, Brent AC, van Erck RPG (2005). Assessing the sustainability performances of industries. *J Cleaner Prod* 13: 373–385.

Landi BJ, Cress CD, Raffaelle RP (2010). High energy density lithium-ion batteries with carbon nanotube anodes. *J Mater Res* 25: 1636–1644.

Lang-Koetz C, Pastewski N, Rohn H (2010). Identifying new technologies, products and strategies for resource efficiency. *Chem Eng Technol* 33: 559–566.

Lee K, Zhang J, Wang H et al. (2006). Progress in the synthesis of carbon nanotube- and nanofiber-supported Pt electrocatalysts for PEM fuel cell catalysis. *J Appl Electrochem* 36: 507–522.

Lu W, Lieber CM (2007). Nanoelectronics from the bottom up. *Nat Mater* 6: 841–850.

Maynard AD (2007). Nanotechnology: The next big thing, or much ado about nothing? *Ann Occup Hyg* 51: 1–12.

McMurray TA, Dunlop PSM, Byrne JA (2006). The photocatalytic degradation of atrazine on nanoparticulate TiO_2 films. *J Photochem Photobiol A* 182: 43–51.

Meyer DE, Curran MA, Gonzalez MA (2009). An examination of existing data for the industrial manufacture and use of nanocomponents and their role in the life cycle impact of nanoproducts. *Environ Sci Technol* 43: 1256–1263.

Moghaddam S, Pengwang E, Jiang Y-B et al. (2010). An inorganic-organic proton exchange membrane for fuel cells with a controlled nanoscale pore structure. *Nat Nanotechnol* 5: 230–236.

Moth-Poulsen K, Bjørnholm T (2009). Molecular electronics with single molecules in solid-state devices. *Nat Nanotechnol* 4: 551–556.

Neetirajan S, Jayas DS (2011). Nanotechnology for the food and bioprocessing industries. *Food Bioprocess Technol* 4: 39–47.

NSTC (2011). *The National Nanotechnology Initiative Strategic Plan*. National Science and Technology Council, Washington DC. Available at http://www.nano.gov/sites/default/files/pub_resource/2011_strategic_plan.pdf. Accessed 21 October 2012.

OECD (2008). Working Party on Manufactured Nanomaterials, Guidance for the use of the OECD Database on Research into the Safety of Manufactured Nanomaterials, Ver.1. Available at http://www.oecd.org/science/nanosafety/44033847.pdf. Accessed 21 October 2012.

Pak C, Kang S, Choi YS et al. (2010). Nanomaterials and structures for the fourth innovation of polymer electrolyte fuel cell. *J Mater Res* 25: 2063–2071.

Park H, Afzali A, Han S-J et al. (2012). High-density integration of carbon nanotubes via chemical self-assembly. *Nat Nanotechnol* 7: 787–791.

Pendergast MTM, Hoek EMV (2011). A review of water treatment membrane nanotechnologies. *Energy Environ Sci* 4: 1946–1971.

Presting H, König U (2003). Future nanotechnology developments for automotive applications. *Mater Sci Eng C* 23: 737–741.

Reddy ALM, Gowda SR, Shaijumon MM et al. (2012). Hybrid nanostructures for energy storage applications. *Adv Mater* 24: 5045–5064.

Rickerby DG, Morrison M (2007). Nanotechnology and the environment: A European perspective. *Sci Technol Adv Mater* 8: 19–24.

Robinson D, Saejova G (2010). ObservatoryNANO briefing No. 1: Agrifood. Biodegradable food packaging. Available at www.observatory-nano.eu. Accessed 23 November 2012.

Roome NJ (1998). *Sustainability Strategies for Industry: The Future of Corporate Practice.* Island Press, Washington DC.

Rosenhahn A, Ederth T, Pettitt ME (2008). Advanced nanostructures for the control of biofouling: The FP6 EU integrated project AMBIO. *Biointerphases* 3: IR1–IR5.

Sakintuna B, Lamari-Darkrim F, Hirscher M (2007). Metal hydride materials for solid hydrogen storage: A review. *Int J Hydrogen Energy* 32: 1121–1140.

Sanchez F, Sobolev K (2010). Nanotechnology in concrete—A review. *Construct Build Mater* 24: 2060–2071.

Sarkis J (2001). Manufacturing's role in corporate environmental sustainability—Concerns for the new millennium. *Int J Oper Prod Man* 21: 666–686.

Scardino AJ, Zhang H, Cookson DJ et al. (2009). The role of nano-roughness in antifouling. *Biofouling* 25: 757–767.

Schubert EF, Kim JK, Luo H et al. (2006). Solid-state lighting—A benevolent technology. *Rep Prog Phys* 69: 3069–3099.

Seager TP, Linkov I (2008). Coupling multicriteria decision analysis and life cycle assessment for nanomaterials. *J Ind Ecol* 12: 282–285.

Sekhon BS (2010). Food nanotechnology–An overview. *Nanotechnol Sci Appl* 3: 1–15.

Şengül H, Theis TL, Ghosh S (2008). Towards sustainable nanoproducts: An overview of nanomanufacturing methods. *J Ind Ecol* 12: 329–359.

Serrano E, Rus G, García-Martínez J (2009). Nanotechnology for sustainable energy. *Renew Sust Energ Rev* 13: 2373–2384.

Serventi AM, Rodrigues IR, Trudeau ML et al. (2012). Microstructural and electrochemical investigation of functional nanostructured TiO_2 anode for Li-ions batteries. *J Power Sources* 202: 357–363.

Shatkin JA (2008). Informing environmental decision making by combining life cycle assessment and risk analysis. *J Ind Ecol* 12: 278–281.

Sichel C, Blanco J, Malato S et al. (2007). Effects of experimental conditions on *E. coli* survival during solar photocatalytic water disinfection. *J Photochem Photobiol A* 189: 239–246.

Siró I, Plackett D (2010). Microfibrillated cellulose and new nanocomposite materials: A review. *Cellulose* 17: 459–494.

Sitja R (2010). ObservatoryNANO briefing No. 6: Transport. Nano-enhanced automotive plastic glazing. Available at www.observatory-nano.eu. Accessed 23 November 2012.

Soutar AM, Chen Q, Raja Khalif RE (2008). Evaluation of commercially available transparent hard coating for polycarbonate. *SIMTech Tech Rep* 9: 161–165.

Sozer N, Kokini JL (2009). Nanotechnology and its applications in the food sector. *Trends Biotechnol* 27: 82–89.

Stafford N (2007). Catalytic converters go nano. RSC Chemistry World. Available at http://www.rsc.org/chemistryworld/News/2007/October/10100701.asp. Accessed 23 November 2012.

The Royal Society and The Royal Academy of Engineering (2004). *Nanoscience and Nanotechnologies: Opportunities and Uncertainties.* The Royal Society and The Royal Academy of Engineering, London. Available at http://www.nanotec.org.uk/report/Nano%20report%202004%20fin.pdf. Accessed 21 October 2012.

Theis TL, Bakshi BR, Durham D et al. (2011). A life cycle framework for the investigation of environmentally benign nanoparticles and products. *Phys Status Solidi RRL* 5: 312–317.

Thiam HS, Daud WRW, Kamarudin SK et al. (2011). Overview on nanostructured membrane in fuel cell applications. *Int J Hydrogen Energy* 36: 3187–3205.

Van Vliet K, Pellenq R, Buehler MJ et al. (2012). Set in stone? A perspective on the concrete sustainability challenge. *MRS Bull* 37: 395–402.

Vasiliadis H (2011). ObservatoryNANO briefing No. 23: Nanotechnology in automotive tyres. Available at http://www.observatory-nano.eu. Accessed 23 November 2012.

Vasiliadis H (2012). ObservatoryNANO briefing No. 32: Nanotech in next-generation electric car batteries: Beyond Li-ion. Available at http://www.observatory-nano.eu. Accessed 23 November 2012.

von Gleich A, Steinfeldt M, Petschow U (2008). A suggested three-tiered approach to assessing the implications of nanotechnology and influencing its development. *J Cleaner Prod* 16: 899–909.

von Helmholt R, Eberle U (2007). Fuel cell vehicles: Status 2007. *J Power Sources* 165: 833–843.

Wadia C, Alivisatos AP, Kammen DM (2009). Materials availability expands the opportunity for large-scale photovoltaics deployment. *Environ Sci Technol* 43: 2072–2077.

Wakefield G, Wu X, Gardner M et al. (2008). Envirox™ fuel-borne catalyst: Developing and launching a nano-fuel additive. *Technol Anal Strateg Manage* 20: 127–136.

Walter P (2012). Nanocellulose has paper potential. RSC Chemistry World. Available at http://www.rsc.org/chemistryworld/News/2012/February/cellulose-nanomaterial-paper-environment.asp. Accessed 23 November 2012.

Wouters M, Rentrop C, Willemsen P (2010). Surface structuring and coating performance novel biocide free nanocomposite coatings with antifouling and fouling-release properties. *Prog Organ Coat* 68: 4–11.

Wright M, Uddin A (2012). Organic–inorganic hybrid solar cells: A comparative review. *Sol Energ Mater Sol Cells* 107: 87–111.

Zhou Y, Fuentes-Hernandez C, Khan TM et al. (2013). Recyclable organic solar cells on cellulose nanocrystal substrates. *Sci Rep* 3: 1536. doi:10.1038.

Zhu W, Bartos PJM, Porro A (2004). Application of nanotechnology in construction. *Mater Struct* 37: 649–658.

2 Nanotechnology in Electronics

Larry Larson, Seongsin Margaret Kim, Patrick Kung, Quigkai Yu, Zhihong Liu, Deb Newberry, and Walt Trybula

CONTENTS

2.1 INTRODUCTION

This chapter addresses nanotechnology in electronics. While there are many discussions about what classifies something as nanotechnology, the approach employed in this chapter is to address the emerging areas that appear to have the potential for enhancing the performance of electronics. The realm of nanotechnology is providing significant developmental efforts to produce superior devices. While developments and research for various types of devices will be presented, this chapter does not predict the application of any of these developments to actual devices. Beyond this beginning, development stage of the technology, the volume manufacturing of devices will still hold known challenges.

Many people consider the semiconductor industry as the pioneer nanotechnology industry because of the early manufacturing of devices in the sub-100-nm range. The dimensional shrinkage of features that has been progressing for decades follows a path defined from the rules known as Moore's law. The increase in density has followed a reduction in dimensions of approximately 70% every 18–24 months. This doubling of density (70% width times 70% length yields 49% area reduction) has permitted enhanced functionality, higher speed computations, and further miniaturization of the devices. As this shrinkage continues, new challenges must continuously be addressed. Copper lines for device power with sizes less than 50 nm improve some properties and manufacturing processes; smaller volumes can have additional variability of conductivity due to crystallization and grain boundaries. As the size of the transistors continues to diminish, there are concerns about the reduced number of electrons available within a given volume to change the state of a transistor. New devices are entering the realm of dimensions where the properties that are exhibited are between bulk properties and atomistic properties. Understanding the phenomena at this size scale is only beginning. Some projections for semiconductors are discussed in Section 2.2.

One of the most fundamental elements, carbon, has provided the many surprises with potential to impact electronics. The discovery of carbon 60 (aka buckyball) in 1985 opened the door to the discovery and study of various forms of carbon. The carbon nanotube (CNT) and multiple variations were discovered in 1991. Variations include single-walled and multiwalled tubes as well as tubes with different arrangements of the individual carbon atoms. While the first application of CNTs was increasing strength of materials, intriguing electronic properties were also found. Depending on the chirality of the nanotube, it can be either conducting or semiconducting. This discovery led to further research in producing nanotube transistors. In the course of this work, graphene was discovered. Graphene is a sheet of carbon atoms and is often described as a two-dimensional material. An interesting feature of the material is that the properties are significantly different if the actual structure has one atomic layer, two layers, or three layers. Quantum dots (QDs) are another nanoelectronics development. While some of the "dots" are actually in excess of 100 nm, the devices are still considered in the nanorange. These devices offer potential for electronics, communication, and other industries; however, the materials used to create QDs are often hazardous or toxic. There are some recent studies that show CdSe QDs can degrade in the environment and separate into its individual elements. QDs are an example of new nanoscience-driven developments where the benefits and risks must be considered and weighed.

More efficient sensors due to nanoscale enhancements in structure will positively impact photovoltaic applications. Another area that is receiving attention is the fundamental material of the semiconductors. One issue that exists with solar energy is that the basic form of electricity generated is direct current. To efficiently send the power generated to other locations, a conversion to alternating current is beneficial. The size of the required conductor is significantly reduced when the power is converted. The issue that arises is most applications of solar energy have high temperatures involved in their surroundings. The current production of semiconductor devices is mainly on silicon, which has a failure mode at high temperatures. There are a number of other materials, called III-V from their position in the periodic chart, that are employed for very specialized applications. The high temperature requirements of automotive under-the-hood applications have spurred development in silicon-carbide (SiC) substrates. While there has been progress, the issue of controlling micropipes (defects in the crystal structure) is the limiting item to large-scale production of devices on this material.

This chapter provides an overview of some of the materials under investigation and potential future applications. Due to the rapidly expanding nature of the field, the future devices and their applications are left to the imagination of the researchers.

2.2 SEMICONDUCTOR TRANSISTORS

The semiconductor industry was one of the first industries to move into the production of large quantities of devices that have features that are in the double-digit nanometer region. It is fortunate that this industry has long worked at the leading edge of technology.

The development of the International Technology Roadmap for Semiconductors (ITRS)[1] was originally necessitated by the need to keep the equipment suppliers updated on future technology requirements. The time to develop the equipment required for manufacturing smaller and smaller dimensions is normally more than 7 years. The ITRS is formulated by researchers from various leading-edge semiconductor companies. The various elements of the ITRS, lithography, front end processes (FEP), interconnect, and so on, all have working groups that meet and review potential developments that might be applicable to being inserted into the semiconductor manufacturing processes.

What started out as a means of projecting equipment requirements became a means of evaluating the future designs of transistors. As the shrinkage of size continues, the ability to manufacture the precise formations required for various devices becomes almost impossible. Fortunately, the ITRS provided an indication of the coming challenges and the industry expert started to evaluate the various options in the late 1990s.

In 2011, the vast majority of devices were produced using planar bulk complementary metal-oxide-semiconductor (CMOS) devices. The FEP chapter of the ITRS claims the end of planar bulk CMOS is becoming visible within the next several years. They note that as a consequence one can expect the emergence of CMOS technology that uses nonconventional metal-oxide-semiconductor field-effect transistors (MOSFETs) or alternatives such as planar fully depleted silicon on insulator devices and dual-gate or multigate devices of vertical geometry.

Planar CMOS: Planar devices (Figure 2.1) are expected to scale from 24-nm gate length in 2011 to 12-nm gate length in 2019, but to be overtaken by other technologies

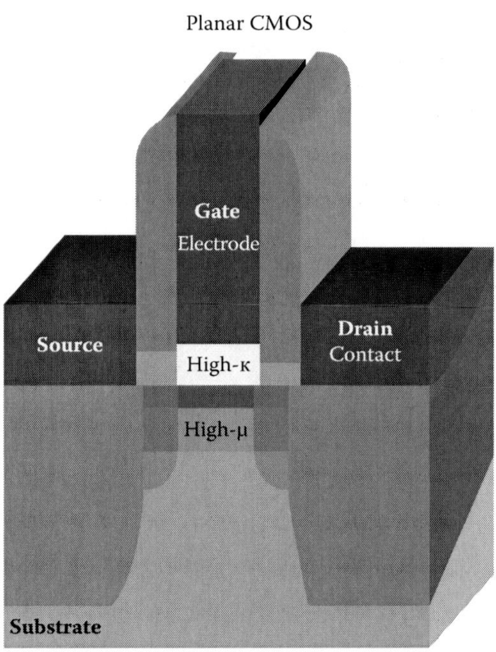

FIGURE 2.1 Planar complementary metal-oxide-semiconductor (CMOS) device.

for use in leading-edge devices. This includes a transition to fully depleted operation in the later part of that time frame.

Nonplanar: The key distinction for planar devices is that the current flow is primarily restricted to the thin area below the gate, essentially one-dimensional current flow. The rest of the structure and detail in the source, drain, and gate of the device is to enable the maximum current flow in a well-designed and controllable manner.

According to the ITRS, nonplanar devices would become mainline at 18-nm gate length in 2014 and extend in some form to the end of the roadmap, which is 5.8 nm in 2026. FinFET devices (Figure 2.2) are the most common of these types of devices in development and can be discussed as a good example of the technology. Their main benefit is that current conduction in this device is three dimensional through the rods (channel) that connect the source and drain through the gate, which controls the switching action. Because of this utilization of the full three-dimensional volume of the device, nonplanar devices are expected to have an advantage of 3–4× in current flow over the equivalent planar device.

Memory: The aforementioned device configurations are the leading candidates envisioned to be used to manufacture logic devices for the next two decades. A similar set of devices are planned for memory use, with implementation time frames based on manufacturing capabilities. For instance, FinFET-like structures have already been demonstrated effectively for dynamic random-access memory (DRAM). The bar-like nature of the channel fits very naturally into the design of DRAM word- or bit-lines. Similarly, process and device structure have been limitations in scaling down NAND Flash Memories to less than 20-nm half-pitch. This makes NAND Flash a leading candidate device technology for various three-dimensional stacked memory architectures that are under development. These architectures consist of stacked bit-lines or word-lines. The implementations are FinFET-like structures, shown in Figure 2.3, which may feature as follows:

- Horizontal channel and gate
- Vertical channel and horizontal gate
- Horizontal channel and vertical gate

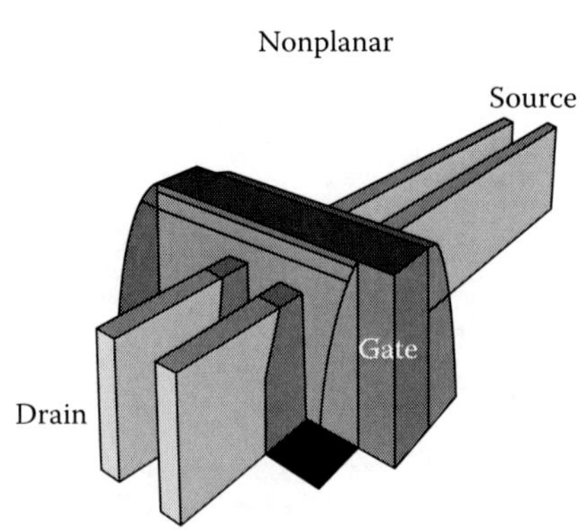

FIGURE 2.2 Nonplanar CMOS device (FinFET).

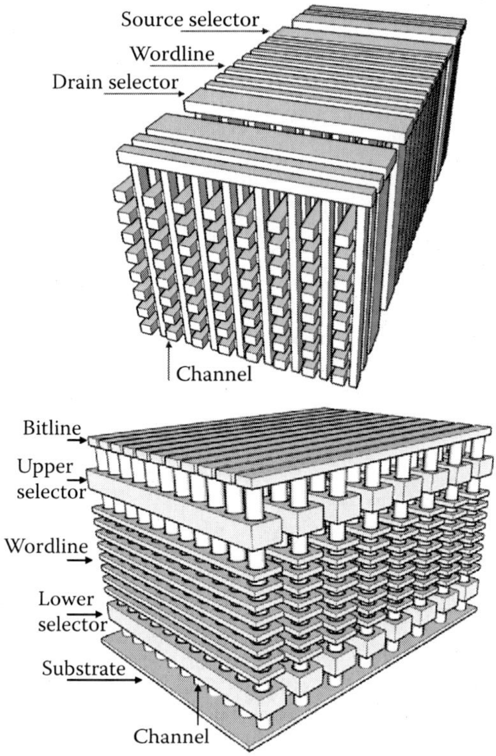

FIGURE 2.3 Proposed structures for three-dimensional NAND Flash Memory. (Data from International Technology Roadmap for Semiconductor [ITRS] http://www.itrs.net.)

Future devices: The investigations of the top researchers in the industry realize that even the novel innovation that is being developed does not provide for the requirements in the next 10 years.

In the farther future, more radical devices are envisioned within the ITRS Working Group called Emerging Research Devices. For logic applications, they have reviewed devices including the following:

- Ferromagnetic (including magnetic quantum-dot cellular automaton)
- Field-effect transistor (FET) extension—one-dimensional structures
- FET extension—channel replacement
- Resonant tunneling
- Molecular (including electric QCA)
- Single-electron transistor (SET)
- Spin transistor

A *ferromagnetic transistor* is a SET with ferromagnetic outer electrodes and a nonmagnetic island. Tunneling current causes nonequilibrium electron–spin distribution in the island. The dependencies of the magnetoresistance ratio on the bias and gate voltages show the dips that are directly related to the induced separation of Fermi levels for electrons with different spins.

FET extension—one-dimensional structures: These are planar transistors that use one-dimensional materials such as graphene as the active channel.

FET extension—channel replacement: These can be FETs of either configuration but using high mobility materials as replacement channels.

Resonant tunneling devices: These are transistors constructed by placing an insulating barrier between two two-dimensional wells. An electron can tunnel through the barrier only if its energy and its momentum in the plane of the well are both conserved. In general, when no voltage is applied to the device, there are no matching states in the two wells, and the device is off. Resonant tunneling happens when the energy level of the electrons in one well is shifted appropriately; the energy states line up opposite each other and tunneling occurs.

Molecular transistors: These are transistors formed from molecules. It has been shown that a benzene molecule attached to gold contacts could behave just like a silicon transistor.

The molecule's different energy states depends on the voltage applied to it through the contacts.

SETs: A SET consists of two tunnel junctions sharing one common electrode with a low self-capacitance, known as the *island*. The electrical potential of the island can be tuned by a third electrode (the *gate*), capacitively coupled to the island. In the blocking state, no accessible energy levels are within tunneling range of the electron on the source contact. All energy levels on the island electrode with lower energies are occupied. When a positive voltage is applied to the gate electrode, the energy levels of the island electrode are lowered. The electron can tunnel onto the island, occupying a previously vacant energy level. From there it can tunnel onto the drain electrode where it inelastically scatters and reaches the drain electrode Fermi level.

Spin transistor: These transistors rely on the energy splitting between electron spin states arising from structural inversion asymmetry. Device concepts involving only *nonmagnetic* materials are especially attractive because they avoid many material issues and unwanted stray magnetic fields associated with the incorporation of magnetic contacts and because their operation relies on applied electric fields only, which may be modulated at considerably higher rates than magnetic fields.

Predictions: As these are all potential future devices, they need considerable development before they are ready for production implementation. Of these devices, the nanofloating gate, SET, and resonant tunneling devices (RTDs) could use many existing processes, but would probably need an engineered dielectric material. The one-dimensional structures (nanotubes, nanowires, etc.) will need new processes to control diameter, location, orientation, and new doping processes. The polymer and molecular devices would require low temperature processing and reliable contacts that are compatible with CMOS integration. The other devices will introduce more radical materials that will require significant work to make them compatible with CMOS processing.

2.3 CARBON NANOSTRUCTURES

Carbon structures have been the object of significant research over the past 15 years and are of great interest for the electronics industry. CNTs are an allotrope of carbon. Most people experience forms of carbon as graphite or to a lesser frequency diamonds. In these cases, the carbon atoms are arranged in a crystalline lattice structure. Carbon can also exist in an amorphous state, which includes coal and carbon black. The discovery of the carbon 60, or as it is now called buckyballs, at the Rice University in 1985 opened up a new world of materials. The discovery of the CNT provided additional emphasis on understanding and applying the material. The CNT is a lattice of carbon atoms that are coupled to create a cylinder. There are two types of CNTs—multiwalled and single-walled.

The multiwalled CNTs are an accurate description, which indicates that there are multiple concentric tubes of carbon. There is a significant application of these types of nanotubes in industry. One advantage is that the nanotubes are very strong. Incorporated into epoxies or plastics, the nanotubes provide a very significant increase in strength of the materials. Automotive bumpers made of materials including nanotubes have strengths that are many times greater than steel at only a fraction of the weight. Reduced weight results in lower transportation costs and also better gas mileage in the lighter weight automobiles.

From the perspective of electronics, the single-walled CNT is very promising. There is difference in the properties of the material that depends on how the tube is connected together. If one considers a plane of hexagons that can be rolled into a tube, the various ways to connect the edges will provide different lattice structures around the circumference of the nanotube. This fact is important because the resulting nanotube can be either conducting or semiconducting. There has been demonstration of the ability to create FETs using CNTs.

Recently, there has been research to evaluate the construction of transistors using CNTs and graphene. This also includes the emphasis on developing novel electronics devices using different forms of carbon. Section 2.4 covers the advantages of graphene for its application in electronics. An "unrolled" CNT results in a single layer of carbon in a sheet called graphene. Because it has only one layer of atoms (thickness), graphene is often called a two-dimensional material. It has unique properties that are different from the bulk form of carbon and have hinted at the possible capability of creating significant novel electronics.

2.4 GRAPHENE ELECTRONICS

Graphene is a two-dimensional material of sp^2-bonded carbon atoms, which has many unique and exciting electronic properties. Specific characteristics of single-layer graphene, such as high carrier mobility, up to 200,000 $cm^2/v/s$, large critical current density (about 2×10^8 A/cm^2), high saturation velocity (about 5.5×10^7 cm/s^1),[1–3] and very high intrinsic thermal conductivity,[2–5] exceeding 3,000 W/m-K, make it appealing for electronic, sensor, detector, and interconnect applications.[6–8]

Several methods have been developed to get graphene on arbitrary substrates. The first free-standing graphene sheet is prepared by mechanical exfoliation of highly oriented pyrolytic graphite.[2] The quality of graphene fabricated by this method is currently the best. The pristine graphene has very low concentration of structural defects, so it is widely used in many laboratories for basic scientific research and for making proof-of-concept devices. However, the flake thickness, size, and uniform thickness location are largely uncontrollable, making it potentially unsuitable for mass production. Several strategies are presently being used to achieve reproducible and scalable graphene on substrates. One example is the conversion of SiC (0001) to graphene through sublimation of silicon atoms at high temperatures.[9] High-quality wafer-scale graphene with switching speeds of up to 100 GHz has been demonstrated using this technique.[10] The drawback of this approach for graphene synthesis is the high price of the initial SiC wafer. Another approach is chemical exfoliation of graphite in liquids,[11–13] and it is considered to be the most economical mass production method. The graphene achieved from chemical exfoliation methods usually has abundant structural defects and functional groups that degrade the quality of graphene. The most promising inexpensive approach for deposition of reasonably high-quality graphene is chemical vapor deposition (CVD) onto transition metal substrate such as Ni[14,15]; PT, Pd, and Co[16]; Ru[17]; Ir[18]; or Cu.[19] Then following a transfer process, the graphene layer can be transferred to any substrate. In particular, recent developments of uniform single-layer deposition of graphene on copper foil over large areas have opened an access to electronics industry. The array of single-crystal graphene has also been achieved by using prepatterned growth seeds methods[20] and it opens a route toward scalable fabrication of single-crystal graphene devices avoiding grain boundaries. Direct growth graphene layer on insulator substrate has also been developed, but the quality of the graphene is not good enough for most of electrical application.[21,22]

Digital logic devices and radio frequency devices are two principal divisions of semiconductor devices. Graphene is potentially well suited to radio frequency applications because of its promising carrier transport properties and its purely two-dimensional structure.[23] A graphene MOS device was among the breakthrough results reported by the Manchester group in 2004.[2] The first graphene MOSFET with a top-gate was reported in 2007.[24] Recently, graphene MOSFETs with gigahertz capabilities have been reported. Mechanical exfoliated graphene,[25] CVD grown graphene,[19] and epitaxial graphene[10] have been used for these large-area transistors channels. The top-gate dielectric has been made with SiO_2, Al_2O_3, and HfO_2. The substrate also affects the performance of the device.[3] The fastest graphene transistor currently is a transistor with a self-aligned nanowire gate that has a cut-off frequency of 300 GHz.[26] The weak point of all radio frequency graphene MOSFETs reported so far is the unsatisfying saturation behavior that has an adverse impact on the cut-off frequency, the intrinsic gain, and other figures of merit for radio frequency devices.

Ambipolar electronics is based on a new family of nonlinear devices that rely on the fact that the conduction and valence bands in graphene meet at the Dirac point. By changing the gate voltage of the FET, the Fermi level can continuously sweep from the conduction band to the valence band. This ambipolar conduction, which is responsible for the poor performance of amplifiers and logic devices based on graphene, actually enables a new class of nonlinear radio frequency electronic devices. Frequency multipliers, high-frequency mixers, and digital modulators[27–29] have been recently demonstrated using this principle.

Because large-area graphene is a semiconductor with zero bandgap, devices with channels made of large-area graphene cannot be switched off and therefore are not suitable for logic applications. However, the band structure of graphene can be modified, and it is possible to open a bandgap in three ways: constraining large-area graphene in one dimension to form graphene nanoribbons,[30–33] biasing bilayer graphene,[34–36] and applying strain to graphene.[37,38] Graphene nanoribbons with narrow widths (less than 20 nm) can generate a bandgap that is dependent on the ribbon width and crystallographic orientation of the edges, and edge functionalization and doping can also affect the bandgap. Several approaches have been developed for producing grahpene nanoribbons, such as CVD,[39] unzipping of CNTs,[40,41] and lithography.[42] To get very narrow nanoribbons with well-defined edges is still a challenge at the moment. A bandgap can be opened when an electric field is applied perpendicular to the bilayer graphene. This bandgap opening was predicted by theory and has been verified in experiments. The size of the bandgap depends on the strength of the perpendicular field. A uniform and large area bilayer graphene is still a challenge. Graphene p-n junction has already been fabricated by chemical doping[43] and electrostatic doping.[44]

Benefiting from the zero bandgap of graphene, which allows them to absorb light from the infrared to the ultraviolet with almost equal strength, graphene photodetectors have already been fabricated at very high frequencies, which cannot be accessed with silicon-based device.[45] Different kinds of chemical and biological sensors have also been developed.[8]

While graphene is extensively investigated, its derivatives,[46–48] such as graphene oxide, graphane (hydrogenated graphene), and graphene fluoride, also attract great attention owing to their distinct properties from their parent material, graphene. The most prominent property of graphene derivatives beyond graphene is that they are semiconductors or insulators, which bring many benefits for the application of two-dimensional materials. For example, at selective locations, the transformation of graphene into graphene fluoride is a convenient method to fabricate graphene devices and circuits, avoiding dangling bonds from the etching process. Although graphite is known as one of the most chemically inert materials, graphene can be transformed to the discussed derivatives without damaging its skeleton structure using certain (even quite mild) chemical environments. For example, a 2-hour processing by cold hydrogen plasma can completely transform graphene into graphane. In contrast to graphene, its derivatives can have several orders of higher resistivity. We can consider some of them as insulators, such as graphene oxide and graphane. The derivatives of graphene can be transformed back to graphene, but the properties cannot be fully recovered, owing to the generation of defects during chemical processing. Relatively, the graphene materials transformed back from graphane and graphene fluoride can largely recover its electrical performance by annealing or chemical processing; for example, quantum Hall effect can reappear for the graphene transformed back from graphane. However, for graphene transformed back from graphene oxide, electrical performance is much worse than the pristine graphene.

Graphene is an amazing material. Huge progresses have been made in the development of various electronic devices in past several years, such as the demonstrations of the graphene MOSFET with a cut-off frequency of 300 GHz. However, the progress has also been accompanied with the appearance of a number of problems. MOSFETs with large-area graphene channels cannot be switched off, making them unsuitable for logic applications, and their peculiar saturation behavior limits their radio frequency performance. Nanoribbon graphene, which does have a bandgap and results in transistors that can be switched off, has serious fabrication issues because of the small widths required and the presence of edge disorder. Graphene-based electronics is still in its toddler stage and more years are needed for graphene to compete with state-of-the-art silicon and compound semiconductor electronics.

2.5 QUANTUM DOTS ELECTRONICS

A quantum structure refers to material structures in which electrons exhibit quantum characteristics, such as their wave-like behavior and tunneling. This phenomenon occurs when their motion, as governed by the Schrödinger equation, is restricted in at least one physical dimension due to the presence of a potential energy barrier along that dimension. As a result, the energy states that electrons are allowed to have in such lower-dimensional structures become discrete in that dimension, while they can still have a continuum of energy in other directions along which their motion is not restricted. In a QD, also called quantum box or zero-dimensional structure, electrons are confined in all three physical dimensions, and the allowed energy states are therefore completely discrete. This phenomenon is best illustrated in the density of states associated with low-dimensional structures as given in Figure 2.4.

The physical dimension when quantum behavior arises occurs when the wave-like nature of electrons becomes predominant. This occurs when dimensions are smaller than the de Broglie wavelength of the particle:

$$\lambda = \frac{h}{p}$$

where h is Planck's constant and p is the momentum of the electron. In semiconductors, this is on the order of 50 nm. Because the motion of electrons is restricted in small volumes, QDs are often referred to as artificial atoms, superatoms of QD

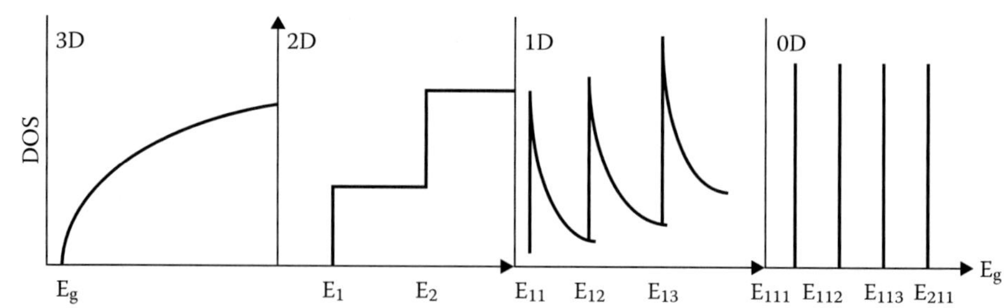

FIGURE 2.4 Schematics of density of states with increasing carrier confinements.

atoms.[49] It was only in the late twentieth century that technology matured enough to manufacture QDs. QDs exhibit a number of unique properties that make them ideally suited for nanoelectronics. These are primarily rooted in the ability to reach the desired electronic quantum states by controlling the physical dimension and shape of the QD.[50–52]

Indeed, the QD size and shape directly determines the number and location of the quantized energy levels, which enables control of the effective bandgap and optical transition energy, as well as the number of electrons that can be confined within the volume. Controlling composition gradients between the QD and the barrier materials at the atomistic level can lead to a rectangular, parabolic, or other type of potential well. Furthermore, the quantization of allowed energy states is associated with the quantization of the allowed values of electron momentum in the first Brillouin zone. As a result, interactions between electrons and phonons are tremendously minimized due to the need for overall energy and momentum conservation. This is called phonon bottleneck[53,54] and has wide impact on the operation and efficiency of electronic devices because phonon relaxation of hot electrons to lower energy states cannot be easily accomplished, which increases electron lifetimes.

In the late twentieth century, the driving force behind the research and development of QDs—including means to manufacture them—was their ability to dramatically yield increased diode laser efficiencies and correspondingly reduced threshold current densities, as shown in Figure 2.5. Controlling QD properties could support the exponentially growing telecom infrastructure industry.[55–59] QD-based lasers soon achieved higher output power than quantum well lasers even though the active region was much smaller. Their temperature stability and the reduced carrier dispersion made it possible to achieve high-speed modulation reaching up to 40 Gb/s, which was crucially important to optical data communication industry.[60–66]

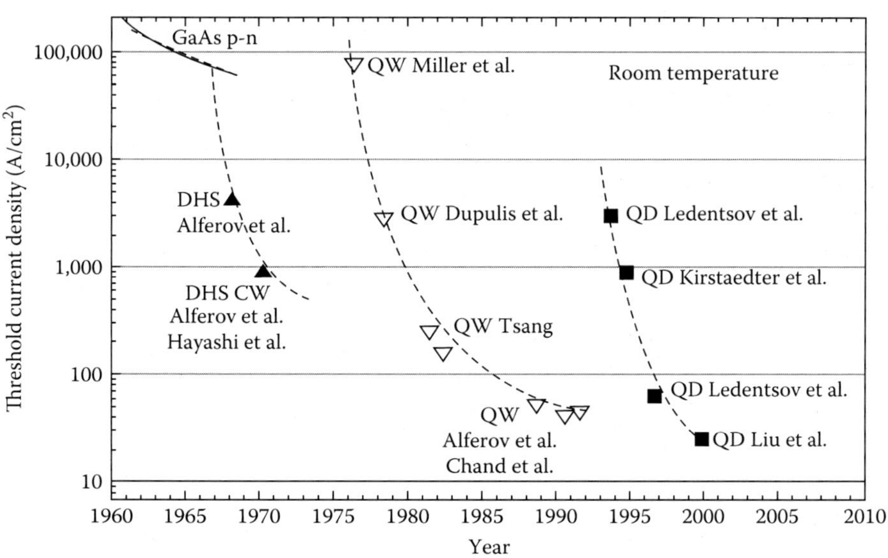

FIGURE 2.5 Development of QD-based laser diodes.

The same fundamental principles that enabled high-efficiency radiative recombination to generate laser photons were subsequently applied to photon absorption in the case of QD infrared photodetectors. Conventional infrared sensors are highly sensitive to thermal noise and usually require cooling to achieve expected nanochannel electroporation. Theoretically, QD-based infrared photodetector should operate with temperature insensitivity and allow for detection of light at normal incidence. Also, intersubband photon absorption can be easily modulated by changing the size of the QDs in the active region, which can lead the multicolor detectors.[67–70]

After decades-long exponential improvements in the size and speed of electronic devices, computing hardware technology has reached a stage where further progress necessitates taking into account the full quantum mechanical nature of the electron, including quantum computing, SET, and spintronics.[71–76] Another device architecture that is being investigated is the QD cellular automata[77] in which binary information is encoded in the electrical charge configuration of the QD and physical interactions between adjacent QDs provide the means for information transport without having the need for electrical current or power flow.[78,79]

Nevertheless, it has always been the fundamental interactions of photons with the discrete electronic states in a QD that have been fueling QD development. In the beginning of the twenty-first century, these unique properties find themselves in resonance with the global and interdisciplinary challenge of *building a sustainable energy future*. Indeed, QDs are being increasingly touted as key to enabling the third and later generation solar cell technologies, which are aimed at overcoming the fundamental efficiency limit established by Shockley and Queisser.[80–85] This limit was determined under the assumption that only one electron-hole pair (exciton) is created per incident photon. However, multiple-exciton generation is fundamentally possible in materials. This is the ability to yield more than one electron-hole pair per photon through a process called impact ionization, in which hot carriers generated from high-energy photon absorption lose some of their excess energy to produce at least one other exciton.[85–90] This phenomenon is most probable to occur in zero-dimensional structures, such as QDs, because of the absence of any continuum of energy states and the resulting phonon bottleneck, which increases the hot carrier relaxation time as approximated by

$$\tau_c \approx \frac{1}{\vartheta} e^{\frac{\Delta E}{k_B T}}$$

where ΔE is the discrete energy level spacing, T is the carrier temperature, k_B is the Boltzmann constant, and ϑ is the phonon frequency.[91] In addition, QDs are more stable compared to organic absorbing materials (Figure 2.6), and the absorption properties of QDs can be tuned both by adjusting the constituent material composition and by controlling the dot size (Figure 2.7), which enables a more complete coverage of the solar spectrum.

Despite these advantages, QDs suffer from difficulties in the extraction of photogenerated charge carriers, due to the inherent need to achieve quantum confinement of carriers within and therefore a surrounding barrier material. Charge transport thus needs to be accomplished through tunneling and Coulomb coupling. A promising

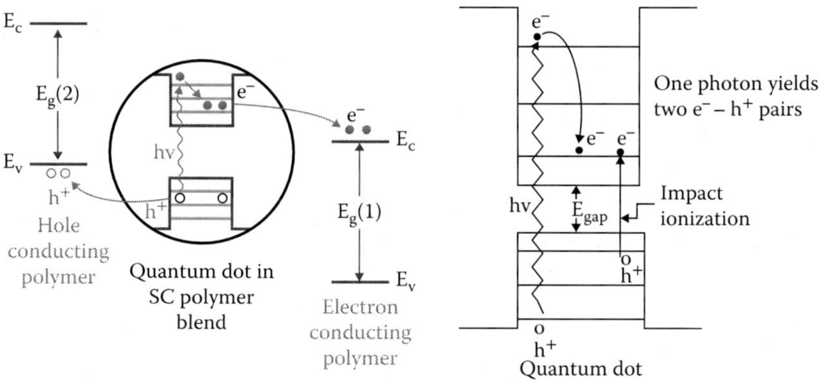

FIGURE 2.6 Schematic diagram of QD-based solar cell carrier transport. Enhanced photovoltaic efficiency is expected from impact ionization. (Data from Nozik, A.J., *Physica* E 2002, 14, 115–120.)

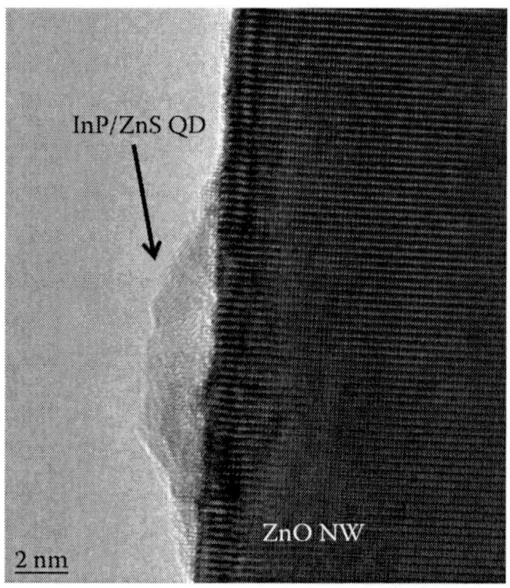

FIGURE 2.7 InP/ZnS QDs attached to ZnO NW for solar cell. (Data from Raghavan, S. et al. *Appl. Phys. Lett.*, 2002, 81, 1369.)

approach to alleviate this challenge is to embed the QDs inside a low conductivity polymer matrix that would sustain the carrier confinement while at the same time providing an electrical conduction path to the external contacts. The polymers most typically used include poly-3(hexylthiophene) and poly(3,4-ethylenedioxythiophene) poly(styrenesulfonate). Such QD-polymer structures are at the foundation of the fourth-generation solar cells, which consist of hybrid structures that combine solid state nanoparticles with organic polymers together to realize a single multispectrum material.

To date, QDs of various materials, including CdSe, PbSe, PbS, PbTe, InAs, Si, and InP, have been investigated,[92–97] as well as QD of varied sizes to efficiently absorb larger portions of the solar spectrum.[98] Although QD-based solar cells have exhibited a power conversion efficiency of up to 8%,[99] QD solar cells have the potential to reach efficiencies upward of 65%,[53] while at the same time offering a much lower cost than other solar cell approaches, thanks to advances in chemistry and nanotechnology manufacturing.

2.6 CONCLUSIONS

Where is nanoelectronics going? That is a hard question to answer. There are many efforts under way to produce nanotechnology-based novel devices in research laboratories and large quantities. To be an effective alternative for the mass production of electronic circuitry, like microprocessors, the methods employed need to be capable of production rates of literally billions of transistors a second. This rate is currently achieved through projection lithography processes. The new developments are not at this stage yet.

There are other efforts attempting to develop new types of transistors that can circumvent the limitations of device sizes where the active constituents consist of small number of individual atoms or electrons. There is currently work being done on entire device systems that will be in the micron size range. As more research and development is accomplished, additional characteristics of novel material will be identified. The future of electronic devices and systems will continue to depend on our understanding of materials at the molecular and atomic level—the nanoscale.

REFERENCES

1. International Technology Roadmap for Semiconductor (ITRS) http://www.itrs.net.
2. K.S. Novoselov, A.K. Geim, S.V. Morozov, D. Jiang, Y. Zhang, S.V. Dubonos, I.V. Grigorieva, and A.A. Firsov, Electric field effect in atomically thin carbon films. *Science* 2004, *306* (5696), 666–669.
3. J.H. Chen, C. Jang, S.D. Xiao, M. Ishigami, and M.S. Fuhrer, Intrinsic and extrinsic performance limits of graphene devices on SiO2. *Nat Nanotechnol* 2008, *3* (4), 206–209.
4. A.A. Balandin, S. Ghosh, W.Z. Bao, I. Calizo, D. Teweldebrhan, F. Miao, and C.N. Lau, Superior thermal conductivity of single-layer graphene. *Nano Lett* 2008, *8* (3), 902–907.
5. A.A. Balandin, Thermal properties of graphene and nanostructured carbon materials. *Nat Mater* 2011, *10* (8), 569–581.
6. K. Banerjee, Y. Khatami, C. Kshirsagar, and S.H. Rasouli, Graphene based transistors: physics, status and future perspectives. *ISPD 2009 ACM International Symposium on Physical Design* 2009, 65–65.
7. A.K. Geim, Graphene: status and prospects. *Science* 2009, *324* (5934), 1530–1534.
8. Y.Y. Shao, J. Wang, H. Wu, J. Liu, I.A. Aksay, and Y.H. Lin, Graphene-based electrochemical sensors and biosensors: a review. *Electroanal* 2010, *22* (10), 1027–1036.
9. K.V. Emtsev, A. Bostwick, K. Horn, J. Jobst, G.L. Kellogg, L. Ley, J.L. McChesney, T. Ohta, S.A. Reshanov, J. Rohrl, E. Rotenberg, A.K. Schmid, D. Waldmann, H.B. Weber, and T. Seyller, Towards wafer-size graphene layers by atmospheric pressure graphitization of silicon carbide. *Nat Mater* 2009, *8* (3), 203–207.
10. Y.M. Lin, C. Dimitrakopoulos, K.A. Jenkins, D.B. Farmer, H.Y. Chiu, A. Grill, and P. Avouris, 100-GHz transistors from wafer-scale epitaxial graphene. *Science* 2010, *327* (5966), 662.
11. S. Stankovich, D.A. Dikin, G.H.B. Dommett, K.M. Kohlhaas, E.J. Zimney, E.A. Stach, R.D. Piner, S.T. Nguyen, and R.S. Ruoff, Graphene-based composite materials. *Nature* 2006, *442* (7100), 282–286.

12. G. Eda, G. Fanchini, and M. Chhowalla, Large-area ultrathin films of reduced graphene oxide as a transparent and flexible electronic material. *Nat Nanotechnol* 2008, *3* (5), 270–274.
13. Y. Hernandez, M. Lotya, D. Rickard, S.D. Bergin, and J.N. Coleman, Measurement of multicomponent solubility parameters for graphene facilitates solvent discovery. *Langmuir* 2010, *26* (5), 3208–3213.
14. Q.K. Yu, J. Lian, S. Siriponglert, H. Li, Y.P. Chen, and S.S. Pei, Graphene segregated on Ni surfaces and transferred to insulators. *Appl Phys Lett* 2008, *93* (11).
15. A. Reina, X.T. Jia, J. Ho, D. Nezich, H.B. Son, V. Bulovic, M.S. Dresselhaus, and J. Kong, Large area, few-layer graphene films on arbitrary substrates by chemical vapor deposition, *Nano Lett* 2009, *9* (1), 30–35.
16. J.C. Hamilton, and J.M. Blakely, Carbon segregation to single-crystal surfaces of Pt, Pd and Co. *Surf Sci* 1980, *91* (1), 199–217.
17. P.W. Sutter, J.I. Flege, and E.A. Sutter, Epitaxial graphene on ruthenium. *Nat Mater* 2008, *7* (5), 406–411.
18. J. Coraux, A.T. N'Diaye, C. Busse, and T. Michely, Structural coherency of graphene on Ir(111). *Nano Lett* 2008, *8* (2), 565–570.
19. X.S. Li, W.W. Cai, J.H. An, S. Kim, J. Nah, D.X. Yang, R. Piner, A. Velamakanni, I. Jung, E. Tutuc, S.K. Banerjee, L. Colombo, and R.S. Ruoff, Large-area synthesis of high-quality and uniform graphene films on copper foils. *Science* 2009, *324* (5932), 1312–1314.
20. Q.K. Yu, L.A. Jauregui, W. Wu, R. Colby, J.F. Tian, Z.H. Su, H.L. Cao, Z.H. Liu, D. Pandey, D.G. Wei, T.F. Chung, P. Peng, N.P. Guisinger, E.A. Stach, J.M. Bao, S.S. Pei, and Y.P. Chen, Control and characterization of individual grains and grain boundaries in graphene grown by chemical vapour deposition. *Nat Mater* 2011, *10* (6), 443–449.
21. K.B. Kim, C.M. Lee, and J. Choi, Catalyst-free direct growth of triangular nano-graphene on all substrates. *J Phys Chem C* 2011, *115* (30), 14488–14493.
22. L.C. Zhang, Z.W. Shi, Y. Wang, R. Yang, D.X. Shi, and G.Y. Zhang, Catalyst-free growth of nanographene films on various substrates. *Nano Res* 2011, *4* (3), 315–321.
23. F. Schwierz, Graphene transistors. *Nat Nanotechnol* 2010, *5* (7), 487–496.
24. M.C. Lemme, T.J. Echtermeyer, M. Baus, and H. Kurz, A graphene field-effect device. *IEEE Electr Device L* 2007, *28* (4), 282–284.
25. Y.M. Lin, K.A. Jenkins, A. Valdes-Garcia, J.P. Small, D.B. Farmer, and P. Avouris, Operation of graphene transistors at gigahertz frequencies. *Nano Lett* 2009, *9* (1), 422–426.
26. L. Liao, Y.C. Lin, M.Q. Bao, R. Cheng, J.W. Bai, Y.A. Liu, Y.Q. Qu, K.L. Wang, Y. Huang, and X.F. Duan, High-speed graphene transistors with a self-aligned nanowire gate. *Nature* 2010, *467* (7313), 305–308.
27. T. Palacios, Graphene electronics thinking outside the silicon box. *Nat Nanotechnol* 2011, *6* (8), 464–465.
28. J.S. Moon, D. Curtis, D. Zehnder, S. Kim, D.K. Gaskill, G.G. Jernigan, R.L. Myers-Ward, C.R. Eddy, P.M. Campbell, K.M. Lee, and P. Asbeck, Low-phase-noise graphene fets in ambipolar RF applications. *IEEE Electr Device L* 2011, *32* (3), 270–272.
29. X.B. Yang, G.X. Liu, A.A. Balandin, and K. Mohanram, Triple-mode single-transistor graphene amplifier and its applications. *ACS Nano* 2010, *4* (10), 5532–5538.
30. M.Y. Han, B. Ozyilmaz, Y.B. Zhang, and P. Kim, Energy band-gap engineering of graphene nanoribbons. *Phys Rev Lett* 2007, *98* (20).
31. X.L. Li, X.R. Wang, L. Zhang, S.W. Lee, and H.J. Dai, Chemically derived, ultrasmooth graphene nanoribbon semiconductors. *Science* 2008, *319* (5867), 1229–1232.
32. L. Yang, C.H. Park, Y.W. Son, M.L. Cohen, and S.G. Louie, Quasiparticle energies and band gaps in graphene nanoribbons. *Phys Rev Lett* 2007, *99* (18).
33. X.T. Jia, J. Campos-Delgado, M. Terrones, V. Meunier, and M.S. Dresselhaus, Graphene edges: a review of their fabrication and characterization. *Nanoscale* 2011, *3* (1), 86–95.

34. E.V. Castro, K.S. Novoselov, S.V. Morozov, N.M.R. Peres, J.M.B.L. Dos Santos, J. Nilsson, F. Guinea, A.K. Geim, and A.H.C. Neto, Biased bilayer graphene: semiconductor with a gap tunable by the electric field effect. *Phys Rev Lett* 2007, *99* (21).
35. T. Ohta, A. Bostwick, T. Seyller, K. Horn, and E. Rotenberg, Controlling the electronic structure of bilayer graphene. *Science* 2006, *313* (5789), 951–954.
36. Y.B. Zhang, T.T. Tang, C. Girit, Z. Hao, M.C. Martin, A. Zettl, M.F. Crommie, Y.R. Shen, and F. Wang, Direct observation of a widely tunable bandgap in bilayer graphene. *Nature* 2009, *459* (7248), 820–823.
37. V.M. Pereira, A.H.C. Neto, and N.M.R. Peres, Tight-binding approach to uniaxial strain in graphene. *Phys Rev B* 2009, *80* (4).
38. Z.H. Ni, T. Yu, Y.H. Lu, Y.Y. Wang, Y.P. Feng, and Z.X. Shen, Uniaxial strain on graphene: Raman spectroscopy study and band-gap opening (vol 2, pg 2301, 2008). *ACS Nano* 2009, *3* (2).
39. J. Campos-Delgado, J.M. Romo-Herrera, X.T. Jia, D.A. Cullen, H. Muramatsu, Y.A. Kim, T. Hayashi, Z.F. Ren, D.J. Smith, Y. Okuno, T. Ohba, H. Kanoh, K. Kaneko, M. Endo, H. Terrones, M.S. Dresselhaus, and M. Terrones, Bulk production of a new form of sp(2) carbon: crystalline graphene nanoribbons. *Nano Lett* 2008, *8* (9), 2773–2778.
40. M. Terrones, Sharpening the chemical scissors to unzip carbon nanotubes: crystalline graphene nanoribbons. *ACS Nano* 2010, *4* (4), 1775–1781.
41. A.L. Higginbotham, D.V. Kosynkin, A. Sinitskii, Z.Z. Sun, and J.M. Tour, Lower-defect graphene oxide nanoribbons from multiwalled carbon nanotubes. *ACS Nano* 2010, *4* (4), 2059–2069.
42. L. Tapaszto, G. Dobrik, P. Lambin, and L.P. Biro, Tailoring the atomic structure of graphene nanoribbons by scanning tunnelling microscope lithography. *Nat Nanotechnol* 2008, *3* (7), 397–401.
43. D.B. Farmer, Y.M. Lin, A. Afzali-Ardakani, and P. Avouris, Behavior of a chemically doped graphene junction. *Appl Phys Lett* 2009, *94* (21).
44. M.C. Lemme, F.H.L. Koppens, A.L. Falk, M.S. Rudner, H. Park, L.S. Levitov, and C.M. Marcus, Gate-activated photoresponse in a graphene p–n junction. *Nano Lett* 2011, *11* (10), 4134–4137.
45. F.N. Xia, T. Mueller, Y.M. Lin, A. Valdes-Garcia, and P. Avouris, Ultrafast graphene photodetector. *Nat Nanotechnol* 2009, *4* (12), 839–843.
46. D.R. Dreyer, S. Park, C.W. Bielawski, and R.S. Ruoff, The chemistry of graphene oxide, *Chem Soc Rev* 2010, *39*, 228–240.
47. J.T. Robinson, J.S. Burgess, C.E. Junkermeier, S.C. Badescu, T.L. Reinecke, F. Keith Perkins, M.K. Zalalutdniov, J.W. Baldwin, J.C. Culbertson, P.E. Sheehan, and E.S. Snow, properties of fluorinated graphene films, *Nano Lett* 2010, *10*, 3001–3005.
48. D.C. Elias, R.R. Nair, T.M.G. Mohiuddin, S.V. Morozov, P. Blake, M.P. Halsall, A.C. Ferrari, D.W. Boukhvalov, M.I. Katsnelson, A.K. Geim, and K.S. Novoselov, Control of graphene's properties by reversible hydrogenation: evidence for graphene, *Science* 2009, *323* (5914), 610–613.
49. R.C. Ashoori, Electrons in artificial atoms, *Nature* 1996, *379*, 413–419.
50. Y. Masumoto, and T. Takagahara, *Semiconductor Quantum Dots*, Springer-Verlag, Germany, 2002.
51. L. Jacak, P. Hawrvlak, and A. Wois, *Quantum Dots*, Springer-Verlag, Germany, 1998.
52. D. Bimberg, M. Grundmann, and N.N. Ledentsov, *Quantum Dot Heterostructure*, Wiley, Chichester, 1999.
53. J. Urayama, T.B. Norris, J. Singh, and P. Bhattacharya, Observation of phonon bottleneck in quantum dot electronic relaxation, *Phys Rev Lett* 2001, *86*, 4930.
54. X. Li, H. Nakayama, and Y. Arakawa, Phonon bottleneck in quantum dots: role of lifetime of the confined optical phonons, *Phys Rev B* 1999, *59*, 5069.

55. G. Liu, A. Stintz, H. Li, K. Malloy, and L.F. Lester, Extremely low room temperature threshold current density diode lasers using InAs in an In0.15Ga0.85As quantum well, *Electron Lett* 1999, *35*, 1163.
56. N. Kirstaedter, N.N. Ledentsov, M. Grundmann, D. Bimberg, V. Ustnov, S. Ruvimov, M. Maximov, P. Kop'ev, Zh.I. Alferov, U. Richter, P. Werner, U. Gosele, and J. Heydenreich, Low threshold, large T_0 injection laser emission from (InGa)As quantum dots, *Electron Lett* 1994, *30*, 1416.
57. N. Kirstaedter, O. Schmidt, N.N. Ledentsov et al., Multiphonon-relaxation processes in self-organized InAs/GaAs quantum dots, *Appl Phys Lett* 1996, *68*, 361.
58. Z.I. Alferov, S.V. Zaitsev, N.Y. Gordeev et al., The properties of low-threshold heterolasers with clusters of quantum dots, *Semiconductors*, 1997, *31*, 455–459 (from Russian Semiconductors 1996).
59. D.L. Huffaker, G. Park, Z. Zou, O. Shcheikin, D.G. Deppe, 1.3 μm room-temperature GaAs-based quantum-dot laser, *Appl Phys Lett* 1998, *73*, 2564.
60. R.L. Sellin, Ch. Ribbat, M. Grundmann, N.N. Ledentsov, and D. Bimberg, Close-to-ideal device characteristics of high-power InGaAs/GaAs quantum dot lasers, *Appl Phys Lett* 2001, *78*, 1207.
61. F. Heinrichsdorff, Ch. Ribbat, M. Grundmann, and D. Bimberg, High-power quantum-dot lasers at 1100 nm, *Appl Phys Lett* 2000, *76*, 556.
62. M.V. Maximov, Yu. Shernyakov, A. Tsatsulnikov, A. Lunev, A. Sakharov, V. Ustinov, A.Yu. Egorov, A. Zhukov, A. Kovsh, P.S. Kop'ev, L.A. Asryan, Zh. Alferov, N.N. Ledentsov, D. Bimbert, A. Kosogiv, and P. Werner, High power continuous wave operations of a InGaAs/AlGaAs quantum dot laser, *J Appl Phys* 1998, *83*, 5561–5563.
63. S. Mikhrin, A. Kovsh, I. Krestnikov, D. Livshits, N.N. Ledentsov, Yu.M. Shernyakov, I. Novikov, M. Maximov, V. Ustinov, and Zh. Alferov, High power temperature-insensitive 1.3 μm InAs/InGaAs/GaAs quantum dot lasers, *Semicon Sci Tech* 2005, *20*, 340–342.
64. S.M. Kim, Y. Wang, M. Keever, and J.S. Harris, High-frequency modulation characteristics of 1.3-μm InGaAs quantum dot lasers, *IEEE Photon Tech Lett* 2004, *16*, 377–379.
65. P. Bhattacharya, and S. Ghosh, Tunnel injection In0.4Ga0.6As/GaAs quantum dot lasers with 15 GHz modulation bandwidth at room temperature, *Appl Phys Lett* 2002, *80*, 3482.
66. M. Kuntz, G. Fiol, M. Lammlin, D. Bimberg, M. Thompson, K.T. Tan, C. Marinelli, R. Penty, I. White, V. Ustinov, A. Zhukov, Yu.M. Shernyakov, and A. Kovsh, 35 GHz mode-locking of 1.3 μm quantum dot lasers, *Appl Phys Lett* 2004, *85*, 843.
67. S. Kim, H. Mohseni, M. Erdtmann, E. Michel, C. Jelen, and M. Razeghi, Growth and characterization of InGaAs/InGaP quantum dots for midinfrared photoconductive detectors, *Appl Phys Lett* 1998, *73*, 963.
68. S. Raghavan, P. Rotella, A. Stintz, B. Fuchs, S. Krishna, C. Morath, D. Cardimona, and S. Kennerly, High detectivity InAs quantum dot infrared photodetectors, *Appl Phys Lett* 2002, *81*, 1369–1371.
69. H. Liu, M. Gao, J. McCaffrey, Z.R. Wasilewski, and S. Fafard, Quantum dot infrared photodetectors, *Appl Phys Lett* 2001, *78*, 79–81.
70. S.M. Kim, and J.S. Harris, Multispectral operation of self-assembled InGaAs quantum-dot infrared photodetectors, *Appl Phys Lett* 2004, *85*, 4154–4156.
71. L. Zhuang, L. Guo, and S. Chou, Silicon single-electron quantum-dot transistor switch operating at room temperature, *Appl Phys Lett* 1998, *72*, 1205–1207.
72. D. Klein, R. Roth, A. Lim, A. Alivisatos, and P. McEune, A single-electron transistor made from cadmium selenide nanocrystal, *Nature* 1997, *389*, 699.
73. M. Saitoh, N. Takahashi, H. Ishikuro, and T. Hiramoto, Tunneling barrier structures in room-temperature operating silicon single-electron and single-hole transistors, *Jpn J Appl Phys* 2001, *40*, 2010.

74. D. Loss, and D. DiVincenzo, Quantum computation with quantum dots, *Phys Rev A* 1998, *57*, 120.
75. A. Imamoglu, D. Awschalom, G. Burkard, D. DiVincenzo, D. Loss, M. Sherwin, and A. Small, Quantum information processing using quantum dot spins and cavity QED, *Phys Rev Lett* 1998, *83*, 4204.
76. D. Awschalom, and M. Flatte, Challenges for semiconductor spintronics, *Nature Phys* 2007, *3*, 153–159.
77. C.S. Lent, and P.D. Tougaw, A device architecture for computing with quantum dots, *Proc IEEE* 1997, *85*, 541.
78. C.S. Lent, P.D. Tougaw, W. Porod, and G.H. Bernstein, Quantum cellular automata, *Nanotechnology* 1993, *4*, 49.
79. G. Toth, C.S. Lent, P.D. Tougaw, Y. Brazhnik, W. Weng, W. Porod, R.W. Liu, and Y.F. Huang, Quantum cellular neural networks, *Superlattice Microstruct* 1996, *20*, 473–478.
80. W. Shockley, and H. Queisser, Detailed balance limit of efficiency of p-n junction solar cells, *J Appl Phys* 1961, *32*, 510–519.
81. A.J. Nozik, Quantum dot solar cells, *Physica E* 2002, *14*, 115–120.
82. V.I. Klimov, Mechanisms for photogeneration and recombination of multiexcitons in semiconductor nanocrystals: implications for lasing and solar energy conversion, *J Phys Chem B* 2006, *110*, 16827–16845.
83. P.V. Kamat, Quantum dot solar cells. Semiconductor nanocrystals as light harvesters, *J Phys Chem C* 2008, *112*, 18737.
84. G. Hodes, Comparison of dye- and semiconductor-sensitized porous nanocrystalline liquid junction solar cells, *J Phys Chem C* 2008, *112*, 17778.
85. R.D. Schaller, M. Sykora, J.M. Pietryga, and V.I. Klimov, Band–band impact ionization and solar cell efficiency, *Nano Lett* 2006, *6*, 424.
86. P.T. Landsberg, H. Nussbaumer, and G. Willeke, Seven excitons at a cost of one: redefining the limits for conversion efficiency of photons into charge carriers, *J Appl Phys* 1993, *74*, 1451.
87. R.D. Schaller, and V.I. Klimov, High efficiency carrier multiplication in PbSe nanocrystals: implications for solar energy conversion, *Phys Rev Lett* 2004, *92*, 186601.
88. R.J. Ellingson, M.C. Beard, J.C. Johnson, P.R. Yu, O.I. Micic, A.J. Nozik, A. Shabaev, and A.L. Efros, Highly efficient multiple exciton generation in colloidal PbSe and PbS quantum dots, *Nano Lett* 2005, *5*, 865–871.
89. R.D. Schaller, M.A. Petruska, and V.I. Klimov, The effect of electronic structure on carrier multipcation efficiency: a comparative study of PbSe and CdSe nanocrystals, *Appl Phys Lett* 2005, *87*, 253102.
90. J.E. Murphy, M.C. Beard, A.G. Norman, S.P. Ahrenkiel, J.C. Johnson, P.R. Yu, O.I. Micic, R.J. Ellingson, and A.J. Nozik, PbTe colloidal nanocrystals: synthesis, characterization, and multiple exciton generation, *J Am Chem Soc* 2006, *128*, 3241.
91. A.J. Nozik, Spectroscopy and hot electron relaxation dynamics in semiconductor quantum wells and quantum dots, *Ann Rev Phys Chem* 2001, *52*, 193–231.
92. W.A. Tisdale, K.J. Williams, B.A. Timp, D.J. Norris, E.S. Aydil, and X.Y. Zhu, Hot-electron transfer from semiconductor nanocrystals, *Science* 2010, *328*, 1543–1547.
93. S. Hubbard, and R. Raffaelle, Boosting solar-cell efficiency with quantum-dot-based nanotechnology, *SPIE Newsroom* 2010, February 8. DOI:10.1117/2.1201002.0025534.
94. X.J. Hao, E.C. Cho, C. Flynn, Y.S. Shen, S.C. Park, G. Conibeer, and M.A. Green, Synthesis and characterization of boron-doped Si quantum dots for all-Si quantum dot tandem solar cells, *Sol Energy Mat Sol Cells* 2009, *93*, 273.
95. J.B. Sambur, and B.A. Parkinson, Multiple exciton collection in a sensitized photovoltaic system, *J Am Chem Soc* 2010, *132*, 2130.

96. K.S. Leschkies, R. Divakar, J. Basu, E. Enache-Pommer, J.E. Boercker, C.B. Carter, U.R. Kortshagen, D.J. Norris, E.S. Aydil, Photosensitization of ZnO nanowires with CdSe quantum dots for photovoltaic devices, *Nano Lett* 2007, *7*, 1793–1798.
97. N. Harris, J. Brewer, G. Shen, D. Wilbert, L. Butler, N. Dawahre, E. Baughman, S. Balci, E. Rivera, P. Kung, and S. M. Kim, Quantum dot functionalized ZnO nanowire/P3HT hybrid photovoltaic devices, *Proceeding of IEEE NANO*. 2011, 4007–4015.
98. A. Kongkanand, K. Tvrdy, K. Takechi, M. Kuno, and P.V. Kamat, Quantum dot solar cells. Tuning photoresponse through size and shape control of CdSe-TiO2 architecture, *J Am Chem Soc*, 2008, *30*, 4007.
99. A. Luque, Will we exceed 50% efficiency in photovoltaics? *J Appl Phys*, 2011, *110*, 031301.

3 Photovoltaics and Nanotechnology
From Innovation to Industry

Sophia Fantechi, Iain Weir, and Bertrand Fillon

CONTENTS

3.1 INTRODUCTION

Europe is well established as the leading worldwide market for photovoltaics (PV) and is also strongly involved in both nanotechnology innovation and industrial applications in the field. Solar PV electricity production is the most obvious energy technology where innovative *nanostructured materials and nanotechnology* are contributing to technology development and to new sustainable, competitive industrial processes.

Although PV has been funded by many programs at the European level for many years, in the EU's Seventh Framework Programme for Research and Technological Development (FP7), various nanotechnology and nanosciences, knowledge-based multifunctional materials, and new production processes and devices (NMP) projects support PV, in particular through various nanotechnology breakthrough applications. As a matter of fact, nanotechnologies and nanomaterials are used in several ways in a variety of PV projects funded by European Commission programmes. Projects are often fragmented and results are not fully shared with other communities. There is, therefore, the need to join forces and identify common research goals for each given technology of industrial relevance to efficiently support PV innovation in the European Research Area.

The PV projects in the NMP, Energy, Information and Communication Technologies (ICT), and Research Infrastructures Programmes, half-way through FP7, were brought together in a workshop,* for the first time, with those of the Intelligent Energy Europe (IEE) programme of the EU's Competitiveness and Innovation Framework Programme (CIP), running in parallel with FP7 for the period 2007–2013. The workshop has enabled the nanotechnology and PV communities in Europe to identify joint collaboration and application areas and gain new contacts and ideas for strategic industrial partnerships. A European PV Clusters initiative has continued since the workshop, supported by individual project participants. Recent activities are summarized in the European PV Clusters website (http://eupvclusters.eu, which also announced the second EU PV Clusters Workshop and General Assembly of November 2013).

This chapter gives an overview and analysis of nanotechnology for PV in Europe based on a portfolio of over 40 European projects (including in particular all the NMP-funded PV projects from 2008 to the end of 2012, to bring forward the global picture of PV research and innovation in Europe and to highlight the impact of nanotechnology in this area as a leading-edge opportunity for the European PV industry.

A strong portfolio of PV projects addressing a number of the key PV technology development and innovation issues is supported by the European Commission through FP7 and CIP-IEE. The total funding for the projects analyzed is EUR 200 million, of which EUR 185.8 million are from FP7 and EUR 14.2 million from CIP-IEE. It is now important that the results of these projects (which represent a total cost of EUR 290 million, with EUR 271.5 million for the FP7 projects and EUR 18.5 million for the CIP-IEE projects) are effectively exploited and turned into real innovation to optimize the economic impact in Europe from this investment—essentially the development of a strong European PV manufacturing industry.

3.2 EUROPEAN PHOTOVOLTAICS CLUSTERS OF PROJECTS

The FP7 and CIP projects supporting PV in the NMP, energy, ICT, people, research infrastructures, and IEE programmes can be grouped in the following *Clusters and Subclusters*:

Cluster 1: Wafer-based PV cells—First-generation semiconductor PV cells: crystalline, wafer-based solar cells mainly made from silicon materials, that is, semiconductor wafer-based Si PV technologies

Cluster 2: Thin film PV cells—Second-generation PV cells: thin film solar cells, such as amorphous or microcrystalline silicon, copper indium gallium (di)selenide (CIGS), and cadmium telluride
- Subcluster 2.1: Innovative or improved PV manufacturing processes
- Subcluster 2.2: Innovative PV materials

* EU PV Clusters Workshop, Photovoltaics and Nanotechnology: From Innovation to Industry, Aix Les Bains, October 2010.

Cluster 3: Third-generation PV cells—PV cells obtained through the application of advanced concepts and materials, such as various nanomaterials (including quantum dots, super lattices, nanoparticles, nanowires, dyes, and organic/polymer materials), hybrid organic-inorganic concepts, biomimetic materials, and combinations of these
Subcluster 3.1: Nanodots- or nanowire-based PV
Subcluster 3.2: Organic PV cells or dye-sensitized solar cells (DSCs)
Subcluster 3.3: Innovative nanostructures
Cluster 4: Concentrator PV cells—PV based on optical concentration and tracking
Cluster 5: Innovative installations and grid interconnections—PV for distribution systems
Cluster 6: Production equipment and processes—Demonstration of high-performance equipment and processes for PV
Cluster 7: Industry support—Cross-cutting issues addressing infrastructure, market, quality, legal and training aspects of PV

The projects are categorized, using project acronyms, by program and cluster in Figure 3.1.

3.3 SCOPE OF THE PHOTOVOLTAICS PROJECTS

The PV projects listed in Figure 3.1 range from the development of nanomaterials and enhancement of specific properties of PV materials to upscaling of processes to industrial pilot lines and analysis of the performance of PV systems in the field. Projects can be mapped against a schematic of the PV supply chain to show how they are spread across the whole manufacturing chain, as shown in Figure 3.2.

This analysis shows the following:

- A strong emphasis on PV material and cell development and characterization—aiming to develop technologies to enhance efficiencies.
- A focus on enhancing efficiencies of established technologies through novel applications of materials, particularly nanostructures.
- Fewer projects focusing on module, array, and systems development—this is to be expected due to the current status of PV developments and the importance of improving the basic conversion efficiencies of PV materials and cells.

There is also significant variability in the maturity of the projects—ranging from projects that have been completed to some that have only recently started in 2012, with potential opportunities for new projects to benefit from the results of those projects that have finished.

Participation of hundreds of organizations from around Europe is also seen in these projects. Germany provides the largest number of partners, which probably reflects the scale and maturity of the German industry while the majority of participants from Eastern Europe are from academia, which may highlight a lack of industrial development in PV in these countries.

Cluster		Subcluster		EC Program							
				FP7							CIP
				NMP	NMP / Energy R&D (shared)	Energy: R&D	Energy: Demonstration	ICT	People	Research infrastructures	IEE
1	Wafer-based PV cells			HiperSol		HETSI	Ultimate				
2	Thin film PV cells	2.1	Innovative or improved PV manufacturing process	ALPINE NOVA-CI(G)S	R2RCIGS ScaleNano	HIGH-EF ThinSi R2M-Si		SUGAR			
2	Thin film PV cells	2.2	Innovative PV materials		Fast Track	PolySiMode SILICON-Light HELATHIS hipoCIGS					
3	Third-generation PV cells	3.1	Nanodots- or nanowire-based PV cells	SNAPSUN NanoPV NASCEnT AMON-RA		ROD-SOL IBPOWER		LIMA			
3	Third-generation PV cells	3.2	Organics PV cells or DSC	SANS	INNOVASOL EPHOCELL	ROBUST DSC		HIFLEX SUNFLOWER X10D ROTROT	Establis		
3	Third-generation PV cells	3.3	Innovative nanostructures	NanoSpec NanoCharm	SOLAMON			PRIMA			
4	Concentrator PV cells					APOLLON NACIR ASPIS					
5	Innovative installations and grid interconnections						MetaPV				
6	Production equipment and processes						SOLASYS PEPPER PV-GUM				
7	Industry support									SOPHIA	PV LEGAL QualiCert PVs in BLOOM PV-NMS-NET PVTRIN
			Total projects	10	6	14	5	7	1	1	5

FIGURE 3.1 Segmentation of projects by program and cluster (2012).

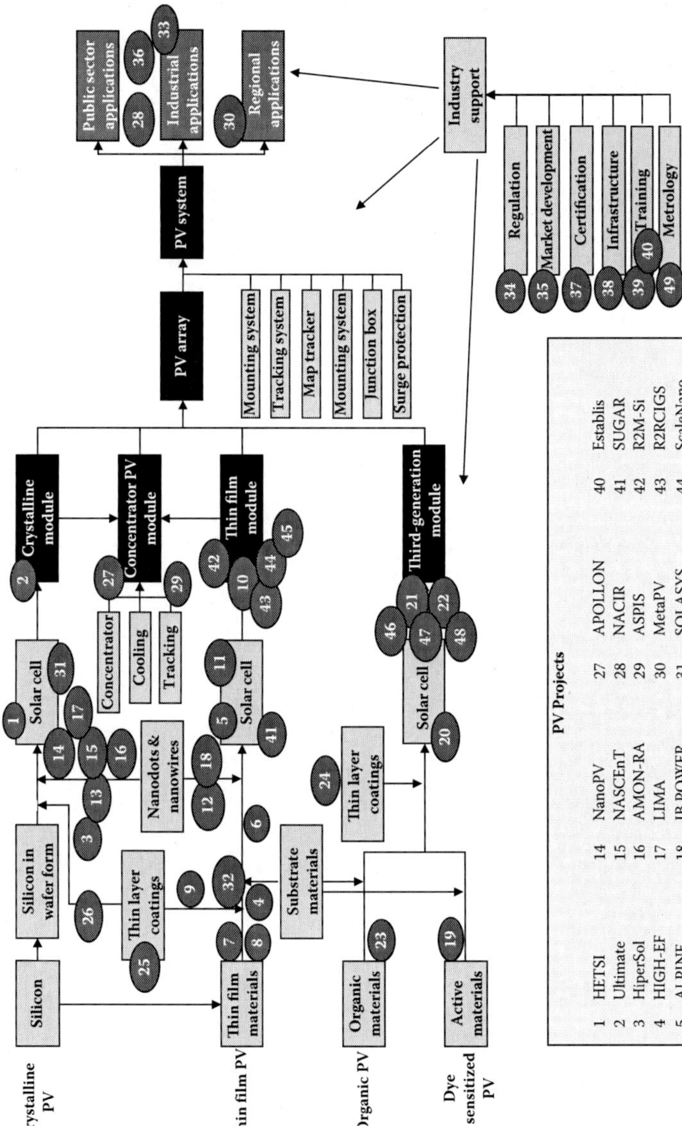

PV Projects

1	HETSI	14	NanoPV	27	APOLLON	40	Establis
2	Ultimate	15	NASCEnT	28	NACIR	41	SUGAR
3	HiperSol	16	AMON-RA	29	ASPIS	42	R2M-Si
4	HIGH-EF	17	LIMA	30	MetaPV	43	R2RCIGS
5	ALPINE	18	IB POWER	31	SOLASYS	44	ScaleNano
6	ThinSi	19	INNOVASOL	32	PEPPER	45	Fast Track
7	NOVA-CI(G)S	20	SANS	33	PV-GUM	46	SUNFLOWER
8	PolySiMode	21	ROBUST DSC	34	PV-LEGAL	47	X10D
9	SILICON-Light	22	HIFLEX	35	PV-NMS-NET	48	ROTROT
10	HELATHIS	23	EPHOCELL	36	PVs in BLOOM	49	NanoCharm
11	hipoCIGS	24	SOLAMON	37	QualiCert		
12	ROD-SOL	25	PRIMA	38	SOPHIA		
13	SNAPSUN	26	NanoSpec	39	PVTRIN		

FIGURE 3.2 **(See color insert.)** Clustering of projects by supply chain activity (2012).

3.4 MAJOR FINDINGS AND POSSIBLE BREAKTHROUGHS

The clusters include a number of projects that have recently started as well as several projects that have been completed—the current maturity of projects is summarized in the following discussion, categorized by funding source (Figure 3.3) and PV cluster (Figure 3.4).

The major results and findings to date have therefore been achieved in the more mature projects. These include, for example, the following:

- HETSI: 20% efficiency achieved in small area heterojunction silicon solar cells
- Ultimate: 18% efficiency achieved in thin silicon solar cells
- ThinSi: Production of silicon powder–based substrates proven
- hipoCIGS: 17.6% efficiency of CIGS solar cell on polyimide (new world record)
- INNOVASOL: Promising initial data for novel excitonic solar cell devices
- ROBUST DSC: 12% efficiency achieved for small-scale DSCs
- NACIR: First stand-alone concentrator PV systems installed
- MetaPV: Demonstration of local PV networks

It is expected that a number of the other projects will, in time, present similarly impressive results. The development of new techniques in a number of projects (e.g., optimized silicon cell efficiencies through controlled nanoparticulate synthesis and bandgap engineering [SNAPSUN], optimized light absorption using plasmonic layers [LIMA], and novel manufacturing techniques [e.g., HIGH-EF]) is expected to contribute to further major breakthroughs and subsequent exploitation.

It is further expected that these results will lead to performance breakthroughs in a range of solar cell technologies, although it is expected that for some of the more innovative projects (e.g., NASCEnT, SNAPSUN, and NanoPV) it could be 10–15 years before industrial applications are developed.

These observations do, however, underline that European PV R&D activities are a strong player in the global market. Projects are addressing the key needs of the industry and European researchers are demonstrating their abilities to produce state-of-the-art results.

3.5 RELEVANCE OF PROJECTS TO GLOBAL PV INDUSTRY NEEDS

The key objectives of EC-funded projects can be compared with global PV trends and the R&D needs specified in key technology roadmaps. In this context, it is important to look beyond the European PV Technology Platform roadmap to seek a global perspective.

Key global PV developers are striving to achieve higher and higher cell conversion efficiencies. Enhanced performance figures are continually being claimed—for all types of PV cell. European participation in several of these achievements (e.g., Oerlikon Solar, ZSW) is a very positive observation, although the number of

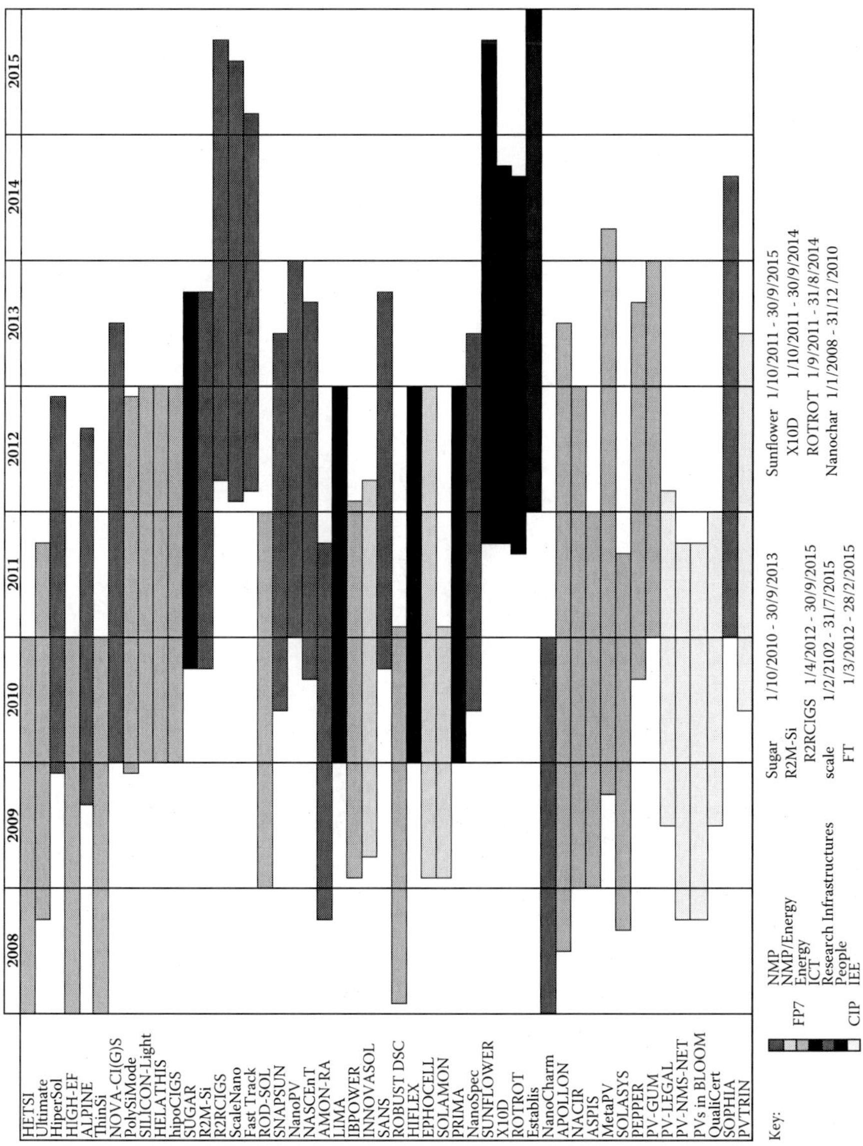

FIGURE 3.3 **(See color insert.)** Timescale and duration of projects by programme (2012).

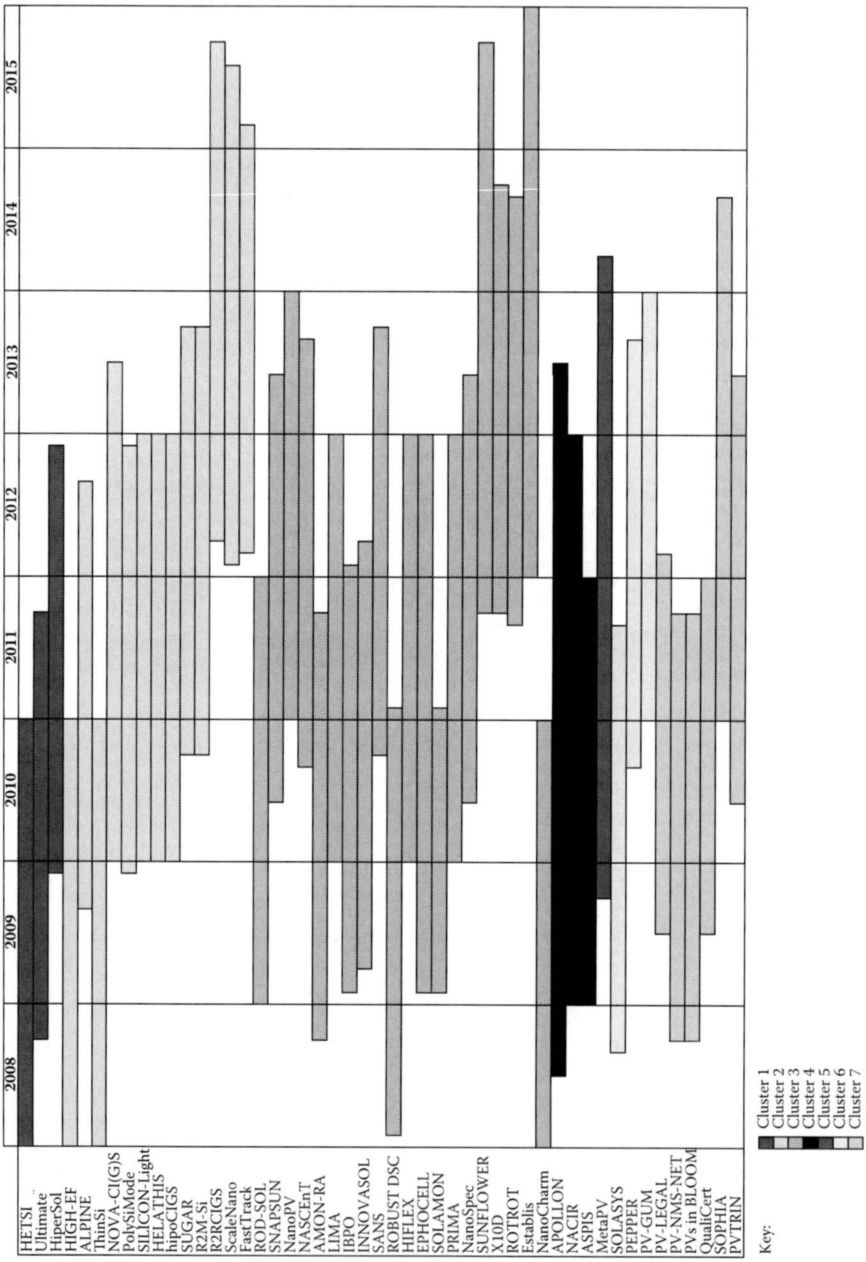

FIGURE 3.4 **(See color insert.)** Timescale and duration of projects by photovoltaic (PV) cluster (2012).

U.S. and Japanese company achievements clearly highlights the progress of the PV industry in these countries.

It is helpful to divide the portfolio of EC-funded projects into two groups.

Clusters 1–3 comprise a number of projects that aim to enhance the properties of materials to increase the efficiency of PV cells and, as a result, reduce the cost per watt of energy production. This is entirely consistent with the goals of global technology roadmaps and research elsewhere.

For example, the International Energy Agency (IEA) technology roadmap for solar PV energy* predicts increasing efficiency rates for all types of PV cells over the period to 2030 and the JRC PV Status Report† highlights the need for reduced materials consumption per solar cell. Furthermore, the IEA roadmap highlights the following key R&D issues:

- Crystalline silicon
    - New silicon materials and processes
    - Cell contacts, emitters, and passivation
    - Improved device structures
- Thin film technologies
    - Large area deposition
    - Improved substrates and transparent conductive oxides (TCO)
    - Advanced materials and concepts
- Emerging technologies
    - Encapsulation of organic-based devices
    - Processing, characterization, and modeling of nanostructured materials and devices

A number, if not all, of projects in clusters 1–3 address at least one of these R&D issues.

Clusters 4–6 focus on concentrator cells, innovative installations, and production equipment and processes. These clusters focus on systems (rather than materials) development and the application of these systems in real situations. Here, we can also note a strong consistency between research activities and those highlighted in the IEA roadmap. For example, the IEA roadmap highlights the need for

- Over 45% efficiency in concentrator PV cells
- Low-cost, high-performance solutions for optical concentration and tracking

We therefore conclude, based on this overview, that the projects being supported by the European Commission are fully consistent with the development activities being pursued by the global industry.

---

* Solar Photovoltaic Energy Technology Roadmap, ©OECD/International Energy Agency 2010. This roadmap can be considered to be representative of global activity as it consulted with, or assessed output from, PV organizations worldwide.

† PV Status Report 2010—Research, Solar Cell Production, and Market Implementation of Photovoltaics, European Commission, Joint Research Centre, Ispra, August 2010.

3.6 RELEVANCE OF NANOTECHNOLOGY PROJECTS TO EUROPEAN PV POLICY TARGETS AND STRATEGIES

The low-carbon economy targets for Europe set by the European Commission* establish clear goals for renewable energy generation and for efficient energy use. These targets are as follows:

- A reduction of at least 20% in greenhouse gas emissions by 2020
- A 20% share of renewable energies in EU energy consumption by 2020
- A 20% reduction of the EU's total primary energy consumption by 2020 through increased energy efficiency

It has already been asserted that solar energy generation can make a major contribution to these targets. The European Commission Strategic Energy Technology (SET) Plan technology roadmap† estimates that up to 15% of EU electricity can be generated by solar power by 2020 while others‡ indicate that a figure of 12% is achievable over the same period. Key technology requirements identified are as follows:

- PV systems
  - Increased conversion efficiency, stability, and lifetime
  - High-yield, high-throughout (and low cost) manufacturing
  - Development of advanced concepts and new generation PV systems
- Integration of PV-generated electricity
  - Develop and validate innovative, economic, and sustainable PV applications
  - Develop grid interfaces and storage technologies capable of optimizing the PV contribution to EU electricity generation

These two documents further highlight that achieving these targets will primarily be based on enhancing current (silicon, thin film, and concentrator) technologies and will require an investment in technology and manufacturing development projects together with market development incentives. It is expected that these incentives will result in lower solar energy generation costs and thus further market growth. A dependence on current technologies is also underlined by the IEA technology roadmap, as shown graphically in Figure 3.5.§

This suggests that the current investment in R&D programs focusing on novel and emerging PV concepts, including nanotechnology developments, will have the most

---

* 20 20 by 2020, Europe's climate change opportunity, European Commission, January 23 2008.

† Investing in the Development of Low Carbon Technologies (SET-Plan)—A Technology Roadmap, SEC (2009) 1295, 07/10/2009.

‡ Implementation Plan of the Solar Europe Industry Initiative (SEII)—May 2010—European Photovoltaic Technology Platform Solar Europe Industry Initiative (SEII), Summary Implementation Plan 2010—2012, EPAI and the Photovoltaic Technology Platform, February 2010.

§ Solar Photovoltaic Energy Technology Roadmap, ©OECD/International Energy Agency 2010.

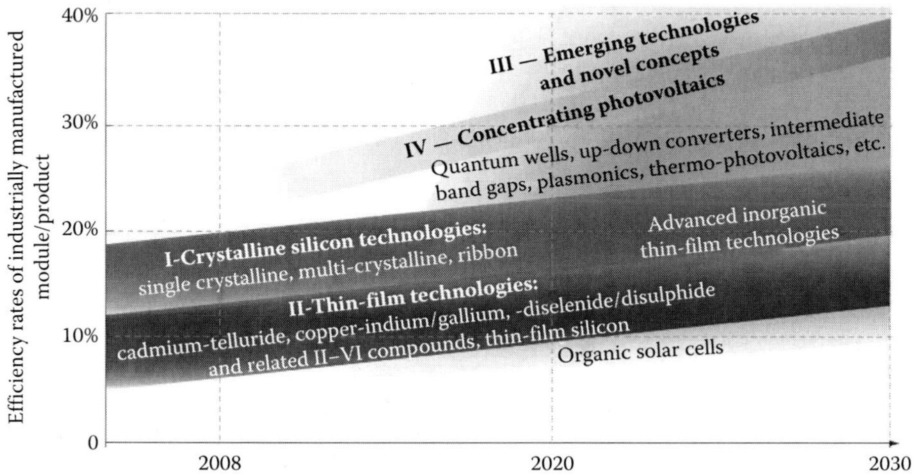

FIGURE 3.5 **(See color insert.)** Expected impact of photovoltaic technologies. (Data from Solar Photovoltaic Energy Technology Roadmap, ©OECD/International Energy Agency 2010.)

significant impact in the longer term. It is possible that the nanotechnology-based process development activities in the recent European Commission–funded projects* (e.g., Fast Track, R2RCIGS, and ScaleNano) may offer a shorter-term impact if projects are successfully completed.

The development in performance of established solar cells designs is likely to make a much higher short-term contribution to PV energy generation through a series of technology and manufacturing process enhancements as highlighted by the Solar Europe Industry Initiative†—essentially continuing the ongoing performance improvements of these cells, as shown in the numerous publications by NREL.‡

There is similar evidence (e.g., Suniva Corporation§) of company-specific roadmaps to achieve optimized efficiency in manufacturing.

The Solar Europe Industry Initiative Implementation Plan¶ defines targets for PV module efficiencies for crystalline silicon, thin films, and concentrators, as shown in Figure 3.6. This implementation plan also identifies a number of project areas to contribute to achieving the targets listed. These targets and proposed projects underline the drive to optimize cell efficiencies and reduce costs of energy generation.

---

* A joint call with the Energy Program.

† Implementation Plan of the Solar Europe Industry Initiative (SEII)—May 2010—European Photovoltaic Technology Platform Solar Europe Industry Initiative (SEII), Summary Implementation Plan 2010—2012, EPAI and the Photovoltaic Technology Platform, February 2010.

‡ For example, Lawrence Kazmerski, National Renewable Energy Laboratory (NREL), April 2010, as shown at http://en.wikipedia.org/wiki/File:PVeff(rev110408U).jpg.

§ Developing novel, low-cost high-throughput processing techniques for 20% efficient monocrystalline silicon solar cells, Rohatgi and Meier, Photovoltaics International, November 2010.

¶ Implementation Plan of the Solar European Industry Initiative (SEII)—May 2010—European Photovoltaic Technology Platform.

		2007	2010	2015	2020
Turn-key price large systems (€/Wp)		5	2.5	2	1.5
PV electricity generation cost in Southern EU (€/kWh)		0.30	0.13	0.10	0.07
Typical PV module efficiency range (%)	Crystalline silicon	13%–18%	15%–20%	16%–21%	18%–23%
	Thin films	5%–11%	6%–12%	8%–14%	10%–16%
	Concentrators	20%	20%–25%	25%–30%	30%–35%
Inverter lifetime (years)		10	15	20	>25
Cost of PV + small-scale storage (€/kWh) in Southern EU (grid-connected)		--	0.35	0.22	<0.15
Energy payback time (years)		2–3	1–2	1	0.5

FIGURE 3.6 Future photovoltaic (PV) performance targets.

A number of the projects supported by the European Commission NMP Programme are addressing objectives that are consistent with these targets, albeit at the research or prototype stage. For example, efficiency targets and demonstration activities proposed by some of the more recently started nanotechnology-based PV projects* are as follows:

- NanoPV
  - Crystalline silicon efficiency >20%
  - Thin film silicon efficiency >15%
  - Potential energy cost <EUR 1/watt
  - Demonstrated (by April 2014) at proof of concept level
- NASCEnT
  - Silicon tandem solar cells with efficiency of >30% are claimed to be achievable
  - Demonstrated (by September 2013) at proof of concept level
- NanoSpec
  - Improvement in efficiency of silicon solar cells by 10%–20% by enhanced light harvesting
  - Demonstrated (by July 2013) at a laboratory scale proof of concept level
- SNAPSUN
  - Demonstration of enhanced cell efficiency using band gap engineering
  - Potential energy cost <EUR 0.5/W
  - Demonstrated (by June 2013) at a laboratory scale proof of concept level
- SANS
  - Demonstration of long-life sensitizer–activated nanostructured solar cells
  - Show stability that enables 20 years outdoor operation
  - Achieve commercially viable solar technology

* Presentations at the EU PV Clusters Workshop, Photovoltaics and Nanotechnology: From Innovation to Industry, Aix Les Bains, September–October 2010.

These projects continue the developments achieved in earlier PV-focused nanotechnology projects under NMP,* which include, in order of project commencement, the following:

- AMON-RA
  - Demonstration of dual junction cell on silicon with enhanced efficiency
- EPHOCELL
  - Enhanced down-shifting and up-conversion for PV
- SOLAMON
  - Development of nanocomposite materials for high-efficiency third-generation solar cells
- ALPINE
  - Improvement of fiber laser systems for scribing of PV
- HiperSol
  - Modeling of interfaces to achieve high-performing solar cells
- INNOVASOL
  - Development of radically new nanostructured materials for excitonic solar cells
- NOVA-CIGS
  - Development of novel, low-cost deposition methods for CIGS layers—with the aim of ultimately achieving module costs of <EUR 0.8/W

These projects demonstrate that European Commission investment under the NMP Programme is highly consistent with some of the themes in the Solar Europe Industry Initiative. This was strongly reinforced by the focus on the development and upscaling of innovative PV cell processes and architectures in the 2011 Work Programme,† which has supported of three large projects (Fast Track, R2RCIGS, and ScaleNano), with a total funding of over €34 million.

Similar developments are also being supported by the ICT and Energy Programmes—several of the projects listed in Section 2 (e.g., HETSI, Ultimate, hipo-CIGS, ROBUST DSC, and NACIR) are based on nanotechnology and are funded by the Energy Programme. The three projects funded by the ICT Programme (LIMA, HIFLEX, and PRIMA) also show a fundamental nanotechnology basis.

However, there are numerous opportunities for the NMP Programme to further support European industry to develop competitive PV technologies. The Solar Europe Industry Initiative identifies development needs in all key technology areas. There is a strong emphasis in these requirements on applied development and innovation, that is, a focus on optimizing commercial technologies and associated manufacturing processes, rather than developing longer-term novel technologies. This is extremely logical considering the need to make significant industrial progress by 2020. These requirements essentially define the areas for further work and offer a focus for future NMP funding actions.

---

* EPHOCELL, SOLAMON, and INNOVASOL were jointly funded by the NMP and Energy Programs.

† A joint call with the Energy Program.

This applied emphasis is extremely important to support, as much as possible, the development of a strong European PV industry to address European market demands. It seems to be assumed in some policy documents that a significant European industry will develop to satisfy the need for enhanced PV generating capacity, but this will not necessarily be the case. Already Europe is the major global region for installed solar energy generation capacity* but only holds a small share of the global PV manufacturing capacity (less than 25% in 2010, with projections of a 14.6% share in 2015).† In contrast, by 2015, it is expected that the Far East will manufacture in excess of 50% of the global PV module output and already the United States and Japan are considering initiatives to ensure development of their indigenous industries. Recent developments in the European industry suggest these predictions for Europe are optimistic. Therefore, it is important that the market incentives implemented to develop solar power–generating capacity in Europe are supplemented by technology and industry development initiatives to ensure a strategic advantage and subsequent impact for the European PV manufacturing industry.

3.7 COMMERCIALIZATION OPPORTUNITIES

The analysis of project objectives and scope, especially those projects focused on materials developments and PV cell optimization, shows that there are a number of common issues between projects that are fundamental to achieving effective PV cells. These include the following:

- Improving TCO properties
- Optimization of rear contact cells
- Surface texturing, to optimize light capture
- Understanding behavior at interfaces
- Developing and applying advanced characterization methods

A number of the consortia are structured to include strong industrial participants to exploit these results—there are over 100 companies involved in the portfolio of projects. These industrial representatives include the following:

- Large companies such as ST Microelectronics and Philips
- Specialist PV companies, for example, Isofoton, Oerlikon, and Flisom
- Specialist technology suppliers, for example, Aixtron, Trumpf, SAFC Hitech, Oxford Instruments Plasma Technology, and Dyesol

Thus there are, in the majority of the projects, clear exploitation partners for commercially attractive technologies—as long as the participant companies have an

---

* European Photovoltaic Industry Association, Global Market Outlook for Photovoltaics Until 2016, 2012.

† PV Status Report 2010—Research, Solar Cell Production and Market Implementation of Photovoltaics, European Commission Joint Research Centre, Ispra, August 2010.

interest in further investment and development. This should lead to effective commercialization of attractive project outcomes.

A number of projects are focusing on the early stages in the PV supply chain (e.g., materials or thin film coatings). Successful exploitation of results in these projects will be dependent on companies further down the supply chain being interested in adopting the technology—possibly through licensing. A number of projects have already involved such parties in the consortium (e.g., ST Microelectronics in NASCEnT and SNAPSUN and Isofoton in ThinSi), which should also optimize the commercial potential. Some projects funded more recently (Fast Track, R2RCIGS, and ScaleNano) are focusing on developing pilot production lines for thin film technologies, again supporting exploitation of novel PV systems.

3.8 APPLICATIONS AND INNOVATION PERSPECTIVES

The market for PV is expected to continue to grow significantly over the next 10 years, with growth rates depending on the incentive environment in different regions as well as technology breakthroughs. The European Photovoltaic Industry Association (EPIA)* presents the market growth predictions under different scenarios (Figure 3.7).

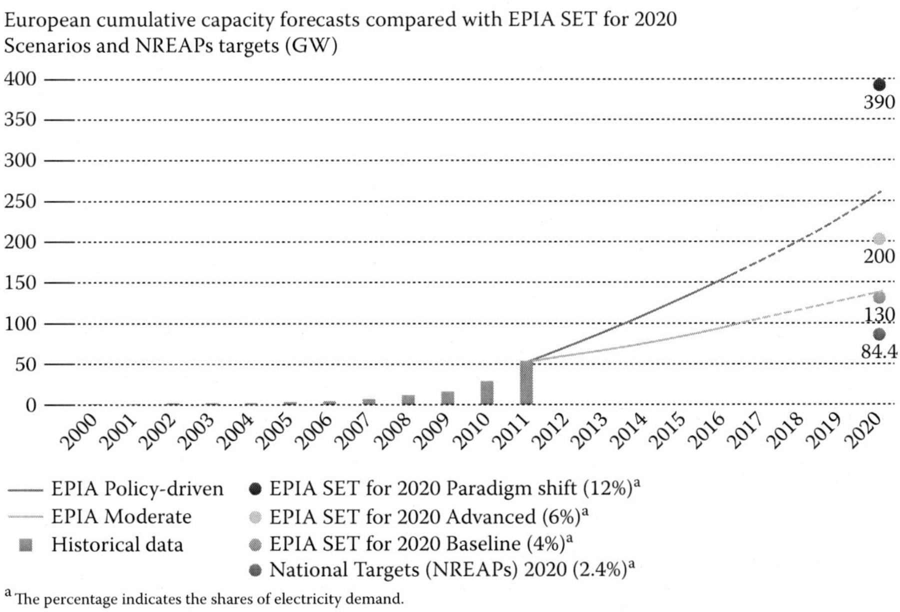

FIGURE 3.7 **(See color insert.)** Alternative futures for the global photovoltaic (PV) market. (Data from European Photovoltaic Industry Association, Global Market Outlook for Photovoltaics Until 2016, 2012.)

* European Photovoltaic Industry Association, Global Market Outlook for Photovoltaics Until 2016, 2012.

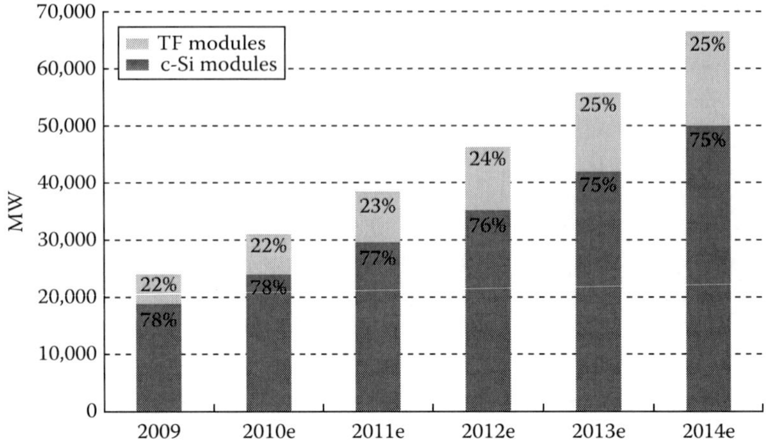

FIGURE 3.8 Photovoltaic market segmentation by technology. (Data from European Photovoltaic Industry Association, Global Market Outlook for Photovoltaics Until 2014, May 2010.)

Evidence to date suggests that strong incentive frameworks on a national basis are the major catalyst for market growth. Furthermore, EPIA* segments the PV market by technology as given in Figure 3.8.

This shows an increase in the thin film share of the market over the period to 2014. This figure also indicates that the market share of other more novel technologies is assumed to be negligible over the period. Other EPIA analyses† suggest that novel PV structures (including organics) will start to achieve market share beyond 2015, as shown in Figure 3.9.

Significant breakthroughs in technologies (from EC funded or other projects) that enhance the performance (efficiency) of different PV systems will influence the growth of the PV market and the market share of different technologies—so it should be assumed that significant project achievements will be adopted by the industry. Already issues relating to technology breakthroughs are influencing market dynamics. Some commentators‡ have indicated that the market growth (and share) of thin film technologies will be lower than predicted due to the difficulties in achieving a technology breakthrough while others§ have highlighted the imperative of enhanced efficiency and associated reduced costs to achieve market share and predicted that silicon will retain its dominant position.

The major challenge for the PV industry, as indicated in Section 3.6, is to optimize the efficiency of cells, modules, and systems and thus minimize the cost of energy generation. As a result, success with these projects, supported by the European Commission, will provide highly relevant technologies for industry to adopt and apply. Successful project results will be highly relevant to industry, provided a series of technology challenges can be overcome.

* European Photovoltaic Industry Association, Global Market Outlook for Photovoltaics Until 2014, May 2010.

† European Photovoltaic Industry Association—Greenpeace International, Solar Generation VI, 2010.

‡ Solar Energy: Growth Opportunities for the Semiconductor Industry, IC Insights, May 2009.

§ Module Cost Structure Breakdown, Lux Research, November 2010.

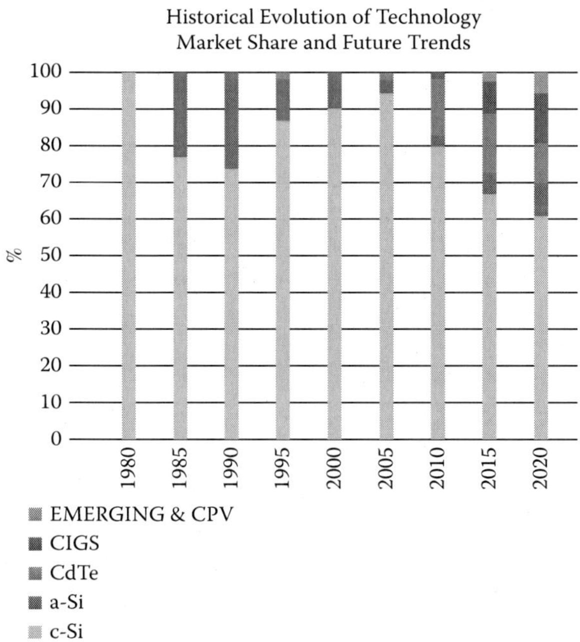

FIGURE 3.9 (**See color insert.**) Long-term photovoltaic market segmentation by technology. (Data from European Photovoltaic Industry Association—Greenpeace International, Solar Generation VI, 2010.)

However, it should be noted that exploitable results from a number of projects, especially those that are focusing on novel materials and structures, are likely to take up to 20 years before they are applied in industry.

3.9 ROLE OF NANOTECHNOLOGY IN ADDRESSING TECHNOLOGY CHALLENGES IN THE PV SECTOR

There have already been a number of nanotechnology developments to address PV technology challenges. These challenges are linked with processes or materials whatever the cluster (first generation, bulk silicon; second generation, thin film technologies; and third generation, organic PV [OPV], nanotextured cells).

First-generation semiconductor PV cells (crystalline, wafer-based solar cells, mainly made from silicon materials) are currently facing issues related to impurities and passivation layers. The HiperSol project, for example, indicated a limited understanding of what parameters determine the passivation quality and interface to obtain a higher efficiency of the cells. Also, the fine-tuning of the surface roughness could be improved, as well as the standardization of back contact solar cell interconnections or alternative material interconnections. The Ultimate and HETSI projects have looked for such development. The TCO material needs further improvement to avoid aging issues and for better resistivity and surface quality.

For second-generation cells, based on thin film technologies, Si epitaxial growth requires enhanced control and understanding. For example, the large stresses that form in the layers due to expansion coefficient differences between glass and the Si layer still represent one of the main bottlenecks. Another challenge is process control and control of layers that can accommodate growth on rough substrates with appropriate crystallinity profiling preventing crack formation. Posttreatment, like rapid thermal annealing and hydrogen passivation, is still in the development phase and needs improvement and better control (project PolySiMode). Laser processes for texturing, scribing, and patterning to replace mechanical processes are still facing issues like shunting induced by the heat-affected zones caused by laser treatment. Advanced beam characteristics or specific pulses need to be optimized to validate their interest.

Low-cost substrates also represent one of the main requirements for thin film technologies. But different types of bottleneck exist, whatever the thin film technology. As an example, even if an Si powder film substrate is used, full compatibility with deposition/crystallization needs to be well controlled. The project ThinSi addressed this challenge to produce low-cost thin film Si cells and modules. This low-cost substrate is also a main target for CIGS thin film technology. Projects hipo-CIGS, NOVA-CIGS, ScaleNano, R2RCIGS, and Fast Track are looking for such a low-cost approach and are facing bottlenecks linked to interface compatibility, high temperature control, and deposition technique. Although these low-cost technologies are being developed for thin film technologies, the described bottlenecks need to be overcome before market launch of such processes. In fact, tape casting, thermal spraying, wet coating, or printing technologies are some of the low-cost approaches that can be realized at high speed, but they need to provide a better reliability, enhanced control of the deposition parameters and of the morphology of produced layers, fewer cracks and less impurities formation.

Also, to increase absorption, either several rounds of Solid Phase Epitaxy or light trapping by the respective structures (still in the optimization stage) are required. Some first results obtained at laboratory scale show good nanotexturization to improve the absorption. Some projects like HIGH-EF, SILICON-Light, SOLAMON, and PolySiMode are looking at such approaches. Fabrication of cells on substrates with high aspect ratio surface texture is in progress. Like first-generation cells, TCO layers are one of the bottlenecks for thin film technologies.

For third-generation cells (OPV, nanotextured cells), there are material and process challenges. For OPV or DSSC, control of thin layers (below 100 nm) and interfaces between different layers to minimize defects needs significant improvement to offer an enhancement in cell efficiency (>10%) and an increase of the durability of the cells (>10 years). The morphology of the polymer or composite needs to be better controlled at the nanoscale to increase the free carrier lifetime, which is the primary goal whatever the generation of solar cell (e.g., bulk Si, thin Si, and CIGS). An improved control of the purity of the polymer, of different layer interfaces, and of barrier properties is still the main challenge for OPV cells. Projects like RotRot, X10D, and Sunflower are developing higher-efficiency OPV cells with a higher stability. The degradation mechanisms are still not well understood. Projects like EPHOCELL are investigating them. Similar to thin film technologies, nanotexturization needs to be better controlled and to become an efficient process to increase the optical absorption of the active layer. Projects like PRIMA and LIMA are looking for the best shape (e.g., cylindrical and spherical) and the best position of the nanoparticles in the cell and also for the best architecture of the final cells.

Third-generation solar cells based on nanotexturization with bandgap engineering might play an important role in achieving the required spectral modification and higher-efficiency solar cells. Functional devices have been demonstrated, but their efficiencies are still far below expectations. Good control of the nanotexturization process is the main issue. A large number of projects are addressing the development of such third-generation high-efficiency solar cells. Nanotexturization could be obtained with wet chemical etching of Si (e.g., ROD-SOL and HIGH-EF) or with the growth of silicon nanowires or nanodots (projects like SNAPSUN, NanoPV, and NASCEnT). Better control of nanoparticle size, density, and temperature deposition will offer improved bandgap engineering and eventually higher solar cell efficiencies (>30%). The development of pilot scale cost-effective processes to implement both enhanced standard solar cells and solar cell based on nanomaterials is now being considered. This development is an important part of the required value chain. A first set of three new projects (ScaleNano, R2RCIGS, and Fast Track) that focus on processes able to produce thin film solar cells with nanotexturation started in 2012. These projects will develop the advanced nanotechnologies required for the second generation of solar cells. But these projects should be seen as the first stage. To reach the targeted 30% efficiency, other pilot scale developments of nanotechnologies will be essential.

However, work to date has highlighted the need for more investment in research and development and enables us to make recommendations on a number of nanotechnology research priorities, which highlight the role of nanotechnology to overcome such barriers. These are summarized in the following for the three different generations of PV technologies.

For first-generation solar cells, which constitute a more mature technology, nanotechnology could offer some improvements. However, and first, to increase the free carrier lifetime, a better understanding of the germination phenomena is required. Second, for improved process control, reduction of impurities during casting/wafering, reduction of lattice defects, and the use of pure/inert crucibles are necessary. A better control of the mechanisms at the nanoscale could offer a cost reduction for the global process chain. In fact, the current processes require a high level of energy input, particularly for the silicon refining steps, so that a key concern is to reduce the energy input per Wp of output. In this respect, new process routes (e.g., plasma, segregation, purification, and fluidized bed reactor) should be investigated to develop potentially low-cost feedstock materials. In parallel, direct epitaxial growth of silicon wafers onto substrates from the gaseous phase (e.g., silanes, trichlorosilanes, polysilianes, or polychlorosilanes) should also be investigated. Actually, epitaxial growth is a nanoscale phenomenon that needs to be better controlled and understood, whatever the generation of Si cells (i.e., bulk or thin film technology).

Also fundamental material behavior, through an understanding of the material interfaces, interdiffusion processes, inhomogeneities, and grain boundary effects, should be further investigated. Higher bandgap width engineering, through new inorganic, organic, or nanomaterials, is a fundamental research topic that needs deeper analysis.

For thin film technologies, on the other hand, interface control and stoichiometry of the absorber layer are areas that require further development. A better understanding of the chemistry of formation of homogeneous absorbers, buffer layers, and processes for doping materials is needed to offer lower-cost and higher-efficiency thin film solar cells. To obtain this low cost, high-throughput deposition processes must be developed. Nonvacuum deposition processes can replace vacuum-based thin layer deposition processes. This development can be applied to active absorber layers, TCO, antireflective coatings, and antisoiling coatings. Also interfaces between layers need to be well controlled. Advanced deposition technique for fast, large area, low-waste, low-energy solar cells will require the development of new equipment, processes, and tooling. The control of these processes at the nanoscale is a key factor to improve efficiency and durability of the final solar cells. A number of these issues are being addressed in the recently funded Fast Track, R2RCIGS, and ScaleNano projects that are all developing pilot production lines that use nanotechnologies to achieve sustainable thin film PV manufacturing.

Light trapping and guidance (up and down) and spectral conversion and exploitation of plasmonic effects are shown to be growing activities in the European PV projects portfolio. In this field, it is expected that there will be a substantial improvement of the performance of existing solar cell technologies in the coming decade, through the inclusion of up- or down-convertors or through the exploitation of plasmonic effects. The application of such effects in PV is definitely still at a very early stage, but the fact that these effects can be tailored to boost existing solar cell technologies by merely introducing modifications outside the active layer represents a valuable approach that would reduce the time-to-market considerably. The development of such technologies would require metallic nanoparticle synthesis with control over size, geometry and functionalization, and better stability of enhancing layer materials.

Finally, third-generation cells, based on quantum wells, quantum wires, quantum dots, and nanoparticles in a host semiconductor, imply the development of (dry, wet) deposition technologies, nanoparticles synthesis techniques, and metallic intermediate band bulk materials (both through nano-based and bulk-like approaches). To develop high-efficiency (>30%), low-cost (<EUR 0.5/Wp) reliable solar cells, morphological and optoelectronic characterization of these key parameters is fundamental and needs to be further understood.

3.10 CONCLUSIONS

Through the NMP, energy, ICT, people, research infrastructures, and IEE programmes, the European Commission is supporting a strong portfolio of PV projects. These projects address key PV research and innovation issues and represent a total funding (from FP7 and CIP-IEE) of EUR 200 million at the end of 2012.

A key challenge is to identify how project results can be introduced to the industry and what barriers to overcome. This requires an in-depth understanding of the industry and how it is structured, so it will be important that these reviews are carried out by individuals with such expertise.

The analysis of the 2012 PV portfolio has highlighted a strong emphasis on developing materials and structures that offer improved light conversion efficiencies (and thus lower cost). It is important to recognize that optimization of modules and systems is also required—a lack of developments in these areas may be barriers to commercialization.

Novel nanostructured materials and nanotechnology-based processes must now be transferred as fast as possible to PV industrial applications, the main bottleneck being the manufacturing and production at industrial scale. The manufacture of efficient PV cells is only one contribution to the overall cost of PV modules and there may be the potential for more support for development of enhanced manufacturing methods.

It is recognized that there is already some support for this, such as support to the development of roll-to-roll production, and the recently started (in 2012) projects focusing on the development and upscaling of innovative nanotechnology-based PV cell processes and architectures to pilot-line scale for industrial application (Fast-Track, R2RCIGS, and ScaleNano).

In particular, these three recent projects respond to the challenge to bring nanotechnology innovation developed at laboratory scale to industry, in full compliance with the SET Plan and its implementation plan. It is expected that they will stimulate and accelerate industrial uptake of promising results in an industrial area where nanotechnology plays a central role.

4 How Nanotechnologies Can Enhance Sustainability in the Agrifood Sector

Frans W.H. Kampers

CONTENTS

4.1 INTRODUCTION

Nanotechnologies provide a toolbox full of new and highly precise tools that can be used in a wide range of application fields. Since biological systems are primarily made up of nanostructured materials and the processes in biology largely depend on functionality arising at the nano level, it is not surprising that there are many opportunities for these new tools in the different fields that directly rely on biological processes. The agrifood sector is one of these application fields for nanotechnologies (Joseph and Morrison 2006; Kuzma and Verhage 2006; Chaudhry et al. 2010; Frewer et al. 2011). However, as with all new technologies, there are also critical notes (ETC Group 2004; Friends of the Earth 2008) pointing out that there are also risks associated with these applications, and these will be discussed in Section 4.6. Most people agree that opportunities are enormous and the potential for benefits to consumers, society, and the environment is large. Here, the potential for improving the sustainability of the agrifood sector is explored.

Unfortunately, the agrifood sector is not a very sustainable sector. It is estimated that about one-third of all the food produced in the world for human consumption is wasted (Gustavsson et al. 2011). The processes to produce agricultural products are quite sustainable in principle, but the need to produce enough food for a growing world population, in combination with an increasing demand for biological products for other purposes, has put such pressure on the system that nonsustainable practices have been adopted. Examples are the use of fertilizers and pesticides in agriculture and the use of antibiotics in animal husbandry. Also, the demand for exotic foods and the behavior of consumers toward food quality have resulted in unsustainable aspects in the sector. They all arise from the fact that humankind has only limited control over the biological processes that lie at the base of both food and biomaterial production and the quality degradation that inevitably comes with it. But since these processes also work at the nano level, the nanotechnology toolbox can possibly also provide parts of the solution here.

4.2 APPLICATIONS IN PRIMARY PRODUCTION

Primary production is the production phase in which biological processes produce the biomass that is used in foods or biomaterials. It is the growing phase that ends when the product is harvested, and it is also referred to as the preharvest phase. Here, three production systems are reviewed in the light of possibilities to exploit nanotechnology solutions to make the systems more sustainable: agriculture, horticulture, and animal husbandry (Chen and Yada 2011).

4.2.1 Agriculture

In the agricultural production system, crops are grown in the open air. This makes the system dependent on external conditions like weather, water supply, and pests. Efforts to become less dependent on these conditions require human interaction that is not always sustainable. Nanotechnologies can be used to remedy this to a certain extent.

4.2.1.1 Fertilizers

Obviously, the amount of nutrients in the soil is an important factor that can determine the quality and quantity of a crop. Since fertile soils with the same inputs in capital and labor produce more and higher quality crops, it became apparent very early that some lands are more suited for agriculture than others. But even the best lands get depleted if the same crops are grown on the same fields each year. Further, while harvesting the crops essential nutrients, which are necessary to grow next year's crop, are removed. These have to be replenished to be able to grow the same crop next year. This can be solved by crop rotation or by replenishing the nutrients with fertilizers. Unfortunately, the common practice of fertilizer application is not very sophisticated and many factors that are out of the farmers' control determine how much of the nutrients supplied end up in the crops. The rest, in the best case, stays in the soil and can be used by next year's crop, but it often gets to compartments where it is useless or even harmful.

Nanotechnology provides methods to encapsulate chemical substances and to release them only if certain triggers are present (Weiss et al. 2006). The encapsulates are usually small containers, on the order of 0.3–1.0 μm, enclosed by a wall of molecules with specific properties. Self-assembly—obviously the basis of all biological organisms and exploited by nanotechnology as a very powerful instrument in the toolbox to create functionality in a cost-effective way—driven by the properties of the molecules creates superstructures that contain the nutrients. Because the encapsulates are much larger than the nutrient molecules, they will not get transported to undesired soil compartments as easily.

The wall can be designed in such a way that it breaks down when a specific trigger is present. Nanotechnology can enable many different triggers, including pH, chemicals, light, temperature, and magnetic fields. It is therefore possible to create a nutrient delivery system that remains intact when the trigger is absent but releases its contents when the trigger is present. Plant roots produce certain chemicals that can be used as a trigger. This would mean that the nutrients are released if there are roots close by, almost guaranteeing the uptake of the nutrients by plants. Whether or not such a system will become available depends on its cost and societal acceptance (DeRosa et al. 2010).

4.2.1.2 Pesticides

In the same way that nutrients can be encapsulated, pesticides also can be put in a small capsule and released when necessary (Knowles 2009). By tuning the trigger to either the chemicals produced by the pest or the combination of circumstances that allow a pest to develop, the effectiveness of the pesticide is improved, resulting in less chemicals being sprayed. By using nanotechnology to improve the *stickiness* of the encapsulates, it is even possible to reduce pesticide losses during rainfall. Of course then a trigger must be included to be able to get rid of unused capsules before the consumption of crop or other use.

4.2.1.3 Precision Agriculture

The conditions that influence production can vary greatly over fields. It can be easily seen that if the soil or its compaction varies, the field is flat or not, or parts of the field are shaded, different amounts of crop per unit square are harvested. Plants that produce less crop do not need the same amount of fertilizer as plants that produce more; plants in the sun need more water than plants in the shade. Variations in conditions therefore require variations in farming. These variations can occur at the scale of meters, and the farming is referred to as precision farming (Gebbers and Adamchuk 2010). Ideally, the soil, soil condition, and growing conditions should be known at these length scales. Soil and soil condition can be determined once or at large intervals; but since growing conditions such as light, temperature, and soil moisture can change rapidly, continuous monitoring of these conditions would be best. Another reason to locally monitor specific parameters is that many pests arise if conditions are favorable. Phytophthora, for instance, is a disease in potato crops that is caused by a fungus. The fungus can develop only if conditions like temperature and humidity are favorable. Monitoring these conditions locally is therefore a good predictor for phytophthora and enables local administration of fungicides.

At the moment, it is impractical and economically infeasible to have many sensors distributed over a field, but with nanotechnologies these units can be reduced in size and mass production can make them very cheap. In the defense industry, a concept called *smart dust* (Link and Sailor 2003) has been suggested to monitor battlefield activity through a wireless sensor network comprising sand grain–sized *motes* that harvest energy at the spot and communicate autonomously via neighboring motes to a base station. This concept can easily be adapted to agricultural applications if the business case for such monitoring of local conditions becomes favorable for certain crops. Currently, the concept is being researched in projects like GoodFood* with less miniaturized and cheaper wireless sensor networks.

The result of these applications of nanotechnologies is that fewer chemicals are used and lesser amounts of the chemicals end up in undesired places, while crop production is optimized. More production with fewer inputs obviously makes the agricultural system more sustainable.

4.2.2 Horticulture

The difference between agriculture and horticulture is that agriculture uses outdoor farming on large fields, whereas in horticulture smaller pastures or greenhouses are used to grow often higher quality crops. Here, we focus on horticulture in greenhouses because these create circumstances that are favorable for high-tech applications such as nanotechnologies. The greenhouse system also allows much more control over the growing conditions and, therefore, monitoring these conditions makes even more sense. These aspects, in combination with practical considerations like the availability of infrastructure for power supply and shielding from the environment, often make the greenhouse a good stepping stone for technologies to enter the agricultural application field.

Greenhouses, especially in soilless systems, allow control over nutrients and water supply to a very high extent (Kläring 2001). Crops like tomatoes, grown on rock wool substrates, are supplied with water that contains the exact amount of nutrients that is necessary for the optimal growth and production of crops. The water

* See http://www.goodfood-project.org.

that is not used by the plants is recycled; nutrients are replenished and fed back to the system. To enable this system to work, it is essential to be able to measure the concentrations of individual nutrients and to detect microorganisms that could infect the rest of the crop if fed back to the plants. The system can only work if these tasks are performed cost-effectively and with a high degree of accuracy and dependability. Sensors based on nanotechnologies can offer these requirements. In this way, closed greenhouses can be created that only use the amount of nutrients equivalent to the crops that are produced in them and can therefore be considered as sustainable. In combination with sophisticated energy storage systems, these greenhouses can even produce energy, making them 100% sustainable.

One of the problems of greenhouses is that if a plague develops it can rapidly spread through the whole greenhouse. Therefore, early plague detection is essential to allow localized measures to control or eradicate the plague. Fortunately, the closed environment of a greenhouse not only reduces the risk of plagues but also allows the detection of volatiles that indicate the development of plagues at an early stage. Biosensors developed through the use of nanotechnology and biological detection principles, often mimicking molecular detection in insects, can sense very low amounts of specific volatiles (Iqbal et al. 2000) that are produced by plants affected by the plague. This allows rapid intervention by the farmer to locally apply a pesticide or remove the affected plants.

4.2.3 Animal Husbandry

In primary animal production, there are three major areas where nanotechnologies can be applied (Jennifer 2010) and can help to make this sector more sustainable. First of all, similar to the applications developed for human nutrition, animal feed can also benefit from nanotechnologies to deliver micronutrients to the animal. Second, monitoring animals through sensors that are combined with radio frequency identification (RFID) systems provides invaluable information on the animals and their production to the farm management system. Third, veterinary science can use the same encapsulation principles as human pharmacology for delivering drugs more effectively to diseased animals. However, in all these applications economic viability is the key issue that determines whether or not it is used in husbandry practice, and economic viability is very much dependent on animal species and macroeconomic fluctuations.

The animal husbandry sector is a very large user of vitamin supplements. These supplements are usually mixed with the feed that is provided to animals. Unfortunately, the delivery of these micronutrients is not very effective, resulting in the loss of expensive supplements that end up in the urine or manure. By exploiting the nutrient delivery systems of human food products (Chen et al. 2006), more effective feeding strategies can be exploited, resulting in more sustainable animal production. By combining it with RFID systems and feeding automation, these strategies can be made even more sophisticated by tailoring them to the needs of the individual animal, derived from its production status and available health information.

Implantable RFID systems have been available for some years. They have been internationally standardized (Kampers et al. 1999) to ensure that animals equipped with an RFID transponder can be identified throughout the world. RFID systems provide a unique identification code that allows recognition of individual animals throughout the chain and links information stored in various databases to the individual animal. However, since these transponders are injected into the animal subcutaneously, they can also be equipped with sensors that measure specific data. Nanotechnologies nowadays have enabled the development of biosensors that can quantify the amount of specific substances accurately, cheaply, and reliably throughout the lifetime of the animal. This allows not only direct monitoring of the reproductive status of the animal but also early detection of infectious diseases, removing the need for large-scale antibiotic administration, which is an important reason for the development of multiresistant bacteria and the problems they cause in human medicine.

Of course, the detection of health problems in animals at an early stage is of great benefit to the sustainability of the sector because of the higher effectiveness of cures deployed at the onset of a disease. Nanotechnologies have been shown to provide whole new possibilities in drug delivery in humans (McNeil 2011) that will also make veterinary medicines more effective (Scott 2007). This in turn will result in a lower use of veterinary drugs and higher health status of animals in husbandry systems, improving the sustainability of animal production even further.

4.2.4 Water Treatment

The primary sector uses a lot of freshwater, often turning it into wastewater in the process. The difference between freshwater and wastewater is that the latter has different substances dissolved in it and/or is contaminated by microbial organisms, making it unfit for both human and animal consumption or use in agricultural systems. Since what is dissolved has molecular dimensions, it is obvious that the nanotechnologies toolbox working at these length scales can be used to separate the unwanted substances from the wanted (water). There are two basic approaches to purify water: filtration and absorption. In the first, a molecular filter, membrane, or sieve that allows water molecules, which are very small, to pass but blocks virtually all other molecules is used. In the second approach, special molecules or receptors that selectively absorb the unwanted molecules are used.

Biology has shown that separation using membranes can be very effective. Most biological organisms can keep substances out of cells using semipermeable membranes. Unfortunately, because of the molecular nature of the separation, membranes do not have a large throughput. It is therefore often more practical to combine several separation technologies, that is, using sieves to filter out large contaminations like bacteria or even viruses (Rijn 2004), receptors to get rid of biology-based contaminations like proteins, and semipermeable membranes for molecular-scale separation.

Apart from helping to get rid of the pollutants, nanotechnology-based sensors and monitoring systems, like the ones discussed earlier, can also help to determine the purity and safety of water. Biosensors can detect biological contaminations (Lin et al. 2009) and in the near future will also be able to detect most heavy metals and organic molecules.

4.3 POSTHARVEST PROCESSING

So far, we have discussed nanotechnology's applications in primary production, that is, before products are harvested. But the food sector predominantly deals with processing after harvest and much of the waste occurs there, making the whole sector unsustainable. These losses occur almost exclusively in industrialized countries. Very little agricultural produce is wasted in developing countries. In history, food products have never been as safe as they are today in industrialized countries. This is largely because of the stringent quality assurance methods that the food industry has put in place over the last hundred years. The toll of this achievement is that food materials that are suspected to be below standard are taken out of the process and are at best diverted to a less demanding application. Sometimes this is animal feed, although the rules for this are becoming more and more strict since the discovery of bovine spongiform encephalopathy–type diseases; quite often, they are simply wasted. This obviously does not contribute to sustainability. Although nanotechnologies cannot solve all the sustainability issues associated with postharvest processing, the applications can contribute to some of the solutions.

4.3.1 Process Innovation

Most food materials need processing before they can be consumed. There are several reasons for this. First of all, many food materials are unfit for consumption or even dangerous. They need to be cooked to counteract the measures that plants take to prevent vital parts from being eaten or to kill the microorganisms that are part of the material. Second, digestibility needs to be improved to benefit from the food. This processing usually takes place just before consumption, but it can also be done at an industrial scale. Third, the materials must be processed or treated to stop or slow down the spoilage processes that are inevitably associated with food materials. Virtually none of our food products are consumed as they are harvested. They are often processed to enhance the sensory experience. Finally, modern consumers appreciate convenience, which often requires processing or mixing of raw materials. Virtually nobody still makes his or her own butter or mayonnaise; we prefer them ready-made by the food industry. Many of these processes can be made more efficient through applications of nanotechnologies. Three processes will be discussed in this respect as examples to demonstrate the concepts.

4.3.1.1 Separation

In Section 4.2.4, wastewater treatment has been discussed as an application in primary production. One of the goals discussed was to separate microorganisms from water. This is often a problem in foods also. Milk, for instance, can be contaminated by certain bacteria that make it dangerous to be consumed directly after milking. In fact, in most industrialized countries it is not allowed to sell fresh cow's milk without having treated it. Milk is therefore pasteurized by heating it to a certain temperature. This is also beneficial to the shelf life of milk since it kills spoilage bacteria. To further increase the shelf life, milk can be sterilized by not only killing the bacteria but also inactivating the spores. Heating large quantities of milk to 72°C naturally requires a lot of energy, making the process highly unsustainable. Micro- and nanotechnologies can be used to revolutionize the pasteurization and even sterilization of milk.

Instead of killing the bacteria in milk, it is also possible to separate them from it (Rijn 2004). With a well-designed microsieve, it is possible to sieve out bacteria while letting the other constituents of milk pass through the membrane. In this way, it is possible to achieve a very high logarithmic reduction of unwanted components like bacteria or yeast cells (beer) and even spores. This process requires very little energy input and, as an added advantage, leaves the original taste of milk intact. This process is often referred to as cold sterilization.

4.3.1.2 Fractionation

Sieving out bacteria from milk is a separation process. This process can be extended to increase the value of milk components considerably by fractionation. The aim of fractionation is to get most of the components of a complex mixture as separate substances. In the case of milk, the aim is to separate casein micelles, milk proteins, milk carbohydrates, and fat globules from bacteria and water. The value of the separate components is much higher than that of the milk itself because they can be used in different processes. By intelligently combining different sieves and separation methods based on micro- and nanotechnologies, this can be achieved in principle (Rijn 2004).

4.3.1.3 Emulsification

The same type of membranes as that used in fractionation can also be used for emulsification (Mittal and Kumar 1999; Rijn 2004). By flowing for instance water on one side across the membrane and pressing oil through the holes oil droplets are created in the continuous phase, creating an emulsion of oil droplets in water. Because of the uniformity of the holes of the membranes that is achieved with microtechnology, the droplets are all the same size. Such a monodisperse emulsion has the advantage that it is more stable than a traditional polydisperse emulsion in which the smaller droplets have the tendency to combine to form larger droplets, reducing polydispersity. The membranes are very thin and have a high porosity. As a result, very little pressure is required to push the oil through the holes. Compared to traditional emulsification, where the mixture of oil and water is strongly stirred, membrane emulsification requires much less energy input. Obviously, membrane systems can also be used for substances other than oil in water. By suitable surface functionalization of membranes, it can even invert the emulsion, turning an oil-in-water emulsion into one with water droplets in oil.

Emulsification can also be done in a microfluidic system where the droplet phase is pushed through the standing channel of a T junction and the continuous phase flows across it (Nazir et al. 2010). The standing channel ends in a plateau where the droplet forms and then *drops* into the channel where the emulsion is formed. The height of the plateau is the only factor that determines the droplet size. This means that the plateau can be very wide, resulting in a very high droplet formation capacity (van Dijke et al. 2010).

4.3.2 New Food Products

Consumers like to be seduced by new products with interesting benefits. Food industry is therefore constantly innovating products to create new functionality. But there are also large societal challenges that require constant innovation of processes and products.

Because of the increase in world population, food security is of paramount importance. But with increasing welfare in various regions of the world, the people living there will also shift their diets toward more protein-rich meals. Consequently, producing high-quality protein products, predominantly meat, is an important challenge. Although it can be argued that providing enough basic foods to every person on the planet is largely a matter of redistributing the available food materials, providing everybody with meat is a sustainability issue that cannot be solved with the current meat production systems. The problem is that meat production is highly unsustainable. An animal requires up to 10 times the amount of plant protein to produce the same amount of meat protein, and in the process it turns a large amount of clean water into wastewater. Apart from a reduction in meat consumption—probably forced upon the Western world through steep inclines of meat prices because of the mismatch between supply and demand—it can be interesting to find new production methods for meat or meat-like products. A good meat replacement product will combine commercial success with improved sustainability.

The success of a good meat replacer is based on the efficient use of the primary source of protein: plants. By turning plant proteins directly into a meat-like product, a sustainability gain of up to a factor of 10 can be achieved. Unfortunately, it is not very easy to do this. Meat has a structural hierarchy that starts at the nano level, with the proteins and the protein aggregates; continues to the micro level, with fibers that are packed together in bundles; and eventually ends in the muscle tissue at the macro level. Recreating this structural hierarchy is of primary importance to a meat replacer and, consequently, it should be started at the nano level to obtain a suitable basis to work from. Nanotechnologies, necessary to structure the plant proteins to form the right basic structure, are therefore the key to the production of a good meat replacement product.

An important application of nanotechnologies in food products is to encapsulate nutrients to form functional ingredients (Weiss et al. 2006). Encapsulation of these nutrients is necessary to avoid interference with the taste of the product to be fortified, protect them from external influences that could break them down, keep components apart that otherwise would react with each other, and improve the bioavailability of the nutrients. These encapsulation and delivery techniques reduce the amounts of functional ingredients in food products because they improve the effectiveness of the components and therefore improve the sustainability, although their effect is obviously small.

4.4 RETAIL AND CONSUMER

Now that the food product has been harvested and processed, it can find its way to the consumer. This usually involves extensive logistics and in-between steps like warehouses, retail stores, and home storage. A substantial amount of food is wasted in this process. Next, we look at some of the solutions that nanotechnologies have to offer to reduce such losses.

4.4.1 Packaging

Most food products nowadays are packaged before they are transported, even over short distances. This happens for obvious reasons: to contain the product (e.g., fluids or powdery substances); protect them from external contamination; maintain the quality as long as possible; shield them from light; and avoid drying (fresh produce), wetting (e.g., cookies), or chemical degradation (e.g., oxidation of beer). The packaging systems nowadays are very sophisticated solutions tailor-made for the product being packaged. They are usually a compromise between effectiveness and consumer convenience and preferences. Nanotechnologies can help to improve this compromise (Brody 2003; Kampers 2011).

4.4.1.1 Barrier Properties

An important property of packaging materials is their ability to function as a barrier between the elements that must be kept separate from the food. In the previous century, packaging was done by simply storing food products in tin cans or glass jars. These solutions, although effective for many products, have been abandoned because of consumer preference (you cannot see the contents of a can) or weight considerations and were replaced with polymer bags and bottles. Unfortunately, many of these materials have an important disadvantage in that they are permeable to certain substances like oxygen or water. In some cases, it was apparently acceptable to the consumer to solve this problem by applying a thin layer of aluminium to the polymer, but this again blocked the sight of the product contained. Nowadays, nanocomposites can combine transparency with good barrier properties, bringing lightweight packaging materials within reach of sensitive products (Paul and Robeson 2008). Nanosized clay platelets force oxygen or water molecules to take a tortuous path, thus substantially elongating the diffusion path length and reducing the amount that can leak through the package. Thanks to these developments, it is now possible to store beer in polyethylene terephthalate bottles. This not only reduces the total weight of the packaged food product but also substantially reduces the amount of packaging material needed to maintain product quality.

These barrier properties are also important when the concept of modified atmosphere packaging is applied. In this process, an essential element, usually oxygen, of the normal atmosphere is eliminated, substantially reducing the biological activity in the package. Spoilage organisms either die or go into a dormant state, and most spoilage processes stop. Obviously, this concept relies on the fact that the packaging material can maintain the modified atmosphere for as long as possible.

4.4.1.2 Antimicrobial Properties

As mentioned Section 4.1, biodegradability of food materials makes them vulnerable to spoilage. If no measures are taken, spoilage organisms start doing their job of degrading the materials as soon as the product loses the defenses of the living organism. The most effective way to counteract these spoilage processes is to kill all biological agents on or in the food product and to package it in such a way that new ones cannot get to it. Unfortunately, in most cases this means that the food product has to undergo processes that can also change the product itself. Consumers nowadays in many cases object to these processes and demand mild conservation processes or no conservation at all. The consequence is that other technologies have to step in to slow down spoilage of packaged food products. Some nanomaterials, such as silver nanoparticles and certain polymer layers, are capable of killing microorganisms that come in contact with them (Cha and Chinnan 2004). These antimicrobial properties can be incorporated in the packaging systems to reduce the microbial pressure inside and to slow down quality degradation. In this way, shelf life is extended and less food products are wasted.

4.4.1.3 Volatile Monitoring

If a packaged food product is only mildly preserved or no preservation is applied to it at all, unless the product is kept at very low temperatures, spoilage will inevitably take place. For many food products, it is known how these processes develop as a function of time and temperature and an estimate can be made as to when the product quality has deteriorated to such a level that consumption is no longer advisable. Based on assumptions of storage conditions, these models form the basis of the *sell-by* or *use-by* date that is printed on the package. Although this system is widely used in industrialized countries and has substantially improved overall food quality, there are two major problems connected with it. The first one is related to the assumption of storage conditions. If the actual conditions under which a product is stored, even only for part of the storage time, substantially differ from the ideal conditions, spoilage steps up and quality deterioration diverges from the model. In effect, the product will be spoiled before the use-by date has expired. If detected before consumption this is not a serious problem, although obviously it contributes to food wastage. If instead the spoiled product is consumed, this can lead to serious health consequences and even hospitalization. To reduce this problem, food manufacturers usually keep a substantial safety margin in the sell-by and use-by dates. This large safety margin leads to the second problem: many perfectly good food products get thrown away just because the use-by date has expired.

Spoilage processes produce volatiles. Our nose is well equipped to detect these volatiles, warning us not to eat a spoiled product. With the use of nanotechnology-enabled biosensors, these volatiles can be detected in the headspace of the closed package and the consumer can be warned of spoiled products (Fonseca et al. 2007). Conversely, if the sensor is not signaling quality deterioration a product can safely be consumed even after its use-by date has expired. The concept of detecting spoilage products is preferable because it is a direct way to monitor product quality, whereas the use-by date system is an indirect system relying on models and assumptions.

Spoilage is caused by microorganisms that proliferate on the biodegradable substrate that food products often have. Especially if a fresh product is damaged, conditions are favorable for the spoilage organisms. Cutting fresh produce for the convenience of consumers is a very damaging operation. Therefore, freshly cut products are vulnerable to spoilage. Packaging them and keeping them under a modified atmosphere at a reduced temperature does help to extend shelf life. The amount of spoilage organisms present at the time of packaging very much determines the speed of the spoilage process. For an accurate estimation of the use-by date, it is therefore necessary to determine the amount of spoilage organisms at the packaging stage. Unfortunately, a classical determination—where a sample of the product is ground, put on a substrate, and put in a stove for several days—is obviously not a feasible alternative because it takes too much time. The product would be spoiled before it could be shipped to the stores. Therefore, food industry is very much in need of devices that can determine the amount of specific microorganisms in the complex matrix of a food product rapidly and cost-effectively and that might be operated by untrained personnel at the production line. Current developments in polymerase chain reaction (PCR) tests and lateral flow assays (Amerongen and Koets 2005; Posthuma-Trumpie et al. 2009; Noguera et al. 2011) will result in rapid methods for the detection and quantification of microorganisms. These technologies will help to make more accurate and reliable estimations of the shelf life of fresh produce and will therefore reduce the amount of good food products that end up in the garbage, although they are still perfectly edible.

Although not directly contributing to the sustainability of the food sector, the aforementioned microorganism measurement and detection technologies can also be used to detect pathogens in food products, making them even safer than they already are today. This is necessary because a number of consumers, even in industrialized countries, are still getting hospitalized each year for eating food products contaminated with various pathogens (WHO 2002). These kinds of devices will hopefully also help people in developing countries to reduce the number of fatalities, especially among children, from drinking spoiled water or eating spoiled food.

4.4.2 Logistics

Because of the biodegradable nature of food products, especially for fresh produce, getting the products from the producer to the consumer is a race against the clock. To cope with this problem, the food sector has organized itself in chains from farm to fork. Many of these chains require a lot of transportation of products, ingredients, and/or base materials. The logistics involved are extensive and have been fine-tuned to reduce the risks of product loss. At certain points in the chains, decisions have to be made, for instance, to either ship the products to more distant markets where they will bring in more money or sell them at local markets that can be reached within the shelf life time window available. Preferably, these decisions are made on the quality state in which the products currently are. Obviously, this again requires the ability to either reliably predict this quality state or measure it at the time of making a decision. Unfortunately, the latter, although technology is advancing quickly, is not yet feasible, as discussed in Chapter 3. Therefore, this sector has to rely on quality predictive models. Including sensors in the packaging of products would enable monitoring of spoilage and therefore would help to verify that the models are correct. If closed containers are used to transport the produce, for example, under a modified atmosphere, determining the amount of specific volatiles that correlate with the ripening or spoilage processes would make this aspect more readily feasible.

The ultimate challenge, of course, is to monitor the quality on an individual product basis and to communicate the quality at various points in the chain, including the retail shop and even the refrigerator of the consumer. Apart from the sensors in the packaging system, this would also require a sophisticated communication system to be available in the packaging of the product. For obvious reasons, this cannot be equipped with batteries but must be passive. RFID systems, which have been developed for animal identification (Kampers et al. 1999) among other uses, not only can satisfy the requirements of food products but also are being proposed by certain retail organizations as a successor of bar codes. With advances made in nanotechnologies, these RFID systems can be further miniaturized and/or implemented in polymer electronics (PolyIC), resulting in a cost-effective communication system that is cheap enough for food applications. However, the simple RFID systems that are used today only to identify an object need to be extended with capabilities to sense the volatiles and/or monitor other important parameters like storage temperature.

RFID systems are also used for tracking and tracing purposes. Knowing where a product comes from and where it has been can be helpful to make the spoilage models that are used to estimate the quality of a product as a function of time more accurate. But the main purpose of tracking and tracing systems is to verify that the product is original, with the corresponding quality assurance.

4.4.3 Convenience

Unfortunately, in developed countries much of the sustainability problem in the food sector is caused by the consumers' demand for convenience. As a matter of fact, the food sector in developing countries is still sustainable. Convenience has resulted not only in food products being produced in advance and packaged in such a way that the quality is maintained for as long as possible but also in trade-offs in food products that do not always favor sustainability. For instance, to combine our desire for freshness or mild preservation techniques with extended shelf life and short preparation times, sophisticated packaging systems have been developed. But any packaging system is compromising sustainability since it is largely wasted. If we were to go back to harvesting the products locally, preparing them as soon as possible after harvesting, and consuming them immediately, we would improve substantially the sustainability of food production. Unfortunately, we cannot envisage a return to the old ways without causing major disruptions in the current systems, which is nothing to wish for. Technologies are therefore implemented to improve the compromise between noncombinable issues in food production, processing, storage, preparation, and consumption. As explained earlier, nanotechnologies are used to improve packaging systems and monitor the quality of the packaged food product, thus extending shelf life and reducing the amount of food waste.

It has even been suggested that nanotechnologies can enable food products of which the consumer can modify the flavor according to current preferences (Greßler et al. 2010) by changing preparation parameters. It is usually explained as a *Magic Pizza*, which can be turned into a pepperoni pizza simply by choosing the right microwave conditions. Although this could improve the sustainability of the sector, it has not been realized so far, if it ever will.

4.5 SOCIETAL ACCEPTANCE OF APPLICATIONS IN AGRIFOOD

The applications of nanotechnologies in the food sector described so far hold interesting promises of benefits to the consumer, the society, the food sector, and also sustainability. However, as has been seen in the case of genetically modified organisms (GMOs), whether or not these benefits will be achieved largely depends on their acceptance by individual consumers and the society as a whole. Moreover, since the food sector is well aware of this situation, it is reluctant to introduce new products with these benefits because it is afraid of the reaction in the marketplace. In this respect, it is again the consumer—who in the first instance is responsible for the sustainability problem in the food sector—who holds the key to implement this technology, which can contribute to improving the sustainability or at least reduce the problem.

To accept the new technology, both the individual consumer and the society as a whole must first be able to weigh the benefits against the perceived risks (Siegrist et al. 2008). This requires that they are objectively informed on the subject. It is up to scientists and the food industry to provide them with information on the benefits and the risks and to also inform them about the uncertainties. Usually, nongovernmental organizations and the press tend to provide information that overemphasizes the risks (ETC Group 2004; Friends of the Earth 2008). But even if provided with good information, it is not easy for consumers to make such a decision. This has to do with the nature of the issue and also the fact that food and eating are highly emotional to individuals (Siegrist et al. 2007).

4.5.1 Risk Perception

People are not very skilled in understanding small chances, and also tend to overestimate them (and underestimate large ones). Risks in the food sector are very small; chances of being negatively affected by food products are extremely small. The emotional perception of risks associated with technologies applied in foods is therefore already biased to begin with. But there is more.

If we discuss the risks of applications of nanotechnologies in foods, we need to distinguish between hazards and exposure. Risk is the product of hazard and exposure. A lion, for instance, is a hazardous animal. But the risks for visitors in a zoo are minimal because cages prevent them from being exposed to the hazard. Hazards of nanotechnologies differ extensively for different applications. It can easily be understood that the hazards of biosensors are not comparable to the hazards associated with certain nanoparticles in food products. It is commonly agreed on that the hazards in food are predominantly associated with persistent engineered nanomaterials (ENMs) that have certain characteristics (Seaton et al. 2010; Brayner 2008; Stern and McNeil 2008; McNeil 2009). Because they are small there is a chance that they can enter the food chain and become systemic, and since they are persistent they have time to cause a reaction. Research is directed toward understanding the characteristics of ENMs more thoroughly to be able to eliminate risks from food applications.

There is very little information regarding exposure. Again, it strongly depends on the application: consumers will generally not eat the biosensors; therefore, exposure will be limited. If nanoparticles are used as ingredients, the exposure can be more substantial, although it has been shown that uptake by the body of ingested particles is limited. Moreover, it is very difficult to compare exposure to ENMs with exposure to naturally occurring nanoparticles or particles that have already been part of our environment for decades (Card et al. 2011). Research on the exposure to nanoparticles is still largely in the start-up phase.

Technologies applied to food products are regarded as *unnatural*. This especially holds true for nanotechnology, a technology that is largely unknown and of which applications are virtually undetectable in foods. Consumers in general prefer natural foods. This means that their first reaction to applications of nanotechnologies is negative. These applications must therefore bring substantial benefits, preferably for the consumer to become acceptable. Trust is an important factor in the process of acceptance. Regulation can play a role in building trust with consumers.

4.5.2 Regulatory Issues

One of the purposes of regulation is to minimize potential risks for individuals and the society. Risks associated with the consumption of food products are generally regarded as unacceptable. Food must be safe. In the risk-averse society of today, this results in strict rules for new food products, especially in Europe and the United States. Not only do these rules leave little room for innovation but also the associated costs to get new products approved by the authorities often discourage the industry to even start developing products with innovative properties and/or improved sustainability.

At present, specific regulations on the application of nanotechnologies in food products are poor. In most regions, regulation focuses on new products or existing products that have been produced with new processing techniques. Nanotechnologies are seen as one of a series of new technologies that, if applied to a food product, would constitute a new processing technology; consequently, all products containing the new ingredient would be regarded as a novel food and would require approval by the appropriate authority. As a fallback scenario, often umbrella regulations regarding the responsibility of producers that put a product on the market are mentioned to eliminate potential loopholes in applications of these new technologies. However, in the case of nanotechnologies legislative bodies seem to be determined to set up dedicated regulations to reduce the risks. This can be advantageous because well-devised regulations and the implementation of impartial bodies that supervise use in sensitive areas like food can help build trust in these applications.

It is almost inevitable that regulatory bodies lag behind when new technologies emerge and start to be applied in various fields. Setting up regulations can only start after the new technology has developed to a certain stage of maturity, often accompanied by its first applications in the market. In the case of nanotechnologies the scientists developing it—for the first time in history and implementing the lessons learned from the GMO debacle—were willing to discuss potential risks in a very early stage of the development. Although this has helped to create awareness at the legislative level, it has not helped in setting up regulations because the hazard characterization and especially the exposure assessment need further research to be able to support good laws.

In the case of nanotechnologies, there is a further problem. Before regulations can be set up, critical elements must be defined in such a way that all stakeholders know how to interpret them and the regulation can be enforced. In the case of nanotechnologies this is very difficult and, despite the efforts of different public bodies, it is doubtful if a consistent definition can ever be achieved. This is a problem especially for applications in food. Food products in most cases are made up of nanostructured materials that have purposefully been modified to achieve certain goals. It is very difficult to exactly determine where these modifications constitute classical processing and where they become nanotechnology. Instead of trying to define the technology as a whole, it probably makes more sense to define the products of the technology that have potential risks associated with them. This would mean that regulations and the associated definitions would focus on persistent ENMs. This approach not only has the advantage that definitions can be drawn up but also makes the enforcement of regulations achievable and avoids unnecessary extension of the regulatory burden for the food industry.

4.6 DISCUSSION AND CONCLUSIONS

There are many possibilities for applications of nanotechnologies in the agrifood sector to enhance the sustainability of the sector and at the same time improve products for consumers and the society as a whole. And with the technologies still in their infancy, many more opportunities will develop in the near future. Of course, nanotechnologies cannot solve all the problems of the sector. It is one of many new technologies that can make contributions to solutions. But because of its fundamental nature and wide application field, it is extremely promising. Many future developments will substantially improve the quality of our life and reduce our ecological footprint, helping us to hand over our planet to future generations in a good state, without having to compromise on our current way of life. Because of the generic nature of the technology and the unprecedented application width, it is difficult to predict what promises future applications of nanotechnologies hold. Some new applications that researchers in various places in the world are working on have been discussed in the Sections 4.2 through 4.4. Which of them will become reality is largely a question of economics, that is, if the return on investment is large enough for the industry to start developing the technology, incorporate it in new products, and bring them to the market. Unfortunately, improved sustainability is rarely seen as a return on investment. This probably means that many of the applications that could improve sustainability will not get to the market, especially if regulations do not favor the development of these innovations.

As with any technology, nanotechnology applications have positive and negative aspects, and some of the negative effects may be associated with the benefits to be achieved. The exact trade-offs between benefits and risks are not yet fully understood, but they will be important both in the economic evaluations of applications and in the acceptance by the individual consumer and society as a whole. Scientists are working hard to understand the mechanisms behind the risks posed by certain products of nanotechnologies. Hopefully, we will see the results of their efforts in the near future and be able to harness some benefits in a responsible and acceptable way.

At the moment, the trade-off between risks and benefits for applications of nanotechnologies in the food sector is entirely at the benefits' side. Risks in food products are unacceptable no matter what benefits are presented for how many people, even though this rule cannot be applied to existing food products without drastically reducing our current diets. Also, from an ethical point of view it can be questioned if it is acceptable in the long run to deprive large population groups or sustainability from benefits because of risks perceived among groups that already are economically privileged. Unfortunately, without large disruptions or economic instability it will not be feasible to change the principle that a small group of consumers with sufficient purchasing power determines what is good for the rest of the world. Let us hope that the benefits for them will turn out to be benefits for others also and especially for the sustainability of the agrifood sector.

REFERENCES

Amerongen, A. V. and M. Koets. (2005). "Simple and rapid bacterial protein and DNA diagnostic methods based on signal generation with colloidal carbon particles." In: *Rapid Methods for Biological and Chemical Contaminants in Food and Feed.* A. van Amerongen, D. Iarug and M. Lauwaars (eds), Wageningen, Wageningen Academic Publishers, 105–126.

Brayner, R. (2008). "The toxicological impact of nanoparticles." *Nano Today* **3**(1–2): 48–55.

Brody, A. L. (2003). ""Nano, nano" food packaging technology." *Food Technology* **57**(12): 52–54.

Card, J. W., T. S. Jonaitis et al. (2011) "An appraisal of the published literature on the safety and toxicity of food-related nanomaterials." *Critical Reviews in Toxicology* **41**(1): 1–30.

Cha, D. S. and M. S. Chinnan (2004). "Biopolymer-based antimicrobial packaging: A review." *Critical Reviews in Food Science and Nutrition* **44**(4): 223–237.

Chaudhry, Q., L. Castle et al., Eds. (2010). *Nanotechnologies in Food*, London, Royal Society of Chemistry.

Chen, H. and R. Yada (2011). "Nanotechnologies in agriculture: New tools for sustainable development." *Trends in Food Science & Technology* **22**(11): 585–594.

Chen, H. D., J. C. Weiss et al. (2006). "Nanotechnology in nutraceuticals and functional foods." *Food Technology* **60**(3): 30–36.

DeRosa, M. C., C. Monreal et al. (2010). "Nanotechnology in fertilizers." *Nature Nanotechnology* **5**(2): 91.

ETC Group (2004). *Down on the Farm: The Impact of Nano-scale Technologies on Food and Agriculture*, Ottawa, ETC Group.

Fonseca, L., C. Cane et al. (2007). "Application of micro and nanotechnologies to food safety and quality monitoring." *Measurement & Control* **40**(4): 116–119.

Frewer, L. J., W. Norde et al., Eds. (2011). *Nanotechnology in the Agri-Food Sector: Implications for the Future*, Weinheim, John Wiley & Sons.

Friends of the Earth (2008). *Out of the Laboratory and on to Our Plates*, Sydney, Friends of the Earth.

Gebbers, R. and V. I. Adamchuk (2010). "Precision Agriculture and Food Security." *Science* **327**(5967): 828–831.

Greßler, S., A. Gazsó et al. (2010). *Nanoparticles and Nanostructured Materials in the Food Industry. Nano Trust Dossier*, Vienna, Institute of Technology Assessment of the Austrian Academy of Sciences. **004**: 1–5.

Gustavsson, J., C. Cederberg et al. (2011). *Global Food Losses and Food Waste*, Rome, Food and Agricultural Organization.

Iqbal, S. S., M. W. Mayo et al. (2000). "A review of molecular recognition technologies for detection of biological threat agents." *Biosensors and Bioelectronics* **15**(11–12): 549–578.

Joseph, T. and M. Morrison (2006). *Nanotechnology in Agriculture and Food*, U.K., Nanoforum.

Kampers, F. W. H. (2011). "Packaging." In: *Nanotechnology in the Agri-Food Sector*. L. J. Frewer, W. Norde, A. Fischer and F. Kampers (eds), Germany, Wiley-VCH Verlag GmbH & Co. KGaA, 59–73.

Kampers, F. W. H., W. Rossing et al. (1999). "The ISO standard for radiofrequency identification of animals." *Computers and Electronics in Agriculture* **24**(1–2): 27–43.

Kläring, H.-P. (2001). "Strategies to control water and nutrient supplies to greenhouse crops. A review." *Agronomie* **21**(4): 311–321.

Knowles, A. (2009). "Global trends in pesticide formulation technology: The development of safer formulations in China." *Outlooks on Pest Management* **20**(4): 165–170.

Kuzma, J. (2010). "Nanotechnology in animal production—Upstream assessment of applications." *Livestock Science* **130**(1–3): 14–24.

Kuzma, J. and P. Verhage (2006). *Nanotechnology in Agriculture and Food Production*. Project on Emerging Nanotechnologies, Washington, DC, Woodrow Wilson Institute. **4**.

Lin, C.-H., C.-H. Hung et al. (2009). "Poly-silicon nanowire field-effect transistor for ultrasensitive and label-free detection of pathogenic avian influenza DNA." *Biosensors and Bioelectronics* **24**(10): 3019–3024.

Link, J. R. and M. J. Sailor (2003). "Smart dust: Self-assembling, self-orienting photonic crystals of porous Si." *Proceedings of the National Academy of Sciences* **100**(19): 10607–10610.

McNeil, S. E. (2009). "Nanoparticle therapeutics: A personal perspective." *Wiley Interdisciplinary Reviews: Nanomedicine and Nanobiotechnology* **1**(3): 264–271.

McNeil, S. E. (2011). "Unique benefits of nanotechnology to drug delivery and diagnostics." *Methods in Molecular Biology* **697**: 3–8.

Mittal, K. L. and P. Kumar (1999). *Handbook of Microemulsion Science and Technology*, New York, Marcel Dekker.

Nazir, A., K. Schroën et al. (2010). "Premix emulsification: A review." *Journal of Membrane Science* **362**(1–2): 1–11.

Noguera, P., G. Posthuma-Trumpie et al. (2011). "Carbon nanoparticles in lateral flow methods to detect genes encoding virulence factors of Shiga toxin-producing *Escherichia coli*." *Analytical and Bioanalytical Chemistry* **399**(2): 831–838.

Paul, D. R. and L. M. Robeson (2008). "Polymer nanotechnology: Nanocomposites." *Polymer* **49**(15): 3187–3204.

PolyIC. from http://www.polyic.com/. Accessed November 25, 2011.

Posthuma-Trumpie, G., J. Korf et al. (2009). "Lateral flow (immuno)assay: Its strengths, weaknesses, opportunities and threats. A literature survey." *Analytical and Bioanalytical Chemistry* **393**(2): 569–582.

Rijn, C. J. M. v. (2004). *Nano and Micro Engineered Membrane Technology*, Amsterdam, Elsevier.

Scott, N. (2007). "Nanoscience in veterinary medicine." *Veterinary Research Communications* **31**(0): 139–144.

Seaton, A., L. Tran et al. (2010). "Nanoparticles, human health hazard and regulation." *Journal of The Royal Society Interface* **7**(Suppl 1): S119–S129.

Siegrist, M., C. Keller et al. (2007). "Laypeople's and experts' perception of nanotechnology hazards." *Risk Analysis* **27**(1): 59–69.

Siegrist, M., N. Stampfli et al. (2008). "Perceived risks and perceived benefits of different nanotechnology foods and nanotechnology food packaging." *Appetite* **51**(2): 283–290.

Stern, S. T. and S. E. McNeil (2008). "Nanotechnology safety concerns revisited." *Toxicological Sciences*. **101**(1): 4–21.

van Dijke, K. C., G. Veldhuis et al. (2010). "Simultaneous formation of many droplets in a single microfluidic droplet formation unit." *AIChE Journal* **56**(3): 833–836.

Weiss, J., P. Takhistov et al. (2006). "Functional materials in food nanotechnology." *Journal of Food Science* **71**(9): R107–R116.

WHO (2002). *Who Global Strategy for Food Safety: Safer Food for Better Health*, Geneva, World Health Organization.

5 Biological Production of Nanocellulose and Potential Application in Agricultural and Forest Product Industry

Nadanathangam Vigneshwaran, Prasad Satyamurthy, and Prateek Jain

CONTENTS

5.1 INTRODUCTION

Cellulose, one of the most abundant organic compounds found in nature, is chemically a homopolymer of β-1,4-linked D-glucose molecules in a linear chain. These linear chains are hydrogen bonded to each other giving cellulose its extended three-dimensional structure. Among various sources of cellulose that include algae, marine creatures, and bacteria, plants form a major source where it is present as the primary cell wall component. From the industrial point of view, the major sources of cellulose are wood and cotton that have a cellulose content of 40–50% and 90%, respectively (Ververis et al. 2004). It is used mainly in the production of textiles, paperboard, and paper and also converted into wide variety of derivative products such as cellophane and rayon.

The cellulose polymer chains consist of highly ordered crystalline regions interspaced by disordered amorphous portion. Cellulose polymer can be hydrolyzed using various acids or hydrolytic enzymes to produce glucose, which can be further used for the production of ethanol (Olsson and Hahn-Hagerdahl 1996), organic acids (Luo et al. 1997), and other fine chemicals (Cao et al. 1997). In the presence of strong acids or even mechanical forces, native cellulose breaks down into microsized and further into nanodimension cellulose, having morphology in the form of whiskers. Acid hydrolysis of cellulose is a well-known process where concentrated sulfuric acid (65% w/v) hydrolysis removes the amorphous regions and enables the isolation of crystallites of cellulose (Krishnamachari et al. 2012). These crystallites of cellulose, also known as nanocrystalline cellulose (NCC) has attracted a lot of interest over the past decade due to their excellent mechanical properties such as very high Young's modulus and mechanical strength. The Young's modulus of NCC is as high as 134 GPa, whereas the tensile strength of the crystal structure is estimated to be in the range of 0.8–10 GPa (Prasad et al. 2011). Thus, NCC is a novel class of cellulosic material that combines the properties of biodegradability, nontoxicity, and renewable nature of cellulose along with enhanced mechanical properties. Even though the most reported method for the production of NCC is the acid hydrolysis, its commercial exploitation is limited from the environmental point of view due to the release of toxic waste products as effluents. Also, the acid hydrolysis of cellulose using sulfuric acid results in the surface modification of cellulose by the introduction of sulfate groups on the surface of cellulose (Araki et al. 1998). The other methods of production of NCC such as the chemomechanical and sonochemical processes are highly energy intensive and require a huge energy input for the production of NCC. Hence, an ecofriendly, cost-effective, and energy-efficient biological process for the production of NCC would be of utmost commercial importance. NCC can be produced by a biological process wherein an agent of biological origin such as microbe itself or an enzyme produced by the microbe hydrolyses cellulose from its original size down to nanosize.

5.2 BIOLOGICAL PRODUCTION OF NANOCRYSTALLINE CELLULOSE

A biological process for the production of NCC involves the use of a biological agent for the production of NCC. The major reason for the hydrolysis of cellulose by microorganisms is to use it as a source of carbon/energy for its growth. The focus is toward controlling the microbial growth so that the reaction could be stopped once the nanosize of cellulose is formed.

5.2.1 Microbial Hydrolysis of Cellulose

The microorganisms secrete a group of enzymes, which are known as cellulases for the hydrolysis of cellulose into its final end product of glucose that can be used as a source of energy for its growth and metabolism. Bacteria and fungi are among the major group of organisms that secrete cellulase enzyme as they are unable to directly use the insoluble cellulose particles; aerobic fungi secrete their cellulases extracellular, and anaerobic bacteria produces cellulosomes attached to their outer surface to facilitate hydrolysis (Wilson 2011).

Among the various microorganisms, fungi are the most potent source for the production of the cellulose-degrading enzyme, cellulase. The filamentous fungus *Trichoderma reesei* (anamorph of *Hypocrea jecorina*) is one of the most efficient producers of extracellular cellulase enzyme. Cellulases are produced as multicomponent enzyme system comprised usually of three components that act synergistically in the hydrolysis of cellulose, namely endoglucanases (EC 3.2.1.4), cellobiohydrolase (CBH) (EC 3.2.1.91), and cellobiase (β-glucosidase, EC 3.2.1.91). The extracellular cellulolytic system of *T. reesei* is composed of 60–80% CBHs, 20–36% of endoglucanases, and 1% of β-glucosidases. The first two components act directly on cellulose yielding oligosaccharides namely cellobiose, cellotriose, or cellotetraose as the reaction products. Further, hydrolysis of oligosaccharides is taken care of by cellobiase (Zaldivr et al. 2001).

5.2.2 Production of NCC by Controlled Microbial Hydrolysis of Cellulose

A microbial process for the synthesis of nanocellulose was described previously wherein NCC was synthesized from microcrystalline cellulose (MCC) by controlled hydrolysis using the filamentous fungi *T. reesei* (Prasad et al. 2011). In this process, the raw material (cotton fiber) was initially converted into MCC. This MCC was further hydrolyzed microbially to produce NCC. For the production of NCC, the 24 h inoculum of the fungus *T. reesei* (ATCC 13631) was prepared in potato dextrose broth by the inoculation of spore suspension (~3×10^6 spores/mL). The optimized concentration of inoculum was added in Mandel's medium containing MCC as the sole carbon source and was incubated at 25°C under shaking condition at 150 rpm.

Unlike other fungus that also can degrade cellulosic biomass like the brown rot fungi that constitutively produce peroxidases in combination with cellulase, *T. reesei* does not produce peroxidases but only produces cellulases resulting in the degradation of cellulose (Martinez et al. 2005). This makes *T. reesei* a better candidate for the production of NCC; also it uses the free enzyme mechanism for the degradation of cellulose wherein the cellulase enzyme is secreted extracellular into the medium and these cellulases possess the carbohydrate-binding module (CBM) joined by a flexible linker to one end of the catalytic domain and the cellulases that are present as mixture of multicomponent enzyme systems act synergistically to degrade highly crystalline cellulose (Wilson 2011). This extracellular production of cellulases is affected by many external factors such as pH, temperature, and substrate concentration that can be manipulated for the controlled hydrolysis of cellulose.

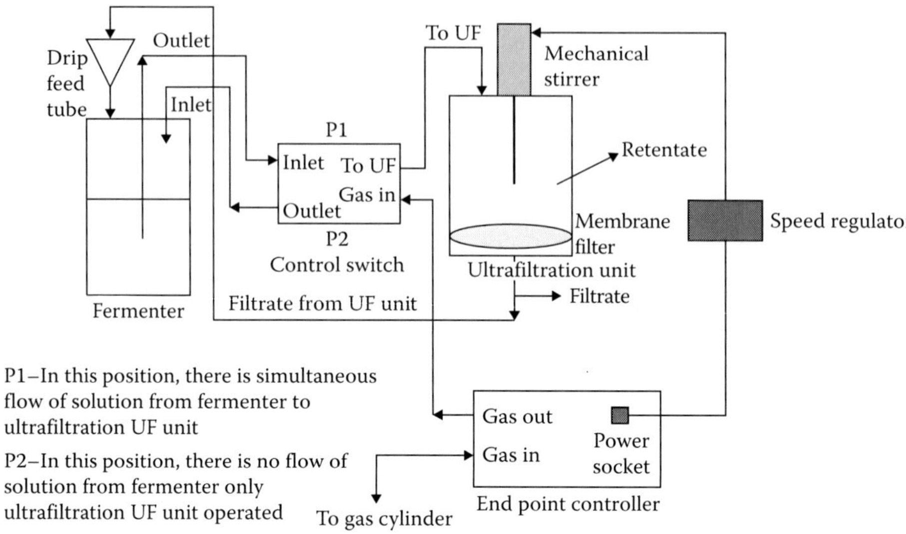

FIGURE 5.1 Schematic representation of continuous fed-batch system for nanocellulose production.

5.2.3 Continuous Batch Process for the Production of Nanocellulose

The major objective of the cellulose-degrading microbes is the formation of the end product glucose that has to be used for its growth. Thus, under optimal conditions the yield of nanocellulose by microbial hydrolysis is very low. So, one of the strategies for increasing the yield of nanocellulose is the continuous removal of nanocellulose as and when it is formed thus increasing the yield of nanocellulose. A batch fermentation system (Figure 5.1) for the continuous removal of nanocellulose and its purification is developed, which is a combination of a fermenter coupled with a filtration system with a flow control valve that works on the principle of positive pressure for movement of liquid in the system. In this system, the supernatant from the fermenter containing the cellulose nanocrystals is transferred to the filtration system wherein the cellulose nanocrystals gets trapped in the membrane retentate and the filtrate containing the other soluble media components is fed back into the fermenter thus enabling the continuous separation of cellulose nanocrystals as they are formed preventing their further hydrolysis into the end product of glucose.

5.2.4 Purification and Characterization of NCC Produced by Microbial Hydrolysis

The nanocellulose formed during the biological process is mixed along with many other impurities such as the biological agents like microbes, enzymes, and many soluble media components. Thus, purification of NCC produced by biological process is an essential step in the production of NCC by biological processes. In the production of NCC by microbial hydrolysis of MCC by the fungus *T. reesei*, after fermentation, the broth was subjected to differential centrifugation for the sedimentation of all particles of size greater than 1 μm. The resultant supernatant was filtered through 100 kDa ultrafiltration membrane by vacuum suction wherein water and lower molecular weight solutes pass through the membrane, and the NCC that was retained on the surface of membrane was removed with a jet of ultrapure water (Prasad et al. 2011). Thus, differential centrifugation combined with ultrafiltration was reported to be a good method for the purification of NCC, which may also be adopted for the purification of any other nanoparticle synthesized by the biological route.

The NCC thus obtained after isolation and purification was subsequently characterized by the techniques such as atomic force microscopy (AFM), particle size analysis based on dynamic light scattering, zeta potential measurement, Fourier transform infrared (FTIR) analysis, and degree of polymerization measurements. Tapping mode AFM, using a silicon nitride cantilever probe to gently oscillate and tap the sample surface, is very well suitable for organic materials like NCC at low forces. The cantilever is vibrated at a resonant frequency of oscillation, and the sample surface is scanned generating an image based on the morphology of the sample. The AFM micrograph of NCC produced by the microbial hydrolysis shown in Figure 5.2 revealed that the morphology of the NCC produced is rod like, also referred to as cellulose nanowhiskers (CNW).

The particle size distribution and zeta potential of CNW in suspension were measured using particle size analyzer based on the dynamic light scattering principle (Figure 5.3).

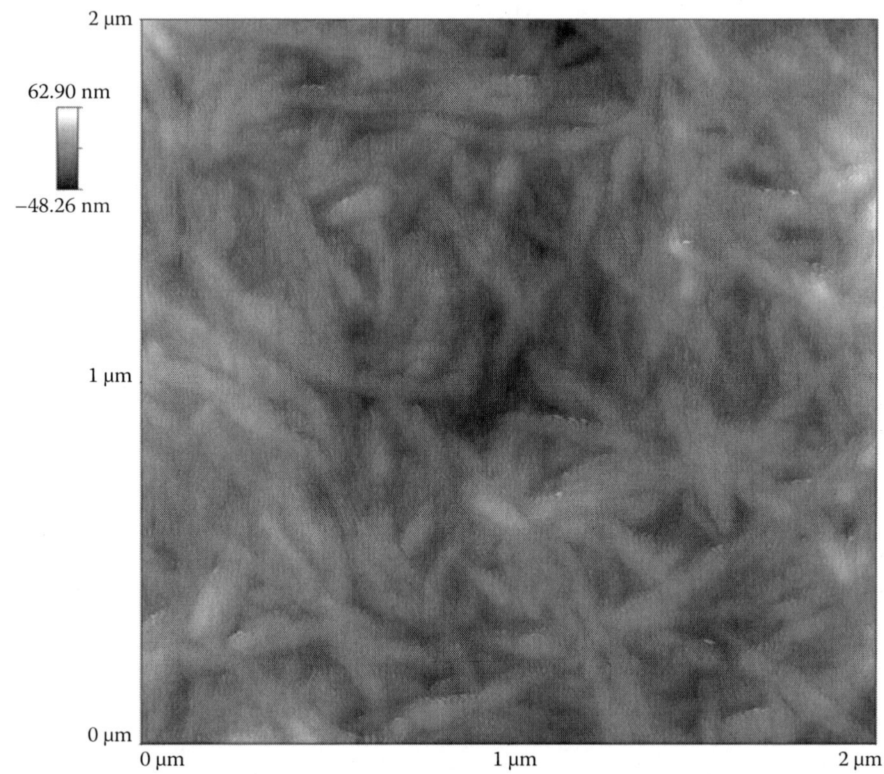

FIGURE 5.2 Atomic force microscopy of nanocrystalline cellulose (NCC) produced by microbial hydrolysis of microcrystalline cellulose in tapping mode.

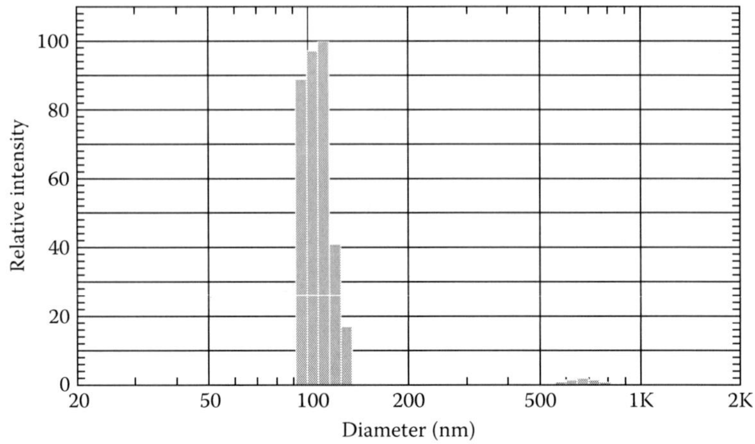

FIGURE 5.3 Particle size distribution of NCC after purification.

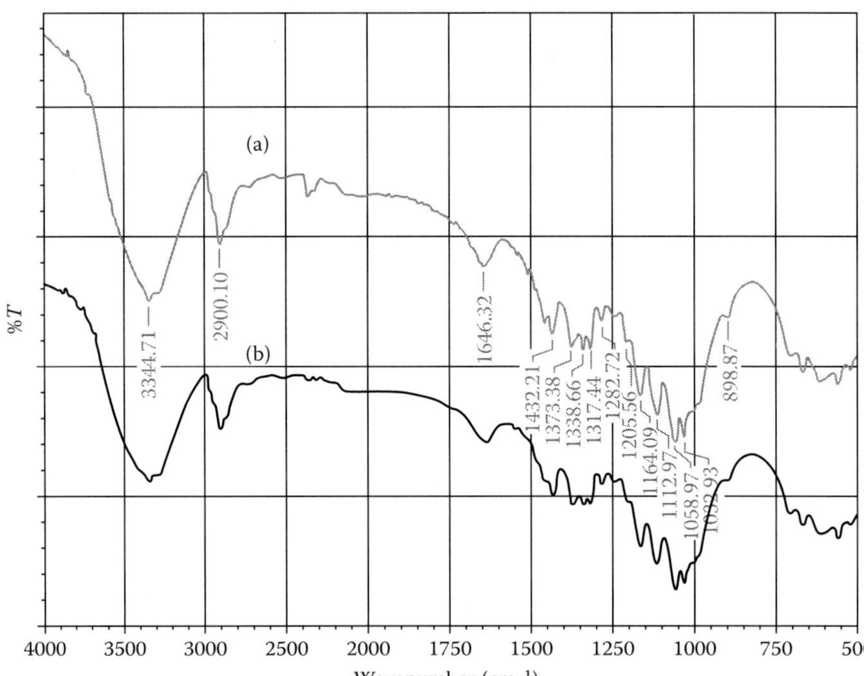

FIGURE 5.4 Fourier transform infrared spectra of (a) NCC prepared by microbial hydrolysis and (b) cellulose derived from cotton.

A bimodal size distribution was obtained with NCC of size 100 ± 30 nm being predominant and NCC of size 620 ± 90 nm was also observed. The average zeta potential of NCC prepared by acid hydrolysis is −69.7 mV and that of microbial enzyme hydrolyzed is −14.6 mV. The high negative charge on the surface of CNW prepared by acid hydrolysis may be due to the attachment of sulfate groups on its surface. The zeta potential of microbial prepared CNW was very close to that of pristine and enzymatically treated cotton as given in the literature (Buschle-Diller et al. 2005). Figure 5.4 shows the FTIR spectra of CNW and pristine cotton cellulose. The peaks match well in both the spectra and no significant difference could be observed, suggesting that there is no chemical modification in the NCC that is produced.

5.2.5 Production of NCC by Enzymatic Hydrolysis of Cellulose

There are various ways of producing cellulose in nanoform, which can be broadly classified as mechanical, chemical, and biological. Among them, mechanical methods that are necessarily top-down approaches, although afford better yields, are very expensive and energy intensive. Biological methods (Bismarck et al. 2011) involving using cultures of bacteria and fungi are very time consuming. A more feasible route is chemical hydrolysis of cellulose by acids, bases, and oxidation reagents to produce cellulose nanoparticles. However, the use of these routes has a number of critical drawbacks such as corrosivity, surface modification of cellulose, and environmental incompatibility. Enzyme-based hydrolysis of cellulose is potentially an efficient way to produce nanocellulose. Such process will generate negligible effluents and is suitable for producing nanocellulose at low energy cost.

Several enzymes act as natural catalysts for the modification of cellulosic materials. Today, several commercial preparations are available for biostoning and biofinishing of cotton fabrics that use enzymes. Cellulase is a typical example of one such enzyme that can specifically hydrolyze the glycosidic bonds in cellulose. They can act on macromolecular cellulose chains and under controlled conditions shred it down to nanodimensional size. Cellulases are multicomponent enzyme system consisting of at least three classes of enzymes, working together to degrade cellulosic material. Endo-1,4-β-d-glucanases (EC 3.2.1.4) hydrolyze internal β-1,4-glucosidic bonds (Figure 5.5) in the cellulose chain, presumably acting mainly on the amorphous or disordered regions of cellulose. Exo-1,4-β-d-glucanases, also called CBHs (EC 3.2.1.91), cleave off cellobiose units from the ends of cellulose chains. Hydrolysis to the final product is accomplished by 1,4-β-d-glucosidases (BG, EC 3.2.1.21) (Andersen et al. 2008), which hydrolyze cellobiose to glucose and also cleave off glucose units from the various soluble cello-oligosaccharides. BG activity has often been found to be rate limiting in enzyme cocktails during enzymatic hydrolysis of cellulose. Each of these components consists of a catalytic or active site that acts on the substrate and carries out the hydrolytic reaction (Figure 5.5). Apart from the active site, there is also a CBM that helps the enzyme to tether itself onto the substrate surface.

The long cellulose polymer chains are depolymerized by the combined action of three enzyme components. The hydrolysis rate is dependent on a lot of factors such as temperature, pH of medium, enzyme concentration, and substrate concentration. Available literature studies indicate that the cellulase activity (acid cellulases) is highest in the temperature range of 45–50°C and pH maintained at 4.8. Generally, the substrate concentration is in large excess so as to reduce nonproductive binding (Qing et al. 2010) of enzymes on the substrate surface. Also, the initial rate of reaction is the fastest (Lu et al. 2002) and accounts for the maximum reactant utilization. During this period, rapid hydrolysis of cellulose molecules takes place. The extent

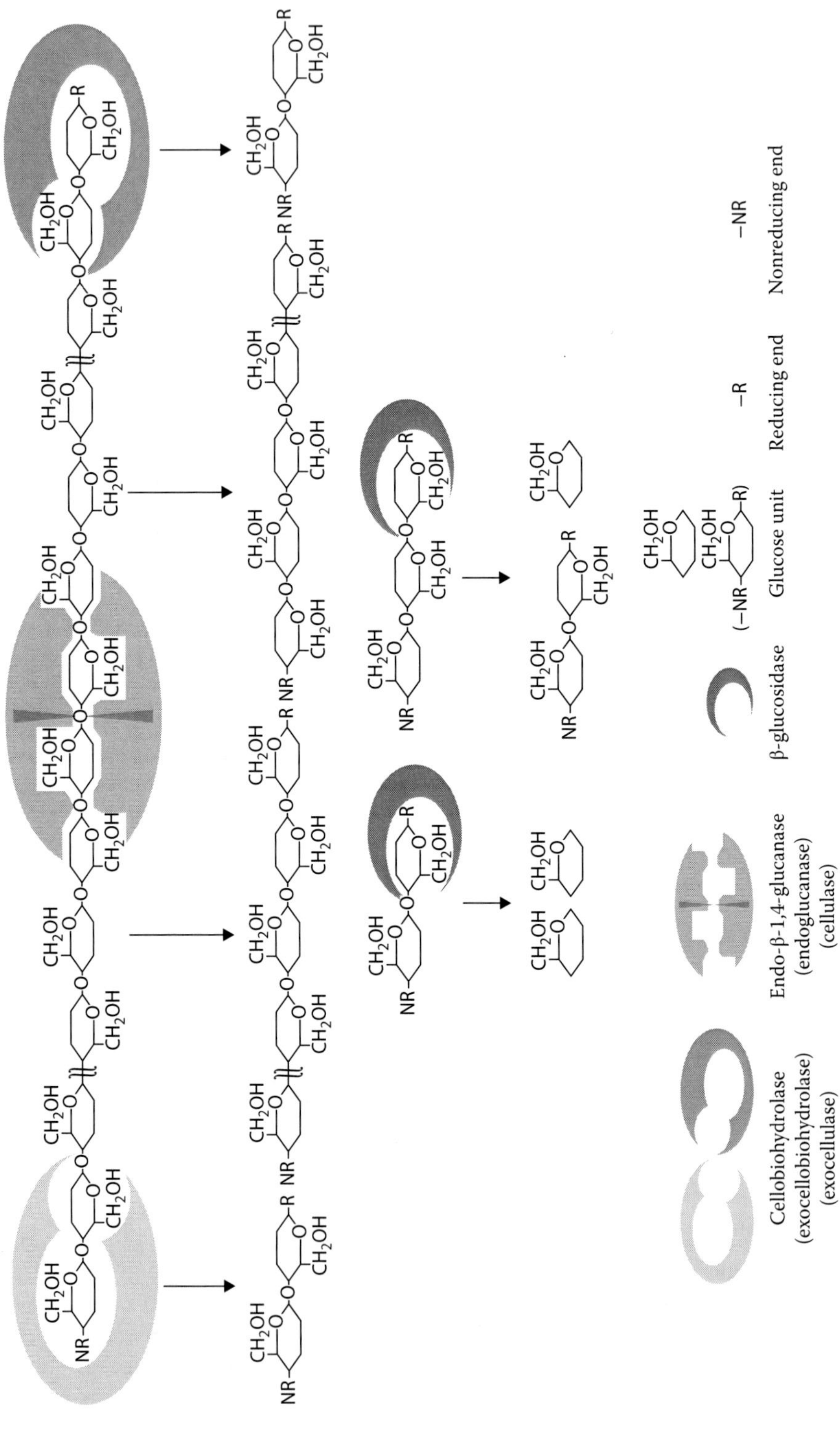

FIGURE 5.5 Cellulose hydrolysis by cellulase enzyme. (Data from Watanabe, H., and Tokuda, G., *Annu. Rev. Entomol.*, 55, 609, 2010.)

of hydrolysis depends on the reaction time for which the substrate and the enzyme are allowed to interact. The objective of harvesting nanocellulose can be achieved by disrupting the hydrolysis process. This can be achieved by disengaging the enzyme (by denaturation) from the partially hydrolyzed substrate or segregating the partially hydrolyzed cellulose (by ultrafiltration). In this way, cellulose nanoparticles may be obtained by separation and the process may be repeated on the unhydrolyzed cellulosic material to increase the yield.

The enzymatic hydrolysis of cellulosic substrate to generate nanocellulose is a very unique approach. There is a vast amount of literature that focuses on the production of biofuel from cellulosic material by enzymatic hydrolysis and a handful of reports also mention using enzyme cocktails as pretreatments prior to treatment with other reagents. One of the first reports for the preparation of nanocellulose (Paralikar and Bhatawdekar 1984) of ~50 nm diameter was by hydrolysis of MCC using cellulase enzyme. The enzymatic hydrolysis route for generating nanocellulose is not routinely followed since the difficulty lies in maintaining proper conditions for the hydrolysis and periodically harvesting the nanocellulose produced before it gets hydrolyzed further to soluble sugars. Other hurdles, besides high cost of enzyme and low yields, such as loss of enzyme activity due to denaturation and inhibition lead to nonuniform hydrolysis. This results in a broad size distribution of nanocellulose and limits its applications in specialized areas. In spite of these drawbacks, this method is attracting attention mainly due to lesser energy and financial input, and most significantly its ecofriendly nature.

Thus, enzymes can be conveniently adapted for biological synthesis of nanocellulose. And with further optimization of operating conditions, this process can be scaled up for industrial scale production of nanocellulose. The process can further be made more efficient by recycling the spent enzyme. With increased production, low operating costs, and its ecofriendly nature, this process may become the process of choice for the production of nanocellulose.

5.3 APPLICATIONS OF NANOCELLULOSE IN AGRICULTURAL AND FOREST PRODUCT INDUSTRY

Cellulose is a representative of an interesting group of sustainable natural polymeric raw materials characterized by their useful properties such as hydrophilicity and biodegradability, and is highly amenable for chemical modification in surface. Cellulose in its nanoform, referred to here as NCC/nanocellulose, combines both the properties of cellulose and the exciting properties such as increased mechanical strength and large surface area exhibited by materials in the nanolevel. NCC as a material finds application in a variety of fields and some of the applications of nanocellulose are discussed below.

5.3.1 Nanocellulose in Paper and Boards

The potential of nanocellulose in the area of manufacturing of paper and paperboards is a vast area of application that is being investigated currently. The incorporation of nanocellulose in paper can enhance the fiber–fiber bond strength, resulting in a strong reinforcement effect on paper materials. It was reported that cellulose obtained from wood can impart a very high degree of toughness to the paper. Papers of different porosities were prepared from NCC of different molar mass. Despite having a porosity of 28%, the Young's modulus of 13.2 GPa and tensile strength of 214 MPa were obtained for the toughest paper, which are remarkably higher than the conventional microfiber-based papers. The authors also reported that the tensile strength, toughness, and strain-to-failure do correlate with average molar mass of nanocellulose (Marielle et al. 2008).

It is well known that a high linting and dusting tendency of newsprint will almost certainly cause a higher frequency of production interruptions in offset printing. Such stops are connected with production losses, health hazards, and high costs for blanket cleaning. The processes such as surface and internal treatments of thermo-mechanical pulp were to reinforce the surface strength of the paper to decrease the linting and dusting of the paper during printing. Thus, it was reported that the surface treatment of the newsprint with starch along with fibrillated cellulose in its nanoform reduced the linting property of the newsprint significantly (Hainong et al. 2010).

5.3.2 Application in Films and Coatings

Nanofibrillated cellulose typically binds high amounts of water and forms gels with only a few percent of dry matter content. This characteristic has been a bottleneck for industrial scale manufacture. In most cases, fibril cellulose films are manufactured through pressurized filtering but the gel-like nature of the material makes this route difficult. In addition, the wires and membranes used for filtering may leave a so-called *mark* on the film, which has a negative impact on the evenness of the surface. Thus, VTT Technical Research Centre of Finland and Aalto University have developed a method wherein films are manufactured by evenly coating nanocellulose on plastic films so that the spreading and adhesion on the surface of the plastic can be controlled. The films are dried in a controlled manner by using a range of existing techniques. Due to the efficacy in management of spreading, adhesion, and drying, the films do not shrink and are completely even. The more fibrillated cellulose material is used, the more transparent films can be manufactured out of them. Highly transparent films of cellulose along with caroboxymethyl cellulose matrix are also reported wherein the nanocellulose was prepared by homogenization method and incremental steps of homogenization increased the transparency and mechanical properties of the film by up to two times (Siro et al. 2010). Carbon dioxide capture from atmospheric air, referred to as *air capture*, can be coupled with carbon dioxide sequestration in geological formations to reduce the atmosphere's CO_2 content, or with the conversion into CO_2–neutral liquid hydrocarbon fuels using renewable energy. Nanocellulose that has been functionalized with amine group finds application as adsorbent for CO_2 capture from air (Christoph et al. 2011).

5.3.3 Application in Barrier Films

Barrier films are one of the potential areas where nanocellulose can be used as filler in the preparation of composites. Dogan and McHugh (2007) described the preparation of edible films that can serve as a protective coating on the food itself. Fibrillated cellulose in carboxymethyl cellulose composite was also used to impart oxygen and oil barrier properties to composite films (Christian et al. 2010).

5.3.4 Application in Food Packaging

Cellulosic fibers, as paper and paperboard, have traditionally been used in packaging for a wide range of food categories such as dry food products, frozen or liquid foods and beverages, and even fresh food (Kirwan 2003). Thus, nanocellulose could be used as an additive in the preparation of packaging films using biopolymers to enhance its mechanical strength and also to reduce its oxygen permeability. Although chemically modified biopolymers such as cellulose derivatives or thermoplastic starches have been used in packaging, renewable biopolymers are currently of central interest as there is the potential to replace conventional petroleum-derived polymers typically used in food packaging (de Vlieger 2003). One of the examples is starch-based films that are limiting their application due to highly hydrophilic nature and poor mechanical properties. This problem was reportedly overcome by forming a nanocomposite of thermoplastic starch and nanocellulose composite. It was reported that the incorporation of nanocellulose in its fibrillated form in the starch matrix improved mechanical strength by up to 46% and also improved the water vapor permeability of the starch matrix (Savadekar and Mhaske 2012).

The use of 2,2,6,6-Tetramethylpiperidinooxy (TEMPO)-oxidized cellulose whiskers obtained from Whatman filter paper was investigated as barrier membranes in a polyvinyl alcohol (PVA) matrix (Paralikar et al. 2008). The incorporation of nanocellulose reduces the water vapor transmission rate (WVTR) of the membranes. However, the authors also obtained more interesting results when nanocomposites were prepared by mixing PVA matrix with low quantities of poly(acrylic acid) (PAA) and nanocellulose. These results can be ascribed to the fact that cross-linking with PAA reduces the number of hydroxyl groups in the composite and thus its hydrophilicity. The incorporation of cellulose nanocrystals into the composite provides the physical barrier through the creation of a tortuous path for the permeating moisture. Overall, it has been seen that the membrane barrier properties were improved by the addition of cellulosic crystals in combination with PAA.

In another study, the effect of nanocellulose whiskers derived from soft wood on water vapor barrier properties of xylan/sorbitol films has been reported (Saxena and Ragauskas 2009). The nanocomposites were prepared with different amounts of whiskers as filler in xylan/sorbitol matrix. The incorporation of cellulose whiskers reduced the WVTR of xylan/sorbitol composite. It has been seen that at 10 wt% of nanocellulose decreased the WVTR of the composite from 304 $gh^{-1}{\cdot}m^{-2}$ (that of the control films without any filler) to 174 $gh^{-1}{\cdot}m^{-2}$. The authors reported that the high degree of crystallinity of whiskers and its rigid hydrogen-bonded network governed by a percolation mechanism are responsible for the improvement of barrier properties of the films, decreasing the WVTR of the final nanocomposite.

5.4 CONCLUSIONS

Nanocellulose is an interesting class of sustainable biopolymer that has potential applications in agriculture and forest product industry. Even though the biological method for the synthesis of nanocellulose offers new ecofriendly method for the synthesis of NCC, the major bottleneck in this route is the low yield of nanocellulose compared to the commercially established chemomechanical process. With the scope of producing nanocellulose with low-energy consumption and using nanocellulose in high-value products, the low yield of NCC by biological route can be compensated for. Also, the yield of NCC can be improved considerably by using a continuous system of production. Overall, the biological route for NCC production shows a lot of promise for commercial exploitation for application in diversified fields.

ACKNOWLEDGMENTS

The authors are thankful to Dr. A.J. Shaikh and Dr. S.K. Chattopadhyay of Central Institute for Research on Cotton Technology, Mumbai, India, for their kind suggestions and support for this research work. This research was financially supported by the National Agricultural Innovation Project, Indian Council of Agricultural Research through its subproject entitled "Synthesis and characterization of nanocellulose and its application in biodegradable polymer composites to enhance their performance," code number 417101.

REFERENCES

Andersen N, Johansen KS, Michelsen M, Stenby EH, Krogh KBRM, Olsson L (2008). Hydrolysis of cellulose using mono-component enzymes shows synergy during hydrolysis of Phosphoric Acid Swollen Cellulose (PASC) but competition on Avicel. *Enzyme and Microbial Technology* **42**:362–370.

Araki J, Wada M, Kuga S, Okana T (1998). Influence of dope solvents on physical properties of wet-spun cellulose. *Colloids and Surfaces A: Physicochemical Engineering Aspects* **42**(1):75–82.

Bismarck A, Juntaro J, Mantalaris A, Pommet M, Shaffer MSP (2011). Material comprising microbially synthesized cellulose associated with a support like a polymer and/or fiber. Also published as US 20110021701A1.

Buschle-Diller G, Inglesby MK, Wu Y (2005). Physicochemical properties of chemically and enzymatically modified cellulosic surfaces. *Colloids and Surfaces A: Physicochemical Engineering Aspects* **26**:63–70.

Cao NJ, Xia YK, Gong CS, Tsao GT (1997). Production of 2, 3-butanediol from pretreated corn cob by Klebsiella oxytoca in the presence of fungal cellulase. *Applied Biochemistry and Biotechnology* **63**:129–139.

Christian A, Gallstedt M, Lindstrom T (2010). Oxygen and oil barrier properties of microfibrillated cellulose films and coatings. *Cellulose* **17**:559–574.

Christoph G, Wurzbacher JA, Tingaut P, Zimmermann T, Steinfeld A (2011). Amine-based nanofibrillated cellulose as adsorbent for CO_2 capture from air. *Environmental Science and Technology* **45**:9101–9108.

de Vlieger J (2003). Green plastics for food packaging. In: Ahvenainen R (ed). *Novel food packaging techniques*. New York, NY: CRC Press LLC.

Dogan N, McHugh TH (2007). Effect of microcrystalline cellulose on functional properties of hydroxy propyl methyl cellulose microcomposite films. *Journal of Food Science and Technology* **72**(1):E16–E22.

Hainong S, Ankerfors M, Hoc M, Lindström T (2010). Reduction of the linting and dusting propensity of newspaper using starch and microfibrillated cellulose. *Nordic Pulp and Paper Research Journal* **25**(4):495–503.

Kirwan M (2003). Paper and paperboard packaging. In: Coles R, McDowell D, Kirwan MJ (eds). *Food packaging technology*. Boca Raton, FL: CRC Press LLC, pp. 241–281.

Krishnamachari P, Hashaikeh R, Chiesa M, Gad El Rab (2012). Effect of acid hydrolysis time on cellulose nanocrystals properties: Nanoindentation and thermogravimetric studies. *Cellulose Chemistry and Technology* **46**(1–2):13–18.

Lu Y, Yang B, Gregg D, Saddler JN, Mansfield SD (2002). Cellulase adsorption and an evaluation of enzyme recycle during hydrolysis of steam-exploded softwood residues. *Applied Biochemistry and Biotechnology* **98–100**:641.

Luo J, Xia LM, Lin JP, Cen PL (1997). Kinetics of simultaneous saccharification and lactic acid fermentation processes. *Biotechnology Progress* **13**:762–767.

Marielle H, Berglund L, Isaksson P, Lindstro T, Nishino T (2008). Cellulose nanopaper structures of high toughness. *Biomacromolecules* **9**:1579–1585.

Martinez AT, Speranza M, Ruiz-Duen FJ, Ferreira P, Camarero S, Guille F, Martinez MJ, Gutierrez A, del Rio JC (2005). Biodegradation of lignocellulosics: Microbial, chemical, and enzymatic aspects of the fungal attack of lignin. *International Journal of Microbiology* **8**:195–204.

Olsson L, Hahn-Hagerdahl B (1996). Fermentation of lignocellulosic hydrolysates for ethanol production. *Enzyme and Microbial Technology* **18**:312–331.

Paralikar KM, Bhatawdekar SP (1984). Hydrolysis of cotton fibers by cellulase enzyme. *Journal of Applied Polymer Science* **29**:2573–2580.

Paralikar SA, Simonsen J, Lombardi J (2008). Poly (vinyl alcohol)/cellulose nanocrystals barrier membranes. *Journal of Membrane Science* **320**:248–258.

Prasad S, Prateek J, Balasubramanya RH, Vigneshwaran N (2011). Preparation and characterization of cellulose nanowhiskers from cotton fibers by controlled microbial hydrolysis. *Carbohydrate Polymers* **83**:122–129.

Qing Q, Yang B, Wyman CE (2010). Impact of surfactants on pretreatment of corn stover. *Bioresource Technology* **101**:5941–5951.

Savadekar NR, Mhaske ST. (2012). Synthesis of nano cellulose fibers and effect on thermoplastics starch based films. *Carbohydrate Polymers* **89**(1):146–151.

Saxena A, Ragauskas AJ (2009). Water transmission barrier properties of biodegradable films based on cellulosic whiskers and xylan. *Carbohydrate Polymers* **78**:357–360.

Siro I, Plakette M, Hedenqvist, Ankerfors M, Lindström T (2010). Highly transparent films from carboxymethylated microfibrillated cellulose: The effect of multiple homogenization steps on key properties. *Journal of Applied Polymer Science* **119**(5):2652–2660.

Ververis C, Georghiou K, Christodoulakis N, Santas P, Santas R (2004). Fiber dimensions, lignin and cellulose content of various plant materials and their suitability for paper production. *Industrial Crops and Products* **19**:245–254.

Watanabe H, Tokuda G (2010). Cellulolytic systems in insects. *Annual Review of Entomology* **55**:609–632.

Wilson DB (2011). Microbial diversity of cellulose hydrolysis. *Current Opinion in Microbiology* **14**:1–5.

Zaldivr M, Velasquez JC, Contreras I, Perez LM (2001). Trichoderma aureoviride 7–121, a mutant with enhanced production of lytic enzymes: Its potential use in waste cellulose degradation and/or biocontrol agent. *Electronic Journal of Biotechnology* **4**(3):1–7.

6 Applications of Nanotechnology in Aerospace

Leonard L. Yowell and Padraig G. Moloney

CONTENTS

6.1 INTRODUCTION

During the past 15 years, the promise of technological advancement through the design and manipulation of materials at the nanoscale has led to great excitement in the popular media and to steady increases in government, private, and corporate research and development (R&D) spending. Significant enhancements in materials performance, including high-strength lightweight structures, improved conducting and semiconducting properties, and the extremes of thermal transport properties, have first interested the sectors of our economy most sensitive to advancement from materials research. The aerospace community—along with other high-performance needs sectors such as renewable energy, medicine, and high-end sports equipment—is still early in the experimentation and successful adoption of nanotechnology in its products.

In this chapter, we will provide a survey of current and near-term use of nanomaterials in aircraft and spacecraft across military, civilian, and commercial sectors. We will present the current state of the art across systems—structures, avionics, power, energy storage, propulsion, life support, and thermal management. A survey of scientific and trade literature will be supplemented with interviews and personal communication with academics, government researchers and program managers, and industry representatives. A summary will be made of the most salient of the challenges that remain in the research, development, and deployment of nanotechnology-enabled applications in the aerospace sector.

6.2 SENSORS

Nanoscale materials offer the ability to reduce both weight and power requirements, while enhancing the fidelity of sensors. Numerous studies have shown the applicability of carbon nanotube (CNT) to this goal. Some of this work has utilized the change in electrical resistivity of nanotubes upon mechanical deformation.[1,2] These types of sensors offer advantages not found in conventional strain sensors such as being embedded within the structure and providing isotropic data at a vastly larger number of points. When combined with a polymer in composite form, a clear application is as a low-weight, low-power structural health monitor.[3] In addition to strain, a high applicability to pressure has been found.[2,4,5] Changes in electrical properties can also be harnessed to sense a flow of both liquids and gases, with great potential to aerospace applications.[6,7] Temperature sensors using nanotubes have also been researched.[8] Some have taken the approach of using shifts in the Raman D band with temperature,[4] while others have focused on electrical properties.[9]

CNT sensors at the research level have yielded impressive results in a laboratory setting including parts per billion sensitivity to chlorine,[10] ammonia[11] and nitrogen dioxide.[12] Among the most common approaches to these sensors is through field effect transistors.[13,14] In the literature, the most frequent target analytes for applying CNT sensors appear to be ammonia (NH_3), nitric oxide (NO), nitrogen dioxide (NO_2), and hydrogen (H_2), with most of the remainder focusing on water vapor, carbon monoxide (CO), volatile organics, and more directed chemicals such as nerve gas simulant dimethyl methylphosphonate.

NASA Ames Research Center was an early applier of CNTs to sensing, with small chemical sensors for a variety of applications in space and on the earth.[15–18] The platform is a 32-channel sensor chip (1 cm × 1 cm) with different nanostructured materials for chemical sensing (Figure 6.1). NASA Ames' chemical sensor module was flown as a secondary payload of a Navy satellite (Midstar-1) and was

(a) (b) (c)

FIGURE 6.1 (a) NASA Ames chemical sensor module was on a secondary payload of a Navy satellite (Midstar-1) launched through Atlas V on March 9, 2007. (b) A 32-channel sensor chip (1 × 1 cm) with different nanostructured materials for chemical sensing. (c) The nanosensor module (5″ × 5″ × 1.5″) contained a chip of 32 sensors, a data acquisition board, sampling system, and a tank with 20 ppm NO_2 in N_2. (Courtesy of Center for Nanotechnology, NASA Ames Research Center.)

launched through an Atlas V rocket on March 9, 2007. The sensor data were downloaded for 60 days. Subsequently, the nanosensor technology was integrated into the JPL "electronic-nose" aboard the International Space Station in January 2009 to monitor air quality in the crew cabin, especially formaldehyde.[19] The sensor worked well in space and helped demonstrate the viability of the technology in a space environment. The maturation of this technology has since reached a technology readiness level (TRL) of six as a result of NASA Ames working on homeland security and fire department applications on the earth. In space, the Ames sensors are intended to be used for fire safety and various types of hazardous leak detection applications (Meyyappan, M. Conversation with authors, 2011).

The application of CNTs to sensors is especially susceptible to lack of quality control of the raw materials, with inconsistencies in chirality, length, and diameter of the nanotube affecting performance. The ongoing maturation of the CNT industry continues to improve viability in this area. Sensors using nanotubes for the detection of NO among others have been successfully commercialized. Fam et al.[20] have noted the number of patents awarded on CNT sensor arrays has risen enormously in recent years and surmise that challenges in device fabrication, nanotube consistency, and the understanding of functioning mechanisms are being solved. They also point to the successful integration of CNTs into industry standard CMOS circuitry as significant for fabricating commercial devices.[21,22] Although to a lesser extent than CNTs, nanowires of other materials have received some focus as gas sensors. Wires of In_2O_3, etched silicon, and SnO_2 to test various gases are just a few examples.[23–25]

Electrically conducting polymers applied as thin films or composites have been researched as sensors for volatile gases. Examples include CO and NO_2 sensing using polyaniline-SnO_2/TiO_2 and polyhexylthiophene nanocomposites and high sensitivity for hydrazine using polythiophene and polyaniline.[26–29] Although of more relevance to biological sensing applications, Rajesh et al. have reviewed conductive polymers for nanocomposite applications.[30] Included is a summary that highlights analyte, diameter/size, detection limit, and voltage among other important data.

Nanophotonics is an area of great promise to sensors, using optical phenomena close to or past the diffraction limit. Although still in relative infancy, rapid progress has been made in the past decade on the fabrication of nanometer scale devices using quantum wires and dots, photonic crystals (PhC), and holographic optical elements.[31] Simulations have shown the viability of integrating PhCs with composites, utilizing the large changes in PhC optical properties when stimulated with small changes in dimension.[32] Mechanical forces on the nanoscale have been detected using microcantilevers combined with arrays of optical waveguides, providing a new level of sensitivity.[33] "Label-free detection" of biological substances by the direct sensing of their dielectric properties may enable new classes of low-weight, low-power sensors relevant to micro air vehicles.[34]

The use of optical fiber data cables on aerospace platforms has grown in the past decade, and combined with the increasing amount of data use and generation on these platforms, the need for optical integrated circuits has grown. Nanophotonics as a field has enhanced and produced photonic integrated circuits such as optical switches, multiplexers, laser diodes, and sensors that offer reduced weight and improved performance.[31] Research on the nanoscale continues to shrink the size, weight, and power requirements of such systems and produce new devices, such as microdot lasers, applications of which for aerospace have yet to be explored.

6.3 ELECTROMAGNETIC INTERFERENCE/ ELECTROSTATIC DISCHARGE

6.3.1 Nanomaterials on Deep Space Exploration Spacecraft

CNTs were used on several parts of the Juno spacecraft, built by Lockheed Martin for NASA. Juno's mission is to explore the environment of Jupiter using an array of sensors. Launched on August 5, 2011 (Figure 6.2), this example of

(a) (b)

FIGURE 6.2 (a) NASA's Juno spacecraft is raised out of a thermal vacuum chamber following tests that simulated the environment of space, March 13, 2011. (Courtesy of NASA/JPL-Caltech/Lockheed Martin.) (b) The Juno spacecraft, enclosed in the payload fairing at the top of an Atlas V rocket, begins its journey to Jupiter on August 5, 2011. (Courtesy of NASA/Bill Ingalls.)

CNTs being used in a real spacecraft is the latest example of successful application of nanomaterials in aerospace.[35] Nanocomp Technologies Inc. is the primary developer of the EMSHIELD, a product made of sheets of nanotubes with applicability to electrostatic discharge protection. With the composite materials of the Juno spacecraft traveling through areas of high radiation, electromagnetic protection was needed. Interestingly it was not the nanotubes' high electrical conductivity coupled with lightweight features that made the CNTs preferable to other solutions. The traditional solution would have been an aluminum foil applied to the areas of the spacecraft needing protection. The CNT sheets that were selected had the advantage of being applied in an integral manner with the composite parts themselves.

While the electrical performance of the nanomaterial was good, it is worthy of noting that the central factor in selecting the material came from a manufacturing and processing advantage. In contrast to traditional metal systems, the nanomaterial was easy to integrate into the final composite by co-curing. We see this as being a positive and noteworthy result and likely trend in nanomaterials for the aerospace sector. Lockheed Martin was the primary contractor and integrator for the Juno mission and was principal in qualifying these materials for spaceflight and bringing nanomaterials to a high TRL. Lockheed internally developed CNT-infused fibers for various parts for Juno, primarily for contoured supports or struts to prevent thermal shielding from interfering with the three solar array support mechanisms. The materials also have the added benefit of multifunctionality by providing an electrical discharge path.[36] The use of nanomaterials in critical components for Juno marks the first high TRL or mission condition use of nanotechnology at NASA.

6.4 AVIONICS/ELECTRONICS

6.4.1 Nanomaterials-Based Memory for Spaceflight

The harsh environment of space, temperature extremes, radiation, and power restrictions for spacecraft are concerns for avionics and computer systems using current technologies. As NASA and the international community endeavor to expand space exploration beyond low earth orbit, future technologies will be needed to enable these systems to survive beyond the comparably benign environment of low earth orbit. One such technology based on nanomaterials that shows promise for near-term application to spaceflight is nonvolatile memory developed by Nantero. Nantero's government business was acquired by Lockheed in 2008.[37] Nantero's NRAM™ is based on high-quality and extremely pure CNTs and attains a solid-state nonvolatile memory that offers radiation hardening/tolerance for space application. The technology is based on the physical position of CNT strips stretched across gaps, where the 0 or 1 position is marked by the physical position of the nanostructure, resulting in short start-up times and resistance to faults and radiation-induced errors. An important milestone for the advancement of this technology was in May 2009 when the Space Shuttle Atlantis (STS-125) launched with an NRAM assembly in the aft payload bay with 100 chips to test the spaceworthiness of the memory in launch, on-orbit, and landing configurations.[38] This initial work has proved successful and work is under way to scale up the technology and increase the density and performance of the memory.

6.4.2 Wiring

While not yet fully realized as high maturity technology in aerospace, developments in high electrical conductivity carbon fibers are worthy of highlighting. Much research has concentrated upon and continues on these materials, mainly single-wall CNTs (SWNT), for a host of different reasons including the potential for higher conductivity than copper, high strength to weight ratio, better thermal management, and higher current carrying capacity. Reliable, affordable, lightweight conductive fibers could enable the replacement of heavy copper wiring in all manned and unmanned air and space vehicles. Usage of electrical solutions versus hydraulic systems aboard various aerospace platforms (e.g., Boeing 787) only increases the need for lightweight conductors. Nanocomp is among a number of enterprises and research groups pursuing this goal, having received funding awards from both the U.S. Air Force and Army.[39,40] While not necessitating the high conductivity of copper for power applications, it is notable that CNT cables for data transmission have been flown and successfully tested aboard a satellite.[41] Basic research by the National Institute of Standards and Technology, Boeing, NanoRidge, and Rice University into improving the conductivity of CNTs is ongoing, with recent research claiming resistivity on the order of 10^{-5} Ω·cm, within 1 order of magnitude of copper.[42]

6.4.3 Thermal Management in Electronics

One of the significant challenges in the miniaturization of electronics is the dissipation of heat in today's high-performance microchips. The Defense Advanced Research Projects Agency's Thermal Ground Plane Program seeks to encourage development of high-performance electronic systems at high-power densities and reduced weight and complexity. General Electric, the Air Force Research Laboratory, and universities have worked together to advance the state of the art.[43] The consumer electronics sector is a powerful driver for the global materials R&D community to focus on the challenge, and it is likely that the aerospace sector will be a secondary beneficiary of advancements made in this field. For instance, better thermal management materials could benefit commercial aircraft by reducing the weight of cabin lighting heat sinks. Much of the effort focuses on CNTs with their highly anisotropic thermal properties, graphene, and other nanoparticles.[44,45] NANOPACK is a European project that seeks to speed technological advancement in the field of thermal interfaces and interconnects in the electronics packaging arena. This consortium of industry and academic entities highlights the interest and commitment to address the thermal challenge in electronics and will certainly benefit the aerospace sector in the future. In addition, research driven by both consumer and aerospace industries has already shown the promise of utilizing microelectromechanical systems (MEMS) and nanomaterials toward improved thermoelectrics for both thermal management and energy generation.[46] Thermoelectric generators, despite poor efficiency, have a long history of use on NASA space probes such as Cassini and Voyager.[47]

6.5 STRUCTURES

6.5.1 Primary and Secondary Structures

Given the potential electrical, mechanical, and thermal properties of nanoscale materials, in particular SWNT, improvements in composite material performance through nanotechnology was, and continues to be, anticipated by the aerospace community. Research literature has shown great progress, but high TRL applications for primary structures have been slower to arrive.[48–50] Recent progress in improving the consistency and quality of CNT material has led to more mature aerospace possibilities. The F-35 Lightning II (Figure 6.3), built by Lockheed Martin, is likely the first mass produced aircraft to use CNT reinforced composites in structural applications.[51] The nanomaterials are not used in primary load-bearing structures, but in wingtip fairings, as has been released in the public domain. There are also hundreds of other parts that are candidates for nanocomposite application due to weight, cost, and manufacturability advantages. For example, a wing material with improved mechanicals that is also electrically conductive to incorporate lightning strike protection or deicing capability has been highlighted by industry experts as very attractive (Maguire, R. Conversation with authors, 2011). The public success Lockheed has had with nanomaterials in air and spacecraft is undoubtedly noticed by other prime and secondary contractors, and it is likely that other applications of nanomaterials are being developed by others.

FIGURE 6.3 According to open press, the Lockheed F-35 Lightning II, also known as the Joint Strike Fighter (JSF), contains nanomaterial-reinforced composites in structural applications. (Courtesy of Lockheed Martin.)

6.5.2 Nondestructive Evaluation

Improved quality assurance and nondestructive evaluation techniques will be increasingly important in the acceptance of new structural materials in commercial, military, and civilian aircraft and spacecraft. Recent work by the Wardle group at MIT[52] focused on the inclusion of CNTs within composite materials at a sufficient volume percentage to allow resistive heating to occur at low-power levels. Localized heating of a composite material permits the use of infrared (IR) thermography to then monitor the part and detect defects or cracks in the material at resolution levels that are better than the current state of the art. Traditional IR thermography relies on heating areas of interest through a peripheral source, applied externally or requiring a substantial amount of onboard power and additional equipment. In the aerospace sector, power, weight, and volume are at a premium, so enhancements to this technology should be viewed as promising. Further R&D is needed though, to assure that any enhancements in resolution and reduced power and mass will not cause interference, or complexity that becomes an overall detriment to the system. For a CNT-based IR thermography capability to mature for aerospace, a long lead time is necessary, since CNTs need to be already within the system for the technique to work. Another consideration in using CNTs to enhance IR thermography would be the geometric configuration and location of the material within the system.

Enhancing the conductivity of any part that needs to be evaluated using NDE is an advantage. In particular, the addition of CNTs or carbon nanofibers[53] may improve other materials systems by improving the utility and effectiveness of electromagnetic inspection techniques such as eddy current testing.

6.6 PROPULSION

6.6.1 Gas Turbine Engines

It has been demonstrated for a number of years that nanostructured ceramic systems offer the potential to improve the performance of thermal barrier coatings (TBC) in gas turbine engines. Enhanced performance includes modification and reduction of thermal conductivity and improved mechanical durability and resistance to coating degradation. Even a small improvement in the thermal barrier performance of a TBC on a turbine blade can lead to significant increases in operating temperatures and efficiency in gas turbine engines. Plasma spray and electron beam deposition techniques can be modified to incorporate nanostructured additives or to form nanoscale morphologies that enhance the performance. A cursory review of filed patents and positive results reported in scientific literature hint that engine makers are actively engaged with nanostructured coatings in R&D and potentially in production.[54–56]

6.6.2 Propellants

It has been reported for years that solid rocket motor propellants could be enhanced by the inclusion of nanoscale particles. In the case of nanoscale aluminum, the enhancements come from a higher specific surface area resulting in better reactivity and faster burn rates by a significant degree. A disadvantage with higher surface areas comes from the nature of the geometry itself. Since bare aluminum is the reactive element, any coatings applied for stability or dispersion, and prevention of aggregation, will reduce the fraction of active elements by weight. Also, increased reactivity of a highly energetic material can lead to concerns of safety in handling and preparation. Research in 2005 in Italy/Russia using Russian nanoparticles highlighted the increase in reactivity of nano-Al and reliable faster burn rates. It also showed evidence of hazards associated with increased reactivity and the tendency for nanoparticles to agglomerate and negatively affect performance.[57]

In 2009, a university collaboration between Penn State and Purdue Universities with the support of the Air Force Office of Scientific Research and NASA conducted a flight test of a rocket using a nanoscale aluminum particle (produced by Austin-based NovaCentrix) and water ice as propellant.[58] The novelty of this mixture sheds light on potential for development and production of propellants in remote environments for space exploration. Any amount of materials that could be produced through in situ *resource utilization* (ISRU) would greatly increase the efficiency and sustainability of in-space propulsion technologies. If water or ice could be used as an oxidizer/fuel, then the recent discoveries of water/ice on the Moon and Mars could open up new possibilities and mission architectures for long duration human space exploration. Since 2009, the group has continued to optimize the propellant mixtures and more confidence has been gained in safety and performance.[59] Due to the national security implications of advances in energetic materials, there is a limitation on the public availability and disclosure of information on operational readiness and flight tests.

6.7 POWER AND ENERGY STORAGE

Whether for aircraft or spacecraft, the ready availability of power is crucial to accomplishment of the mission as well as the survival and performance of the crewmembers. All aerospace applications have the constraints of mass and volume as well as system complexity implications. For unique space applications where only a single or limited number of vehicles are produced, cost is not as much of a concern as is the performance when traded against the weight, volume, and complexity of a system. However, for commercial (and to a lesser extent, military) aircraft production, the cost implications of production and operating the systems are crucial. Similar cost-sensitive traits will apply to the nascent commercial crew and cargo launch industry. In this section, we will highlight some issues and state of the art with respect to lithium ion batteries, supercapacitors, fuel cells, and solar cells.

6.7.1 Lithium Ion Batteries

It is likely that at least for the near-term, lithium ion batteries will be used for an increasing number of applications in aerospace that require high energy density and mass/volume efficiency. Lithium ion batteries are composed of a lithium ion–conducting electrolyte between a negative electrode that stores lithium and a positive electrode into which the lithium ions migrate during discharging. A common material for anodes is graphite; however, there are other materials that offer a higher specific capacity, such as alloys and configurations of lithium, silicon, carbon, titanium, and germanium. The geometry and morphology of any proposed system is critical, due to the impacts on mechanical deformation during discharging and charging.

One of the critical issues in lithium ion batteries is the sizeable volume change that occurs with the lithium ion insertion and removal. The large mechanical deformation that takes place in these materials can seriously limit the number of cycles that a material can remain functional. With nanomaterials, there is great potential to adjust the pore sizes and distributions so as to allow room and flexibility for ions to move and be accessible while allowing the structure to accommodate the changes. By using electrodes with higher available surface area, there is potential for new reactions and higher rates of discharge. Intrinsically, when a material is designed at the same size scale as the charge carrier (i.e., the nanoscale), it is easy to imagine wholly different strategies to enhance the advantages of size and mitigate the disadvantages. For instance, the potential for shorter path lengths for electronic transport can enable the use of different modes of performance and electrolyte composition and configuration. However, lithium ion batteries have nontrivial safety considerations, related to operating temperatures and the potential for cell rupture. In an aircraft or on a spacecraft, especially manned vehicles, there is little room for catastrophic failure modes for highly energetic systems.

Some recent work in the United States has shown promise at allowing the high theoretical specific capacities of Si alloys to be realized with a hierarchical approach to anode morphology and configuration.[60] The work centers on the need for mitigation of the large volume changes during operation that limit battery cycling without using exotic materials or configurations that are not easily scaled up in the short term. The work is highlighted in this chapter because it shows that nanomaterials can be used in nearer-term applications in ways that would be very familiar to current battery material manufacturers, which would therefore increase the likelihood of adoption of the technology into aerospace use. The researchers chose a porous Si-C composite to take advantage of a geometry that would allow for the expansion/contraction of the silicon and could take advantage of the carbon for structural integrity and electrical conductivity. The hierarchical approach begins with submicron carbon black nanoparticles that are annealed to remove impurities, for graphitization, and linking of other nanoparticles. The annealed carbon black is then coated with silicon nanoparticles, resulting in Si-C composite nanoparticles. These building blocks can then be aggregated into rigid spherical granules that have the right level and size of interconnected porosity and that are similar in size to milled graphite, which is routinely used in lithium ion battery production. The researchers report that the electrodes produced in this manner exhibited high capacity, stability, and a high-rate capability. This approach is an interesting example of sophistication at the technical economic boundary that seems to be key to the uptake of nanomaterials into real-world applications.

In this work we have mentioned something that is somewhat common across all nanotechnology, which is the need for manufacturing consistency, specificity, and control. In this case, it is the cost of exotic nanomaterial morphologies that is seen as a limiter. Of course, when a nanomaterial shows promise in a segment of our economy with widespread potential for profit—that is, lithium ion batteries—any breakthroughs or significant discoveries would almost certainly be kept as trade secrets until a time arose where the invention were well established and competitive advantage secured.

6.7.2 Supercapacitors

In a Ragone plot (Figure 6.4) of specific energy versus power, supercapacitors or ultracapacitors occupy a unique place between fuel cells and batteries, which have mainly high specific energies, and traditional dielectric capacitors, which have mainly high specific power. Double-layer electrochemical capacitors offer a bridge between the two types of energy devices that are useful in a range of applications where a significant amount of power is needed for a relatively short period of time. While supercapacitors are similar in configuration to batteries (electrolyte between anode and cathode), the power comes from charge on a surface as opposed to the bulk material as within a battery. So, a lot of power is available quickly, but only for a limited amount of time, whereas a battery or fuel cell can hold a relatively greater amount of energy but is limited in the charge/discharge rate. One of the earliest projected uses of nanomaterials in the late 1990s was in supercapacitors due to the inherently high surface areas and conductivities. Various morphologies of carbon nanomaterials have been attempted in the literature—from spheres to tubes to horns to onions. As in the case of lithium ion batteries, there is enormous potential in designing materials within supercapacitors to be configurable to the specific electrolyte and use of the device. Fine-tuning porosity is of key importance as is the availability of low cost and controlled properties of starting carbon material.[61] As supercapacitors gain in the area of energy density and slowly approach that of lithium ion batteries, more attention will be paid to them in aerospace. For instance, the Airbus A380 contains banks of supercapacitors in each of the plane's 16 emergency

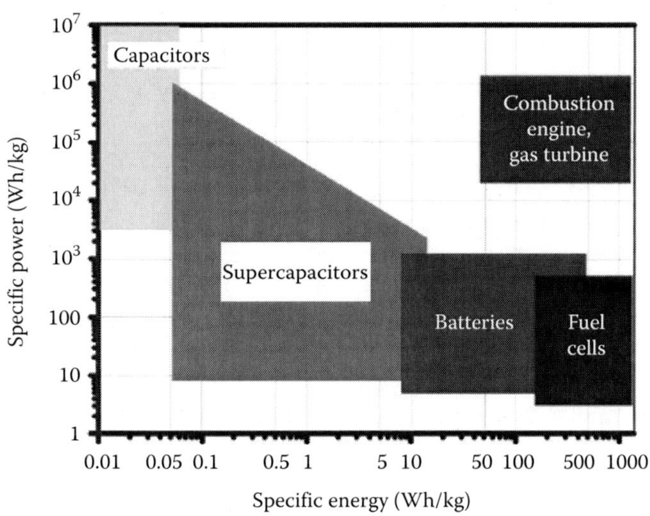

FIGURE 6.4 Simplified Ragone plot showing comparisons of energy storage systems. (From Winter, M. & Brodd, R. J., *Chem. Rev.*, 104, 4245–4270, 2004.[63])

doors, although not with the use of nanomaterials. Also, NASA has experimented with a combination of battery and supercapacitors for an astronaut self-rescue device for extra-vehicular activities in microgravity. We found no information on nanomaterial-enhanced supercapacitors being used on flying aircraft or spacecraft, but given the positive results of research and the lowering of cost of nanomaterials, it is likely that the materials are either flying or are being seriously tested by producers. FastCAP Systems Corp. markets ultracapacitors for the automotive industry that are based on research at MIT[62] using vertically aligned multiwall CNTs, demonstrating a significant maturation of the technology.

6.7.3 Power Generation

While the terrestrial applications of fuel cells are numerous, there is considerable need for efficient and robust fuel cells for use on long duration space missions. Of particular interest for the use of nanomaterials is the proton exchange membrane fuel cell (PEMFC). PEMFCs offer an avenue for ISRU, in particular the use of water on the Moon or Mars as a potential fuel supply. Nanoparticles are of interest as catalyst materials as well as for catalyst support materials. The more that fuel is in contact with the catalyst, the more power can be drawn from the cell. Presenting such catalysts efficiently to the fuel by catalyst support materials is critical. Specifically, CNTs could be used as a porous network of conducting material that supports platinum catalyst and aids in thermal and water management.[64]

The technology maturity level for aerospace-relevant solar cells enhanced with nanomaterials is low and is not covered in depth in this chapter. Research and development is under way in academia, industry, and government labs to advance nanomaterial-based solar cells. Some promising areas of development are in the use of quantum dots, nanoparticles, and nanowire geometries. Specifically in the case of nanowires, it is interesting to consider the potential for nanowires to reduce cost of traditional solar cells by reducing the amount of expensive materials needed and purification requirements and simplifying fabrication processes.[65]

6.8 LIFE SUPPORT SYSTEMS

The greatest challenges for life support technologies surround human exploration of space. Sustaining life and enabling human performance is necessary for both short and medium duration missions in low earth orbit and for projected long duration missions to the Moon, Mars, and beyond. In either case, some of the needs are common—air, water, food, and a functional habitable space. In low earth orbit, the international spacefaring community has learned a great deal about the difficulties on providing a sustainable environment for humans in space. As humankind prepares for missions of multiple years in duration to destinations such as Mars, it is clear that significant advances need to be made in life support systems. For instance, CO_2 removal is essential to providing a nontoxic atmosphere, and current systems are limited by the ability to regenerate and by power.[66,67] While we do not have examples of flying life support systems using nanomaterials, we want to highlight the great potential of high surface area and conductive nanomaterials to act as scaffolds for the reversible adsorption and desorption of carbon dioxide with the potential for low volatility and low consumption of power. The hierarchical bottom-up approach mentioned in Section 6.7.1 has great applicability for air revitalization materials that could enable long duration human space missions. In the area of water filtration and recovery, the challenges in microgravity and in remote locations are many; however, there have been several examples of water filtration systems based on nanomaterials. Carbon fullerenes have shown promise in the purification of water,[68] while NASA-sponsored nanoscale filtration technology has entered the commercial marketplace.[69] With respect to food, nanomaterials stand to impact significantly as packaging materials that enable very long shelf stability.

For long duration missions with limited consumables and logistics support, the resiliency and robustness of these nanomaterial components must be high. Unfortunately, the economic drivers for life support systems used in human space exploration do not lend themselves to large commercial investment. However, since more than half of the world's population live without basic quality "life support" needs such as clean water and air, there is a substantial need for these technologies regardless of the level of commercial investment.

6.9 SUSTAINABILITY: REPAIR AND PROCESSING

On remote human space expeditions, there will be a need for astronauts to repair a wide range of systems and materials. The challenges of materials processing and repair in remote locations will be extreme, since astronauts will be very limited in available power, feedstock, and consumables. CNTs have been observed to absorb microwave energy and convert that energy very efficiently into heat. In research conducted at Rice University and NASA Johnson Space Center, multiwall CNTs were included in low percentages in preceramic materials designed to be used for repair of reinforced carbon–carbon coatings for spacecraft reentry. In this work, it was possible with microwaves directed at preceramic materials and less than 1% CNTs to achieve temperatures of 1150°C in 7 minutes at very low power levels (30–40 W) and without an oven.[70] The systems envisioned would include ceramic or polymeric materials with CNTs included that could be heated to desired temperatures using just a handheld low-power microwave gun. The same type of technique could be used by including small amounts of CNTs into excavated planetary surface materials to enable high-temperature processing and extraction of useful materials for ISRU.

Quality assurance and reparability of structures and materials is also critical for the commercial airline industry. New nano-enhanced materials will face difficulty passing United States Federal Aviation Administration (FAA) or similar certification without appropriate avenues for repair and maintenance (Maguire, R. Conversation with authors, 2011). This area has already received substantial funding and attention by government research agencies.[71]

6.10 RADIATION PROTECTION AND SENSING

Radiation shielding and detection remains one of the most challenging aspects of human exploration beyond the earth's orbit. Current materials and solutions do not provide adequate protection from space radiation, which includes high-charge and energy ions from Galactic Cosmic Rays, at minimum mass and cost. A high-performance shield material should maximize both electrons per unit mass and nuclear reaction cross section per unit mass, while minimizing the production of secondary particles.[72] Liquid hydrogen is seen as a high-performing radiation protection material, but not optimal from a systems or vehicle point of view and has a low neutron absorption cross section. Polymeric materials, in particular polyethylene, have been found to be preferable to aluminum alloys.[73] Nanocomposites incorporating light elements such as boron, nitrogen, carbon, and hydrogen may offer a well-performing solution. To this end, composites of boron nitride nanotubes and polyethylene have been researched as a future shielding composite.[74,75] Boron has been shown as highly suitable for neutron capture.[76] Boron-based composite foams have also been explored by NASA and industry as a multifunctional thermal and radiation shielding material for space applications.[77]

6.11 CHALLENGES IN NANOMATERIALS FOR AEROSPACE

Manufacturing: One of the toughest challenges facing the uptake of nanotechnology into more aerospace applications is the lack of scalable and controlled manufacture of nanomaterials, notably CNTs and graphene (Ajayan, P. M., Conversation with authors, 2011). Many successes have been demonstrated in a range of industries (sporting equipment, automotive, textiles, and cosmetics); however, the advances so far have failed to fully exploit the performance potential of nanostructures. Partially this lack of achievement can be traced to limitations in manufacturing and process control. It would be appropriate for governments to fund more basic research on manufacturing of nanostructures from a host of different technical perspectives. Analytical techniques have kept up fairly well with the advances in nanotechnology, as we have reasonably good capabilities to image and evaluate materials across numerous length scales.

Much research has used atomic force microscope (AFM) type manipulation of nanomaterials, but as of yet we do not have a scaled-up way to do AFM type manipulation. Breakthroughs are needed in this area to leave the current plateau of nanomaterial performance. High-speed AFM offers the possibility of manufacturing, but remains largely in the research domain as an important characterization tool.[78]

Crossing scales: Another limitation to more successful use of nanomaterials in aerospace is the translation of performance at the nanoscale to the predicted and realized performance at the macroscale (Lou, J. Conversation with authors, 2011). For instance, a nanomechanical test on a particular individual nanostructure in a scanning electron microscope and a bending jig cannot be easily converted to the performance of a larger material system containing the same type of nanostructures. The potential solution to this may include a computational approach that can adequately bridge the nanoscale to macroscale with a reasonable time frame of calculation intensity. There may also need to be a better focus on matching experiments to provide the information needed to build models and shed light on performance across size scales. The nanocommunity would be well served to address these challenges head on and focus resources on solutions.

6.12 AFFORDABILITY AND MANUFACTURABILITY

One of the most pleasant and unexpected surprises of writing this chapter was to discover, first, a number of successful applications of nanomaterials in aerospace. Over the authors' history in the field since the late 1990s, the primary focus of advocacy and encouragement for investment has been based on anticipation of dramatic performance enhancements. Implicit in the arguments were assumptions, that while the cost of nanomaterial-based components and structures may be higher than traditional technologies, the enhancements in performance would prove to be worth the higher cost. It is a pleasant surprise, secondly, that some of the nanomaterial successes (Juno and F-35) have been based on manufacturability and reduction of cost as compared to the state-of-the-art and more traditional and conservative approaches. While exciting to think about dramatically lighter, stronger, thinner, and smaller technologies and dramatic new missions enabled, it is far more sustainable to consider the successful uptake of nanomaterials in aerospace as based on affordability and ease of processing. It will be far easier to convince corporate and government executives to use nanomaterials when the cost is lower and no additional risks are introduced into a system. This observation, if true widely, leads now to another challenge in getting nanomaterials more widely introduced into aircraft and spacecraft—reeducation of decision makers.

Throughout this chapter, the need for consistent and scalable manufacturing and supply of nanomaterials has shown itself to be a consistent factor to the adoption of nano-enhanced solutions in aerospace. Since the mid to late 1990s when the attention on nanotechnology increased dramatically, the issue of manufacturing has been of central concern to the R&D community. Much money has been spent over the years and manufacturers have emerged to provide the current levels of bulk nanomaterials for specialty applications and R&D efforts. Progress has definitely been made in this area, with a number of both large and small enterprises producing and ramping up supply of raw materials such as CNTs, in addition to more finished products such as sheeting, yarn, and nanotube-enhanced preimpregnated carbon fiber. The U.S. National Nanotechnology Initiative has identified manufacturing and commercialization as a key focus for its second decade.[79]

6.13 CHALLENGE: MANAGING EXPECTATIONS

It is the opinion of the authors that to achieve more government and commercial R&D funding for the next challenges in nanotechnology development, a push in reeducation needs to be made. Since the promises of nanotechnology were so profound in the early days of the technology advocacy, there is a substantial cadre of aerospace decision makers that are conditioned to expect game-changing performance enhancements from investments in nanotechnology, but have also become conditioned to expect mediocre results. With the exception of the companies and projects who have learned through hard work and experience to now expect affordability, cost savings, and manufacturing advantages, most of the R&D funding decisions for nanotechnology are now still governed by the old paradigm. In our opinion, the true long-term potential of nanotechnology is in the game-changing and revolutionary new materials that have not been produced yet; however, in the short to medium term, it makes sense to advocate investment in the technology based on economically and already-realized advantages of the technology (even if not predicted at the outset).

REFERENCES

1. Dharap, P., Li, Z., Nagarajaiah, S. & Barrera, E. Nanotube film based on single-wall carbon nanotubes for strain sensing. *Nanotechnology* **15**, 379 (2004).
2. Li, C. Y. & Chou, T. W. Strain and pressure sensing using single-walled carbon nanotubes. *Nanotechnology* **15**, 1493 (2004).
3. Kang, I., Schulz, M. J., Kim, J. H., Shanov, V. & Shi, D. A carbon nanotube strain sensor for structural health monitoring. *Smart Materials and Structures* **15**, 737 (2006).
4. Wood, J. R. & Wagner, H. D. Single-wall carbon nanotubes as molecular pressure sensors. *Applied Physics Letters* **76**, 2883 (2000).
5. Stampfer, C. et al. Fabrication of single-walled carbon-nanotube-based pressure sensors. *Nano Letters* **6**, 233–237 (2006).
6. Ghosh, S., Sood, A. & Kumar, N. Carbon nanotube flow sensors. *Science* **299**, 1042–1044, doi:10.1126/science.1079080 (2003).
7. Sood, A. & Ghosh, S. Direct generation of a voltage and current by gas flow over carbon nanotubes and semiconductors. *Physical Review Letters* **93**, 86601 (2004).
8. Arai, F. et al. Ultra-small site temperature sensing by carbon nanotube thermal probes. In *Nanotechnology, 2004. 4th IEEE Conference on* IEEE, pp. 146–148. (2004, August).
9. Fung, C. K., Wong, V. T., & Li, W. J. Towards batch fabrication of bundled carbon nanotube thermal sensors. In *Nanotechnology, 2003. IEEE-NANO 2003*. **2**, pp. 866–869. (2003, August).
10. Gohier, A. et al. Optimized network of multi-walled carbon nanotubes for chemical sensing. *Nanotechnology* **22**, 105501 (2011).
11. Zhang, T., Mubeen, S., Yoo, B., Myung, N. V. & Deshusses, M. A. A gas nanosensor unaffected by humidity. *Nanotechnology* **20**, 255501 (2009).
12. Young, P., Lu, Y., Terrill, R. & Li, J. High-sensitivity NO2 detection with carbon nanotube-gold nanoparticle composite films. *Journal of Nanoscience and Nanotechnology* **5**, 1509–1513 (2005).
13. Bondavalli, P., Legagneux, P. & Pribat, D. Carbon nanotubes based transistors as gas sensors: State of the art and critical review. *Sensors and Actuators: B-Chemical* **140**, 304–318 (2009).
14. Yáñez-Sedeño, P., PingarrÛn, J. M., Riu, J. & Rius, F. X. Electrochemical sensing based on carbon nanotubes. *TrAC Trends in Analytical Chemistry* ***29***(9), 939–953 (2010).
15. Li, J. et al. Carbon nanotube sensors for gas and organic vapor detection. *Nano Letters* **3**, 929–933, doi:10.1021/nl034220x (2003).
16. Li, J. et al. Carbon nanotube nanoelectrode array for ultrasensitive DNA detection. *Nano Letters* **3**, 597–602 (2003).
17. Li, J., Lu, Y. J., Ye, Q., Delzeit, L. & Meyyappan, M. A gas sensor array using carbon nanotubes and microfabrication technology. *Electrochemical and Solid State Letters* **8**, H100–H102 (2005).
18. Lu, Y. J., Partridge, C., Meyyappan, M. & Li, J. A carbon nanotube sensor array for sensitive gas discrimination using principal component analysis. *Journal of Electroanalytical Chemistry* **593**, 105–110 (2006).
19. Li, J., Koehne, J. & Meyyappan, M. Carbon nanotube based nanotechnology for NASA mission needs and societal applications. *Presentation at NSF Carbon Nano Material Workshop (2011).*

20. Fam, D. W. H., Palaniappan, A., Tok, A. I. Y., Liedberg, B. & Moochhala, S. M. A review on technological aspects influencing commercialization of carbon nanotube sensors. *Sensors and Actuators B: Chemical* ***157***(1), 1–7 (2011).
21. Chen, C. L. et al. DNA-decorated carbon-nanotube-based chemical sensors on complementary metal oxide semiconductor circuitry. *Nanotechnology* **21**, 095594 (2010).
22. Liu, Y., Chen, C. L., Agarwal, V., Li, X., Sonkusale, S., Dokmeci, M. R., & Wang, M. L. DNA decorated carbon nanotube sensors on CMOS circuitry for environmental monitoring. In *SPIE Smart Structures and Materials+ Nondestructive Evaluation and Health Monitoring*, pp. 76471W–76471W. International Society for Optics and Photonics. (2010, March).
23. Li, C. et al. Surface treatment and doping dependence of In_2O_3 nanowires as ammonia sensors. *Journal of Physical Chemistry B* **107**, 12451–12455 (2003).
24. Zhou, X. T. et al. Silicon nanowires as chemical sensors. *Chemical Physics Letters* **369**, 220–224 (2003).
25. Kolmakov, A., Zhang, Y., Cheng, G. & Moskovits, M. Detection of CO and O_2 using tin oxide nanowire sensors. *Advanced Materials* **15**, 997–1000 (2003).
26. Ram, M. K., Yavuz, O. & Aldissi, M. NO_2 gas sensing based on ordered ultrathin films of conducting polymer and its nanocomposite. *Synthetic Metals* **151**, 77–84 (2005).
27. Ram, M. K., Yavuz, O., Lahsangah, V. & Aldissi, M. CO gas sensing from ultrathin nano-composite conducting polymer film. *Sensors and Actuators B: Chemical* **106**, 750–757 (2005).
28. McQuade, D. T., Pullen, A. E. & Swager, T. M. Conjugated polymer-based chemical sensors. *Chemical Reviews* **100**, 2537–2574 (2000).
29. Virji, S., Kaner, R. B. & Weiller, B. H. Hydrazine detection by polyaniline using fluorinated alcohol additives. *Chemistry of Materials* **17**, 1256–1260 (2005).
30. Rajesh, Ahuja, T. & Kumar, D. Recent progress in the development of nano-structured conducting polymers/nanocomposites for sensor applications. *Sensors and Actuators B: Chemical* **136**, 275–286, doi:10.1016/j.snb.2008.09.014 (2009).
31. Hunsperger, R. G. Nanophotonics. In *Integrated Optics*, pp. 469–505. Springer, New York (2009).
32. El-Kady, I. & Taha, M. M. R. *Systems, Man and Cybernetics*, 2005 IEEE International Conference, 1961–1966, Vol. 1962.
33. Zinoviev, K., Dominguez, C., Plaza, J. A., Busto, V. J. C. & Lechuga, L. M. A novel optical waveguide microcantilever sensor for the detection of nanomechanical forces. *Journal of Lightwave Technology* **24**, 2132 (2006).
34. Chan, L. L., Gosangari, S. L., Watkin, K. L. & Cunningham, B. T. A label-free photonic crystal biosensor imaging method for detection of cancer cell cytotoxicity and proliferation. *Apoptosis* **12**, 1061–1068 (2007).
35. Dorr, J. Nanocomp reaches new frontiers on NASA's Juno Mission. *Press Release* (2011).
36. Shah, T. Carbon nanostructure fibers boost NASA spacecraft. *Press Release* (2011).
37. LaPedus, M. Lockheed buys Nantero's government unit. In *EE Times* (2008).
38. Lockheed. Lockheed Martin. *Press Release* (2009).
39. Nanocomp. Nanocomp Technologies Awarded Small Business Innovation Research (SBIR) Contract from United States Air Force. *Press Release* (2008).
40. Nanocomp. Nanocomp Awarded Army Wiring Contract. *Avionics Magazine* (2010).
41. Antoinette, P. Scaling carbon nanotube technology commercializing transformational products for defense and industry. *NNI Summit Meeting* (2010).
42. Zhao, Y., Wei, J., Vajtai, R., Ajayan, P. M. & Barrera, E. V. Iodine doped carbon nanotube cables exceeding specific electrical conductivity of metals. *Scientific Reports* **1** (2011).

43. Varanasi, K. K. et al. *Engineered Nanostructures for High Thermal Conductivity Substrates*, Nano Science and Technology Institute (2009).
44. Johan, L., Teng, W. & Campbell, E. E. B. Thermal management technologies for electronics based on multiwalled carbon nanotube bundles. *Nanotechnology Magazine, IEEE* **3**, 16–19 (2009).
45. Balandin, A. A. Thermal properties of graphene and nanostructured carbon materials. *Nature Materials* **10**, 569–581 (2011).
46. Snyder, G. J., Lim, J. R., Huang, C. K. & Fleurial, J. P. Thermoelectric microdevice fabricated by a MEMS-like electrochemical process. *Nature Materials* **2**, 528–531 (2003).
47. Park, G., Rosing, T., Todd, M. D., Farrar, C. R. & Hodgkiss, W. Energy harvesting for structural health monitoring sensor networks. *Journal of Infrastructure Systems* **14**, 64 (2008).
48. Wang, Z., Liang, Z., Wang, B., Zhang, C. & Kramer, L. Processing and property investigation of single-walled carbon nanotube (SWNT) buckypaper/epoxy resin matrix nanocomposites. *Composites Part A: Applied Science and Manufacturing* **35**, 1225–1232 (2004).
49. Zhu, J. et al. Processing a glass fiber reinforced vinyl ester composite with nanotube enhancement of interlaminar shear strength. *Composites Science and Technology* **67**, 1509–1517 (2007).
50. Garcia, E. J., Wardle, B. L. & John Hart, A. Joining prepreg composite interfaces with aligned carbon nanotubes. *Composites Part A: Applied Science and Manufacturing* **39**, 1065–1070 (2008).
51. Trimble, S. Lockheed Martin reveals F-35 to feature nanocomposite structures. *Flight International* May 26 (2011). Available at http://www.flightglobal.com/news/articles/lockheed-martin-reveals-f-35-to-feature-nanocomposite-structures-357223.
52. de Villoria, R. G., Yamamoto, N., Miravete, A. & Wardle, B. L. Multi-physics damage sensing in nano-engineered structural composites. *Nanotechnology* **22**, 185502 (2011).
53. Li, B. & Zhong, W. H. Influence of carbon nanofiber network variability on the AC conductivity of polyetherimide composite films. *Macromolecular Materials and Engineering* **295**, 310–314 (2010).
54. Wei, R. & Gandy, D. W. Nanotechnology coatings for erosion protection of turbine components. *Journal of Engineering for Gas Turbines and Power* **132**, 082104 (2010).
55. Wang, N., Zhao, W., Wang, P. & Wei, Z. To develop nanostructured thermal barrier coatings. *International Journal Modern Physics B* **20** (25n27), 4171–4176 (2006).
56. Tang, F., Ajdelsztajn, L., Kim, G. E., Provenzano, V. & Schoenung, J. M. Effects of surface oxidation during HVOF processing on the primary stage oxidation of a CoNiCrAlY coating. *Surface and Coatings Technology* **185**, 228–233 (2004).
57. De Luca, L. et al. Burning of nano-aluminized composite rocket propellants. *Combustion, Explosion, and Shock Waves* **41**, 680–692 (2005).
58. Wood, T. D. et al. Feasibility study and demonstration of an aluminum and ice solid propellant. In *45th Annual AIAA/ASME/SAE/ASEE Joint Propulsion Conference and Exhibit, Denver, CO. AIAA*, **4890** (2009, August).
59. Kittell, D. E., Pourpoint, T. L., Groven, L. J. & Son, S. F. Further development of an aluminum and water solid rocket propellant. *47th AIAA/ASME/SAE/ASEE Joint Propulsion Conference & Exhibit*, AIAA 2011–6137 (2011). Available at www.arc.aiaa.org.
60. Magasinski, A. et al. High-performance lithium-ion anodes using a hierarchical bottom-up approach. *Nature Materials* **9**, 353–358 (2010).
61. Simon, P. & Gogotsi, Y. Materials for electrochemical capacitors. *Nature Materials* **7**, 845–854 (2008).
62. Signorelli, R., Ku, D. C., Kassakian, J. G. & Schindall, J. E. Electrochemical double-layer capacitors using carbon nanotube electrode structures. *Proceedings of the IEEE* **97**, 1837–1847 (2009).

63. Winter, M. & Brodd, R. J. What are batteries, fuel cells, and supercapacitors? *Chemical Reviews* **104**, 4245–4270 (2004).
64. Moloney, P. et al. PEM fuel cell electrodes using single wall carbon nanotubes. *Materials Research Society Symposia Proceedings* **885** (2005).
65. Garnett, E. C., Brongersma, M. L., Cui, Y. & McGehee, M. D. Nanowire solar cells. *Annual Review of Materials Research* **41**, 269–295 (2011).
66. Moloney, P. et al. Advanced life support for space exploration: Air revitalization using amine coated single wall carbon nanotubes. *Materials Research Society Symposia Proceedings* **851** (2005).
67. Allada, R. K. et al. Nanoscale materials for human spaceflight applications: Regenerable carbon dioxide removal using single-wall carbon nanotubes. *Nanoscale* **1**, 2195 (2006).
68. Pickering, K. D. & Wiesner, M. R. Fullerol-sensitized production of reactive oxygen species in aqueous solution. *Environmental Science & Technology* **39**, 1359–1365 (2005).
69. NASA. Portable nanomesh creates safer drinking water. *STI NASA GOV* (2008).
70. Higginbotham, A. L. et al. Carbon nanotube composite curing through absorption of microwave radiation. *Composites Science and Technology* **68**, 3087–3092, doi:10.1016/j.compscitech.2008.07.004 (2008).
71. IAPETUS. Innovative repair of aerospace structures with curing optimisation and life-cycle monitoring abilities. Conference Presentation ACP7-GA-2008-234333 (2009).
72. Wilson, J., Cucinotta, F., Kim, M. & Schimmerling, W. Optimized shielding for space radiation protection. *Physica Medica* **17**, 67–71 (2001).
73. Zhong, W., Sui, G., Jana, S. & Miller, J. Cosmic radiation shielding tests for UHMWPE fiber/nano-epoxy composites. *Composites Science and Technology* **69**, 2093–2097 (2009).
74. Harrison, C. et al. Polyethylene/boron nitride composites for space radiation shielding. *Journal of Applied Polymer Science* **109**, 2529–2538, doi:10.1002/app.27949 (2008).
75. Harrison, C., Burgett, E., Hertel, N. & Grulkel, E. Polyethylene/boron containing composites for radiation shielding applications. *Nanostructured Materials and Nanotechnology II: Ceramic Engineering and Science Proceedings* **29(8)**, 77–84 (2009).
76. Thibeault, S. A. et al. Shielding Materials Development and Testing Issues. In *Shielding Strategies for Human Space Exploration*, J. W. Wilson, J. Miller, A. Konradi, and F.A. Cucinotta, eds., NASA CP-3360, 397–425 (1997).
77. Kowbel, W., Bruce, C., Withers, J., Ohlhorst, C. & Thibeault, S. Boron based advanced materials for radiation protection. In *Aerospace Conference, 2006 IEEE, 7.*
78. Hansma, P. K., Schitter, G., Fantner, G. E. & Prater, C. High-speed atomic force microscopy. *Science* **314**, 601–602, doi:10.1126/science.1133497 (2006).
79. Alper, J. & Amato, I Report to the President and Congress on the Third Assessment of the National Nanotechnology Initiative. Washington DC, Executive Office of the President (2010).

7 Applications of Nanomaterials in Fuel Cells

Shangfeng Du and Bruno G. Pollet

CONTENTS

7.1 INTRODUCTION TO FUEL CELLS

Although fuel cells were invented in the middle of the nineteenth century (Yeager 1961), they did not find the first application until space exploration in the 1960s with the alkaline fuel cell or the so-called Bacon fuel cell (Whittingham et al. 2004). During the past two decades, a confluence of driving forces has created a sustained and significant worldwide effort to develop fuel cell materials and systems. These driving needs include the demand for efficient energy systems for transportation, the desire to reduce greenhouse gases, CO_2 and particulate emissions and other negative environmental impacts, and the demand for high-energy density power sources for portable electronic applications. Fuel cell technologies are now approaching commercialization, especially in the fields of portable power sources—distributed and remote generators of electrical energy. Due to the high level of interest in fuel cells during the past decade or so, there have been numerous summary articles and symposia focused on the technology state of the art. In this chapter, we present a series of nanomaterials that deal with applications related to fuel cell development.

Fuel cells are direct electrochemical fuel to electrical energy conversion devices and offer higher efficiency (50%–70%) compared with conventional technologies such as internal combustion engines (~35% efficiency). If the waste heat of the fuel cell is also used, fuel efficiencies of 90% are possible. Fuel cells consist of an electropositive anode, where oxidation occurs to fuels (hydrogen, methanol, ethanol, methane, etc.), an electronegative cathode, where reduction occurs (to oxygen, air, etc.), and an electrolyte (e.g., a proton-conducting polymer membrane, ionic conductive doped ceria), where ions carry the current between the electrodes. The scheme of the reactions and processes that occur in the various fuel cell systems is depicted in Figure 7.1. Tables 7.1 and 7.2 list the characteristics for the various fuel cell systems (Brodd et al. 2004).

Fuel cells can be roughly divided into low-temperature (ca. <200°C) and high-temperature (ca. >450°C) fuel cells. Alkaline fuel cell (AFC), polymer electrolyte membrane fuel cell (PEMFC, also known as proton exchange membrane fuel cell under the same acronym), direct methanol fuel cell (DMFC), and phosphoric acid

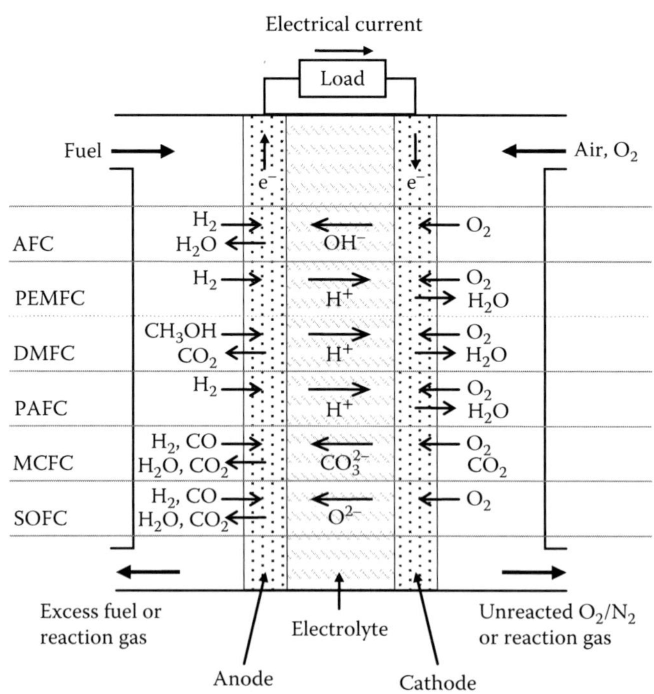

FIGURE 7.1 Operation scheme of typical fuel cells.

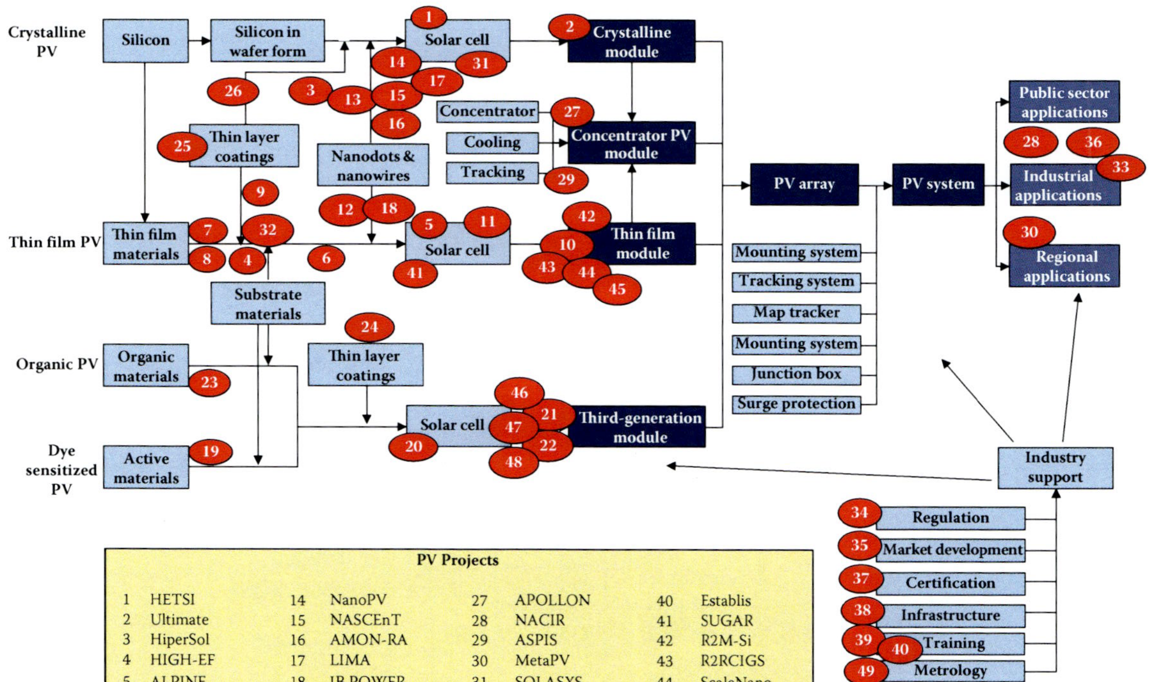

PV Projects

1	HETSI	14	NanoPV	27	APOLLON	40	Establis
2	Ultimate	15	NASCEnT	28	NACIR	41	SUGAR
3	HiperSol	16	AMON-RA	29	ASPIS	42	R2M-Si
4	HIGH-EF	17	LIMA	30	MetaPV	43	R2RCIGS
5	ALPINE	18	IB POWER	31	SOLASYS	44	ScaleNano
6	ThinSi	19	INNOVASOL	32	PEPPER	45	Fast Track
7	NOVA-CI(G)S	20	SANS	33	PV-GUM	46	SUNFLOWER
8	PolySiMode	21	ROBUST DSC	34	PV-LEGAL	47	X10D
9	SILICON-Light	22	HIFLEX	35	PV-NMS-NET	48	ROTROT
10	HELATHIS	23	EPHOCELL	36	PVs in BLOOM	49	NanoCharm
11	hipoCIGS	24	SOLAMON	37	QualiCert		
12	ROD-SOL	25	PRIMA	38	SOPHIA		
13	SNAPSUN	26	NanoSpec	39	PVTRIN		

FIGURE 3.2 Clustering of projects by supply chain activity (2012).

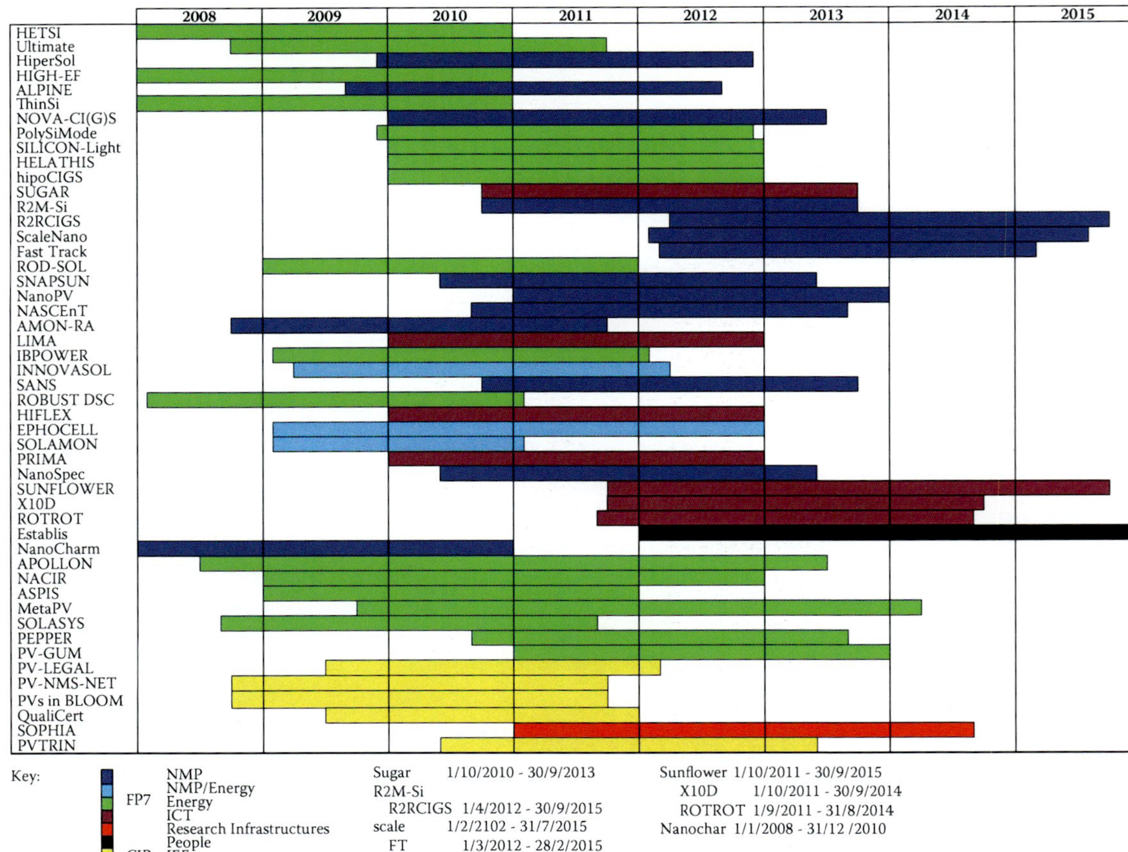

FIGURE 3.3 Timescale and duration of projects by programme (2012).

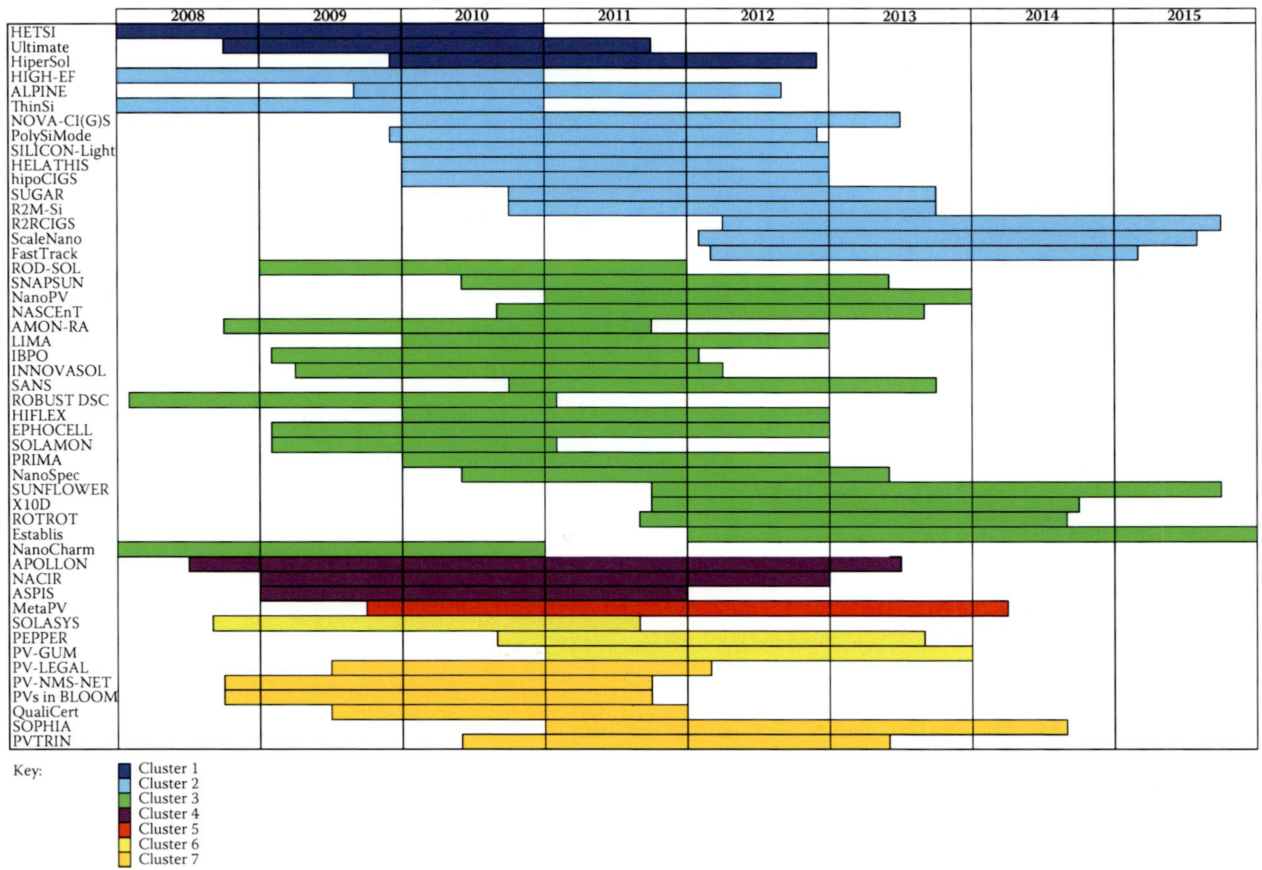

FIGURE 3.4 Timescale and duration of projects by photovoltaic (PV) cluster (2012).

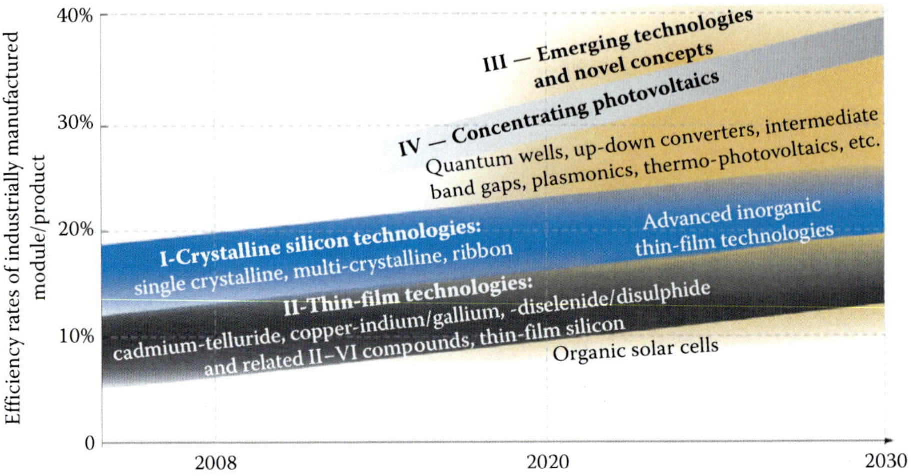

FIGURE 3.5 Expected impact of photovoltaic technologies. (Data from Solar Photovoltaic Energy Technology Roadmap, ©OECD/International Energy Agency 2010.)

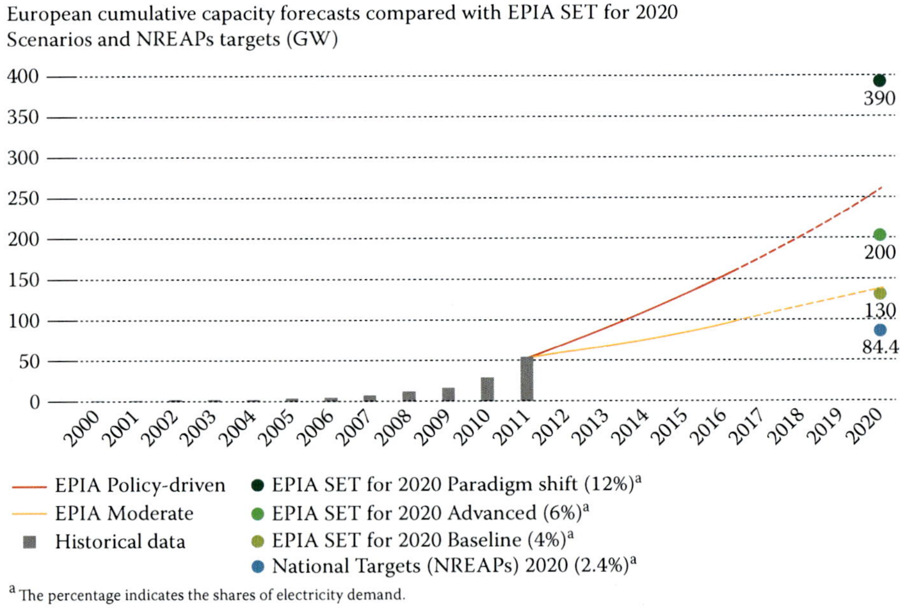

FIGURE 3.7 Alternative futures for the global photovoltaic (PV) market. (Data from European Photovoltaic Industry Association, Global Market Outlook for Photovoltaics Until 2016, 2012.)

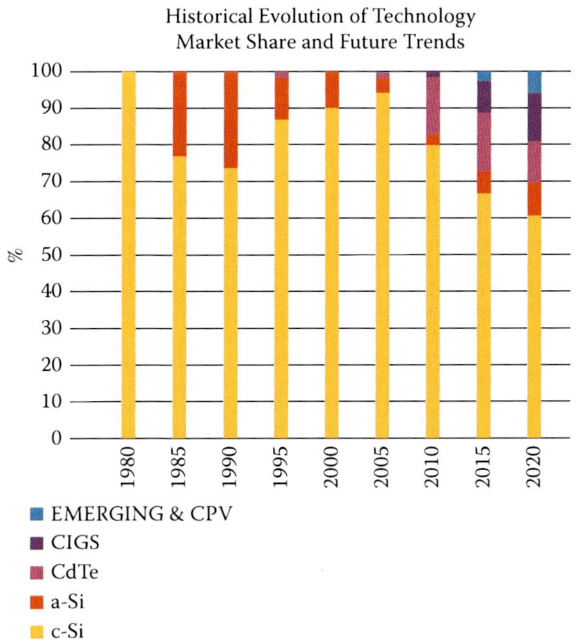

FIGURE 3.9 Long-term photovoltaic market segmentation by technology. (Data from European Photovoltaic Industry Association—Greenpeace International, Solar Generation VI, 2010.)

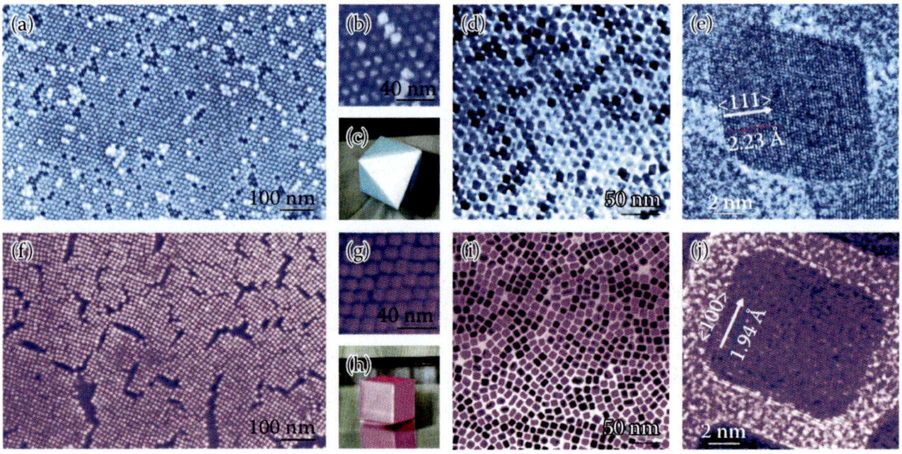

FIGURE 7.4 (a–e) Images for Pt_3Ni nanoctahedra. (f–j) Images for Pt_3Ni nanocubes. (Fang, J.Y. et al., *Nano Letters*, 10, 638–644, 2010.)

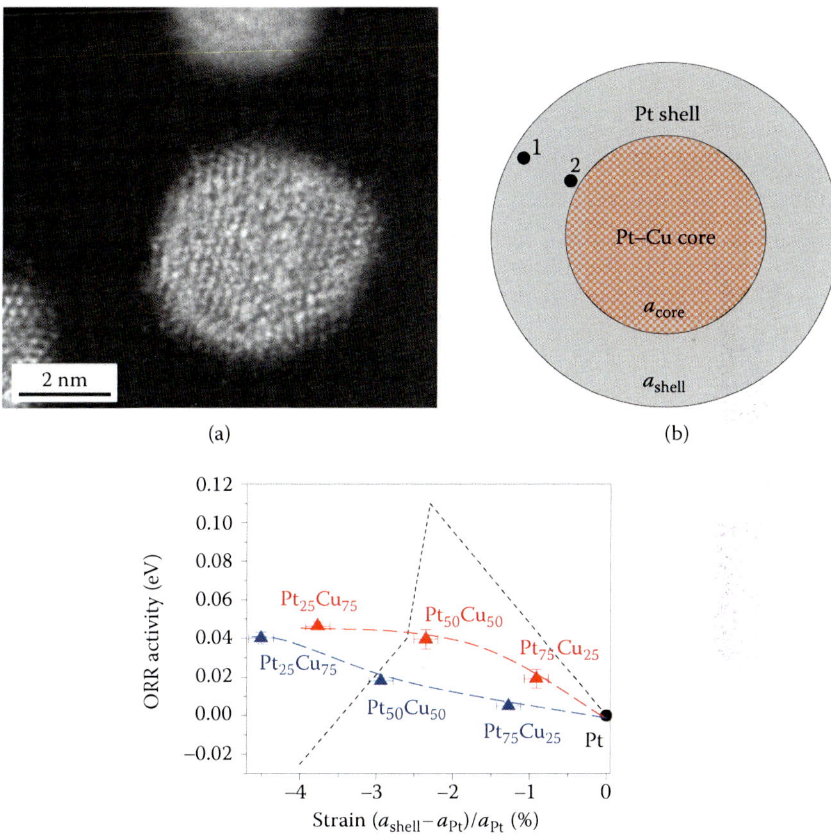

FIGURE 7.7 (a) Subangstrom resolution image of a Pt–Cu dealloyed nanoparticle (~4 nm) (b) Scheme of a simple two-phase structural model for the dealloyed state of a bimetallic particle. (c) Experimental and predicted relationships between electrocatalytic oxygen reduction reaction (ORR) activity and lattice strain (annealing temperatures of 950°C [blue] and 800°C [red]). The dashed line is the predicted density functional theory (DFT) of the ORR activity. (Strasser, P. et al., *Nature Chemistry*, 2, 454–460, 2010.)

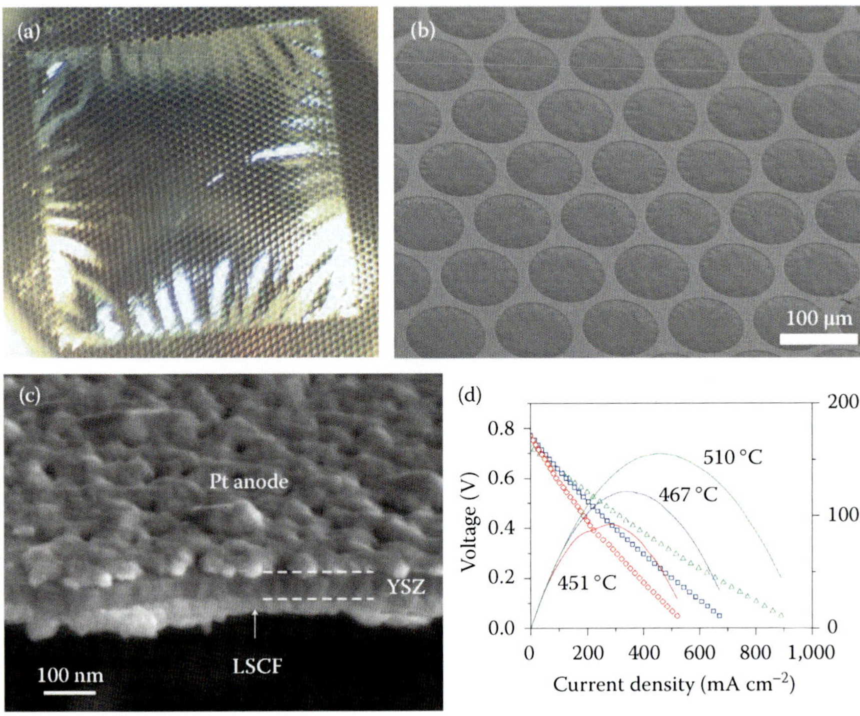

FIGURE 7.12 Platinum-grid-supported 5 × 5 mm micro-solid oxide fuel cell (μ-SOFC): (a) Optical micrographs taken from the cathode side at 480°C; (b) SEM micrograph of slightly buckled freestanding membranes before testing; (c) cross-sectional micrograph of μSOFCs after testing; (d) current voltage sweep of platinum-grid-supported 5 × 5 mm μSOFC at three different temperatures. (Reprinted by permission from Macmillan Publishers Ltd. Tsuchiya [*Nature Nanotechnology*] [M. et al. 2011.], Copyright 2011.)

TABLE 7.1
Types of Fuel Cells

Advantages	Disadvantages	Applications
	Alkaline fuel cell (AFC)	
Mechanically rechargeable; low-cost KOH electrolyte	Limited activated life; pure H_2 only suitable fuel	Military and space
	Polymer electrolyte membrane fuel cell (PEMFC)	
Polymer electrolyte; low temperature; quick start-up	Expensive catalysts; CO a strong poison; oxygen kinetics are slow; intolerant of impurities; high-cost electrolyte; pure H_2 only suitable fuel; limited life; water management essential	Automotive; stationary applications
	Direct methanol fuel cell (DMFC)	
Direct fuel conversion; improved weight and volume; polymer electrolyte	High catalyst loadings; water management essential; low overall efficiency; methanol hazardous	Portable electronic devices
	Phosphoric acid fuel cell (PAFC)	
Higher fuel efficiency	H_2 only suitable fuel; anode CO catalyst poison; O_2 kinetics hindered; high-cost catalysts; limited life	Distributed generation
	Molten carbonate fuel cell (MCFC)	
High efficiency; fuel flexibility; high-grade heat available	Materials problems and life; low sulfur tolerance; high ohmic electrolyte; need to recycle CO_2; limited life	Electric utility; large distributed generation
	Solid oxide fuel cell (SOFC)	
High efficiency; high grade heat available; no electrolyte management; fuel flexibility; tolerant of impurities	High fabrication costs; severe materials constraints	Auxiliary power; electric utility; large distributed generation

Source: Brodd, R.J. et al., *Chemical Reviews*, 104, 4245–4269, 2004.

fuel cell (PAFC) are typical low-temperature fuel cells. Molten carbonate fuel cell (MCFC) and solid oxide fuel cell (SOFC) belong to the high-temperature fuel cell class. In general, low-temperature fuel cells (AFC, PEMFC DMFC, and PAFC) feature a quicker start-up, which makes them more suitable for portable applications, especially PEMFCs have recently gained momentum for application in transportation and as small portable power sources. However, they require as fuel relatively pure hydrogen (minimum 99.999%), and consequently an external fuel processor,

TABLE 7.2
Typical Characteristics of Various Fuel Cell Systems

Type	Anode Composition	Cathode Composition	Electrolyte	Operating Temperature (°C)
AFC	Carbon/platinum catalyst	Carbon/platinum catalyst	Aqueous KOH	Ambient–100
PEMFC	Carbon/platinum catalyst	Carbon/platinum catalyst	Acidic polymer	Ambient–90
DMFC	Carbon/platinum catalyst	Carbon/platinum catalyst	Acidic polymer	60–90
PAFC	Carbon/platinum catalyst	Carbon/platinum catalyst	Phosphoric acid in SiC matrix	150–220
MCFC	Porous Ni	Porous NiO	Molten Li_2CO_3 in $LiAlO_2^-$	550–700
SOFC	Ni-YSZ	Strontia-doped lanthanum manganite perovskite	YSZ	600–1000

Source: Brodd, R.J. et al., *Chemical Reviews*, 104, 4245–4269, 2004.

which increases the complexity and cost and decreases the overall efficiency. They also require a higher loading of the precious metal catalysts (Garcia-Martinez 2010). In contrast to low-temperature fuel cells, MCFC and SOFC are more flexible regarding fuel because they can reform various fuels (methanol, ethanol, natural gas, gasoline, etc.) inside the cells to produce hydrogen, still offering advantages for stationary applications, and especially for cogeneration. They are also less prone to catalyst *poisoning* by carbon monoxide and carbon dioxide (Brandon et al. 2003). However, their slower start-up limits them to more stationary applications (Baxter et al. 2009).

In parallel to the development of classic fuel cells, a new and promising type of fuel cell, the microbial fuel cell (MFC), or biological fuel cell, is currently under intensive research. MFC produces electricity from microbially catalyzed anodic oxidation processes. The greatest potential of MFC lies in the use of wastewater as fuel, which allows combining waste treatment and energy recovery. However, MFC still severely suffers from the short active lifetimes (typically 8 hours to 7 days) (Minteer et al. 2008) and limited power generation (Zhao et al. 2009), which make it far from being used in practical applications. Recently, Barton and Schröder summarized the development and recent work done on MFCs, and more details can be found in their recent reviews (Barton et al. 2004; Harnisch et al. 2010).

Among the various types of fuel cells, PEMFC, DMFC, and SOFC are actively under research and development as they employ solid electrolytes that could make the operation and maintenance of fuel cells easier. However, even like this, their commercialization is still hampered by high cost, poor durability issues, and operability problems that are directly linked to severe materials challenges and systems issues. In low-temperature PEMFCs and DMFCs, the high cost originates from the expensive Nafion® membrane and notable Pt catalyst used; and the poor durability is due to the degradation of membrane and catalyst performance, as well as the instability of catalyst support. While in SOFC, the reason leading to high cost is the high operating temperature, resulting in the expensive interconnect and sealing materials used; and the microstructure decline, carbon deposition, and sulfur poisoning to catalysts are the main factors affecting SOFC durability. These are the main drawbacks limiting the commercial exploitation of these devices. To solve these problems, the bottleneck on materials must be fixed. So, fuel cell progress always attracts a lot of interests in the development of high performance materials with novel design and preparing technologies, in which nanomaterials have played a critical role. For example, in low-temperature fuel cells, nearly half of the cost of the fuel cell is linked to the electrocatalyst cost. To reduce the cost, in PEMFCs and DMFCs, Pt catalyst with novel nanostructure and high performance have been developed to reduce the loading amount of Pt, or directly by using less expensive alternative nanostructured electrocatalysts, such as N-doped carbon nanotube (CNT) (Dai et al. 2009) or iron-based catalysts (Dodelet et al. 2009). Regarding high-temperature SOFCs, to decrease the operating temperature to intermediate range (450–600°C) for a low system cost and an improved durability, nanocomposite electrodes and electrolyte were developed, such as samaria doped ceria (SDC) nanowires (NWs)/Na_2CO_3 nanocomposites electrolyte (Ma et al. 2010) or gadolinium doped ceria (GDC) anode with ionic NW nanoarchitectures (Laberty-Robert et al. 2007).

In this chapter, we discuss fuel cell materials that have been recently achieved by nanotechnology, especially focusing on their applications in electrodes and electrolytes in PEMFCs, DMFCs, and SOFCs.

7.2 BENEFITS OF NANOMETER SIZE EFFECTS

Nanomaterials are becoming increasingly important in the field and hence have attracted great interest in recent years. The variety of nanometer size effects in fuel cell materials can be divided into two types: (1) *trivial size effects*, which rely solely on the increased surface-to-volume ratio and (2) *true size effects*, which also involve the strong influence of surface properties on overall behavior, and the possibility of tailoring electrochemical and other properties when the dimensions of fuel cell systems are confined. There is a significant effect of spatial confinement and the surface on the physicochemical characteristics due to small particle size. New nanomaterials may allow fundamental advances in fuel cell performance.

7.2.1 Nanomaterials for Fuel Cell Electrodes

Nanostructured fuel cell electrodes increase the surface area (per unit weight) of catalysts and enhance the contact between fuels and catalysts, which leads to improved system efficiency. The preparation of nanoscale electrocatalysts for fuel cells typically starts from colloidal nanomaterial precursors, for example, colloidal platinum (Pt) sols.

7.2.1.1 Electrocatalysts in Low-Temperature Fuel Cells

The concern about the limited resources of Pt and its high price (Berger 1999) has made the research on the improvement of Pt catalysts a quite hot research topic for the past 20 years in low-temperature PEMFCs and DMFCs (Chen et al. 2010; Jaouen et al. 2011). To reduce the costs, the Pt loading must be decreased while maintaining or improving fuel cell performance. Already, nanomaterials are having more and more impacts on processing methods in the development of low-temperature fuel cells, the dispersion of precious metal catalysts, the development and dispersion of nonprecious catalysts, as well as the fabrication of membrane electrode assemblies (MEAs).

At the moment, Pt-based catalysts are still the most active materials for low-temperature fuel cells fed with hydrogen, reformate, or methanol (Chen et al. 2010). The oxygen reduction reaction (ORR) limits the fuel cell performance, especially in PEMFCs. A few routes are being actively investigated to improve the electrocatalytic activity of Pt-based catalysts. They consist mainly of alloying Pt with transition metals (e.g., Ni and Co) or tailoring the Pt particle size. One of the present approaches to increase the catalyst dispersion involves the deposition of Pt nanoparticles on a carbon black (CB) support. A typical high-resolution scanning electron microscopy image of Pt/C catalyst is shown in Figure 7.2. Kinoshita (1990) observed that the mass activity and specific activity for oxygen electroreduction in acid electrolytes varies with the Pt particle size according to the relative fraction of Pt surface atoms on the (111) and (100) faces (Figure 7.3). The mass-averaged distribution of the surface atoms on the (111) and (100) planes passes through a maximum (~3 nm), whereas the total fraction of surface atoms at the edge and corner sites decreases rapidly with an increase of the particle size. On the other hand, the surface-averaged distribution for the (111) and (100) planes shows a rapid increase with the particle size, which accounts for the increase of the experimentally determined specific activity with the particle size.

Both CO_2 and CO are present in hydrogen streams obtained from reforming. These molecules are known to adsorb on the Pt surface under reducing potentials. Adsorbed CO-like species are also formed on Pt-based anode catalysts in DMFCs. Such a poisoning of the Pt surface reduces the electrical efficiency and the power

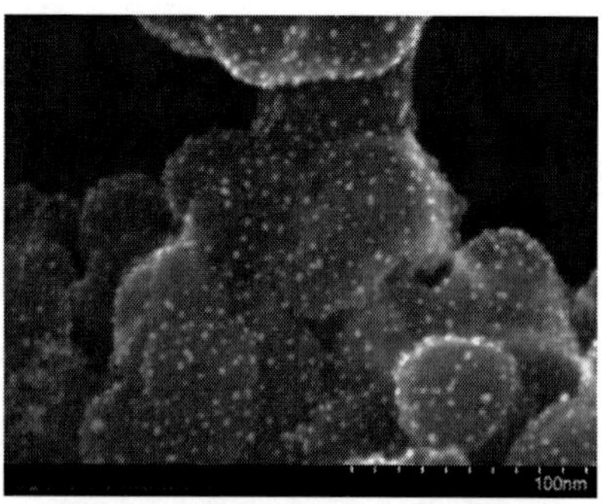

FIGURE 7.2 A typical high-resolution scanning electron microscopy image of platinum (Pt) catalysts deposited on Vulcan XC-72 carbon black. (Fang, B. et al., *Journal of the American Chemical Society*, 131, 15330–15338, 2009a.)

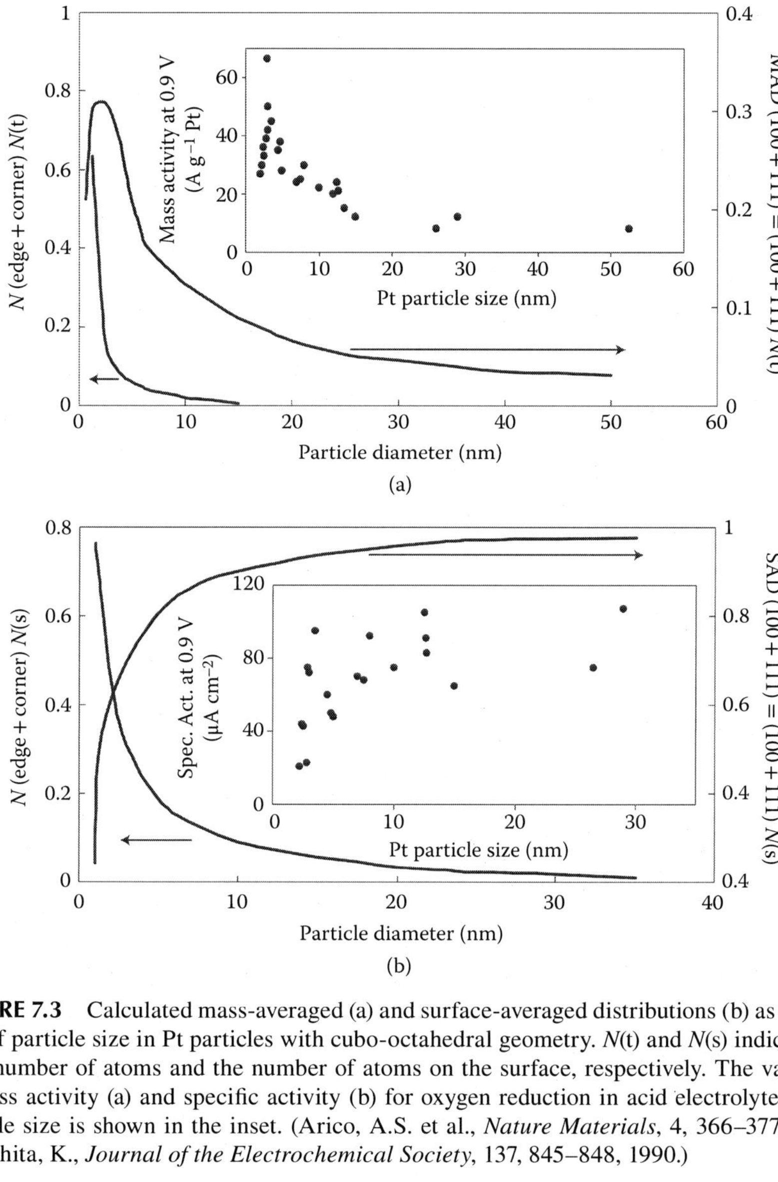

FIGURE 7.3 Calculated mass-averaged (a) and surface-averaged distributions (b) as a function of particle size in Pt particles with cubo-octahedral geometry. *N*(t) and *N*(s) indicate the total number of atoms and the number of atoms on the surface, respectively. The variation of mass activity (a) and specific activity (b) for oxygen reduction in acid electrolyte versus particle size is shown in the inset. (Arico, A.S. et al., *Nature Materials*, 4, 366–377, 2005; Kinoshita, K., *Journal of the Electrochemical Society*, 137, 845–848, 1990.)

density of the fuel cell (Devanathan 2008). The electrocatalytic activity of Pt against, for example, CO_2/CO poisoning is known to be promoted by the presence of a second metal, such as Ru, Sn, or Mo (Chang et al. 2010; Schmidt et al. 1997). The mechanism by which such synergistic promotion of the H_2/CO and methanol oxidation reactions is brought about has been much studied and is still debated. Nevertheless, it turns out that the best performance is obtained from Pt–Ru electrocatalysts with mean particle size 2–3 nm. As in the case of ORR, the particle size is important for structure-sensitive reactions such as CH_3OH and CO electrooxidation.

Therefore, we cannot keep improving the catalytic performance of electrocatalysts by continually reducing the nanoparticle size. In this case, to lower the loading amount of expensive Pt, novel nanostructured electrocatalysts, such as M@Pt core-shell structures and M core decorated with Pt cluster are good options and have attracted many efforts. In addition to the generally recognized small size, recent trends reveal that shape also plays an important role in enhancing the electrocatalytic activity of Pt nanomaterials (Tian et al. 2007; Wang et al. 2011b). For instance, zero-dimensional (0-D) electrocatalysts Pt nanocube (Fang et al. 2009b; Murray et al. 2010), polyhedron and truncated cube (Fang et al. 2010; Herrero et al. 2010), and one-dimensional (1-D) electrocatalysts nanowires (Du 2010; Sun et al. 2011a; Xia et al. 2009) and nanotube (Alia et al. 2010; Gorzny et al. 2010) have all been synthesized for fuel cell applications.

7.2.1.1.1 Pt-Based 0-D Nanoelectrocatalysts

Wang et al. synthesized Pt nanocube, polyhedron and truncated cube by reaction of platinum(II) acetylacetonate with oleic acid and oleylamine in the presence of trace amount of $Fe(CO)_5$ (Wang et al. 2008). They found that 7-nm Pt nanocubes with rich (100) facets were much more active than the other shaped Pt nanoparticles in H_2SO_4 solution, which was due to different adsorptions of sulfate ions on Pt (111) and (100) facets. To eliminate the impact from sulfate ions, You et al. (2009) measured the electrocatalytic activities in perchloric acid ($HClO_4$) for three planes, (100), (111), and (110), which are prepared by utilizing electron beam lithography. The data suggest a *division of labor* between (100) and (111) nanofacets. For example, the (110) facets adsorb oxygen well but cannot reduce efficiently, while the (111) facets can reduce oxygen but cannot easily absorb it. Another important finding was given by Tian et al. (2007), who found that single-crystalline Pt THH (tetrahexahedral) nanospheres enclosed by 24 high-index facets such as (730), (210), and (520) surface have a large density of atomic steps and dangling bonds, which exhibited much enhanced specific activity and stability for electrooxidation of small organic fuels than the commercial E-TEK Pt/C catalysts.

Alloying Pt with transition metals also enhances the electrocatalysis of O_2 reduction. In low-temperature fuel cells, Pt–Fe, Pt–Cr, and Pt–Cr–Co alloy electrocatalysts were observed to have high specific activities for oxygen reduction as compared with that on Pt (Stamenkovic et al. 2007; Wang et al. 2011b). This enhancement in electrocatalytic activity has been ascribed to several factors such as interatomic spacing, preferred orientation, or electronic interactions. The state-of-the-art Pt–Co–Cr electrocatalysts have a particle size of 6 nm (Mukerjee et al. 1995; Wang et al. 2011b). The synthesis of nanocubes with oleic acid and oleylamine has also been expanded to the Pt-based alloy catalysts, for example, (100)-terminated Pt_3M nanocubes (M = transition metals Mn, Co, Fe, and Ni). Generally, they show a higher catalytic activity than their spherical counterparts toward oxygen reduction, formic acid oxidation, and methanol oxidation. By modifying the synthesis procedure, Fang et al. also prepared (111)-terminated Pt_3Ni nanoctahedra (as shown in Figure 7.4). The electrochemical experiments demonstrated that ORR activity strongly depends on the terminal facets of Pt_3Ni nanocubes. The {111}-facet-terminated nanoctahedra are significantly more active than the {100}-bounded nanocubes.

FIGURE 7.4 **(See color insert.)** (a–e) Images for Pt_3Ni nanoctahedra. (f–j) Images for Pt_3Ni nanocubes. (Fang, J.Y. et al., *Nano Letters*, 10, 638–644, 2010.)

More research work is needed to explain the complexities of facet effect. Nonetheless it is clear that controlling the shape of nanocatalyst is an effective way to improve ORR activity and to develop highly active electrocatalysts used under real conditions.

7.2.1.1.2 Nonprecious 0-D Nanoelectrocatalysts

Another solution to overcome the predicament of Pt catalysts is to replace them with nonprecious catalysts. The research on these materials has been studied for over 50 years, but they were insufficiently active for the high efficiency and power density needed for fuel cell applications (Binder et al. 1969; Bohm 1970). Recently, several breakthroughs occurred that have increased the activity and durability of nonprecious electrocatalysts, which can now be regarded as potential competitors to Pt-based catalysts (Jaouen et al. 2011).

Recently, MeN_xC_y catalysts have been receiving increased attention due to their reasonable activity and remarkable selectivity toward ORR. For example, a breakthrough of volume-specific activity from 2.7 to 99 A/cm^3 was achieved by the Dodelet group in Canada (Lefèvre et al. 2009). They produced microporous carbon–supported iron-based catalysts with active sites believed to contain iron cations coordinated by pyridinic nitrogen functionalities in the interstices of graphitic sheets within the micropores. Very recently, by replacing microporous carbon with zeolitic-imidazolate framework, they successfully improved the catalyst performance and reached a power density of 0.75 W/cm^2 at 0.6 V, tested at cathode in H_2–O_2, which is a meaningful voltage for PEMFC operation, comparable with that of a commercial Pt-based cathode tested under identical conditions (Proietti et al. 2011). However, the stability for such a catalyst was still much lower than the corresponding Pt catalysts. A remarkable improvement of stability was achieved by Zelenay et al. (2011). The most stable nonprecious metal catalysts derived from polyaniline, iron, and cobalt showed an 18 μA/h loss in average current density at 700-hour fuel cell performance test at a constant cell voltage of 0.4 V. The presence of the graphitized carbon phase in an active catalyst was suggested to play a role in hosting ORR-active sites and enhancing stability of the polyaniline-derived catalysts. Despite the promising attributes of these catalysts, both their activity and stability are insufficient to satisfy practical requirements. Undoubtedly, the combination of a better understanding of the nature of the active site and introduction of new materials and synthetic methods will be beneficial in improving the performance of this kind of catalyst.

7.2.1.1.3 1-D Nanostructures

The high surface energy brings contemporary 0-D nanoparticles better catalytic activity, but also relatively low stability, especially under the acidic conditions of the fuel cell, such as the presence of Ru in DMFC anode catalyst makes it less stable to permanent oxidation and solubilization in operating DMFCs (Yu et al. 2007a). In light of these shortfalls of contemporary 0-D nanoparticles, the use of 1-D nanomaterials in the development of electrocatalysts has become an excellent choice with which to solve many of the technological challenges that face the development of fuel cells.

From a structural perspective, the asymmetry of 1-D structure suppresses physical ripening processes and thus these nanomaterials are inherently stabilized from dissolution and Oswald ripening, which are known as one of the three primary *breakdown* mechanisms of carbon-supported 0-D Pt and Pt alloy nanoparticles (Chen et al. 2007; Garbarino et al. 2009). And, anisotropic 1-D nanostructures also possess advantageous path-directing effects, which greatly enhance their electron transport properties (Yan et al. 2004). A previous study of metallic NWs in devices has shown that single-crystalline NWs can maintain resistances commensurate with those of their bulk counterparts (Tian et al. 2005).

In recent years, many techniques have been proposed in the preparation and characterization of Pt-based NWs, but only several of them have actually been used for practical fuel cells, for which the history and development has been discussed in our and Wong's early reviews in detail (Du 2012; Koenigsmann et al. 2011). Template preparation has been the earliest technique explored for synthesizing Pt-based NWs for fuel cell catalysts, ever since the use of SBA-15 template nanoreactor from 2003 producing the NW network for anode materials in DMFCs (Choi et al. 2003). Since then, a range of templates have been investigated on this preparation method for fuel cells; from porous silicate (Park et al. 2008; Sun et al. 2006) to anodized aluminum oxide (AAO) (Liu et al. 2004a, 2009b; Lux et al. 2006; Ponrouch et al. 2010; Zhang et al. 2009; Zhao et al. 2006, 2007; Zhong et al. 2008) and then polymer template (Choi et al. 2008), even the fuel cell electrolyte Nafion membrane (Liang et al. 2010b, 2007). Generally, the preparation of NWs using these templates involves three steps: infiltration of the pores with an appropriate precursor through a vapor- or solution-based approach, conversion of this precursor to the desired material, and recovery of the NWs by selectively removing the template. If a sacrificial component (e.g., Cu or Ag) was simultaneously used with Pt to generate NW, after removing this sacrificial component, a porous NW could be obtained. For example, Figure 7.5 shows Pt porous NW array prepared by electrodepositing Pt-Co alloy NWs in AAO and then dealloying the Cu component (Zhang et al. 2009). In case of a 1-D nanostructure, for

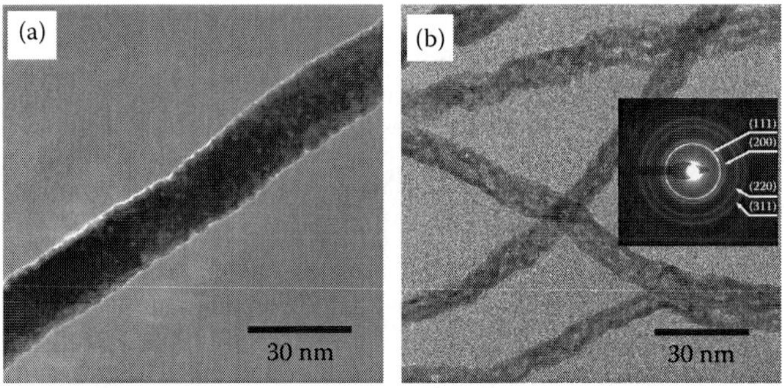

FIGURE 7.5 TEM micrographs showing the crystalline features of as-prepared nanoporous Pt-Co alloy nanowires by (a) 5-minute and (b) 15-hour dealloying, respectively. The dealloying was carried out in 10 wt% H_3PO_4 at 45°C. (Liu, L.F. et al., *Nano Letters*, 9, 4352–4358, 2009b.)

example, silver NWs or tobacco mosaic virus used as template, a porous Pt-based nanotube could also be obtained (Alia et al. 2010; Chen et al. 2007; Gorzny et al. 2010). In both cases, the high specific area from the porous nanostructure, together with the anisotropic 1-D feature, could lead to relative higher activity and durability than contemporary 0-D nanoparticle electrocatalysts. In addition to a template method, an electrospinning method has also been used for preparing Pt-based NWs (Kim et al. 2008, 2009a,b, 2010). However, irrespective of template-based method or electrospinning technique, only polycrystalline nanostructures were obtained for the produced Pt-based NWs. Although the excellent catalytic performance was obtained due to their 1-D feature, the existing lattice boundary and defect sites still result in a relative low catalytic stability, because these regions evince surface Pt atoms with lower coordination numbers and thus are more susceptible to decomposition and prone to irreversible oxidation (Shao et al. 2007; Yu et al. 2007b). Moreover, the template-based and electrospinning methods are technically complicated, owing to the requirement for template or ligand removal to obtain pure products. This limits their applications to some extent.

Apart from the advantages of the 1-D feature as that from polycrystalline NWs, single-crystalline Pt-based NWs give more benefits. First, the anisotropic growth of 1-D nanostructures typically results in the preferential display of low-energy crystal facets to minimize the surface energy of the systems (Cademartiri et al. 2009). In terms of Pt, the low energy (100) and (111) facets are most active for oxygen-reduction and thus the preferential display of these low-energy facets would be highly advantageous for catalysis (Subhramannia et al. 2008). Furthermore, single-crystalline Pt-based NWs possess largely pristine surfaces with long segments of crystalline planes (Seal et al. 2007) and maintain fewer surface defect sites (Williams 2001), leading to much high stability as compared with 0-D morphologies or polycrystalline 1-D nanostructures. More recently, wet chemical methods were successfully employed to grow single-crystalline Pt NWs for fuel cell applications. In 2007, Xia and coworkers (Lee et al. 2007, 2008) demonstrated the direct growth of single-crystalline Pt <111> NW arrays onto a gauze made of Pt or W for methanol oxidation by modified polyol process, combining with the decreased reaction rate obtained by the introduction of a trace amount of Fe^{2+} or Fe^{3+}. Poly(vinyl pyrrolidone) (PVP) was used as a surfactant, and the reaction was carried out at 110°C without any heat treatment, which is typically used for carbon-supported Pt catalysts. TEM images show these NWs had a diameter of ~5 nm, a length up to 200 nm and the growth direction along <111> axis (Figure 7.6). However, the large amount of PVP (molar ratio of PVP to Pt larger than 5) used was very difficult to be removed completely from the sample, especially in a highly dense Pt-NW array, which could hinder the accessibility of the catalytic surface by the fuel molecules and therefore cause a decrease in the catalytic performance. To avoid the influence of PVP, Zhang et al. and Gong et al. replaced ethylene glycerol with oleylamine as the reducing agent, and Pt-Fe and Pt-Mn ultrathin NWs with a diameter of 2–3nm were successfully prepared (Gong et al. 2011; Zhang et al. 2011a).

Another important improvement was given by Sun et al. (2007, 2008a,b, 2009a, 2010, 2011a). To decrease the reaction rate, the reaction was changed from 80°C to room temperature with formic acid used as reducing agent. Single-crystalline Pt <111> NWs were successfully grown in aqueous solution at room temperature, without using any templates, protecting groups, inducing-growth catalysts, or organic solvents. The as-prepared Pt NWs had diameters of approximately 4–10 nm and lengths up to a 100 nm. Using carbon spheres as support, the packed arrays of Pt NWs were densely covered on the surface of carbon spheres. These Pt-NW/C nanostructures were measured as cathode catalysts in PEMFCs. The results showed, for ORR, a 50% higher mass activity of 120 A/g Pt and threefold better specific activity of 275 μA/cm^2 Pt than those of the state-of-the-art Pt nanoparticle catalysts (PRIMEATM 5510 from W.L. Gore and Associates) (Sun et al. 2008a). The 1-D single-crystalline

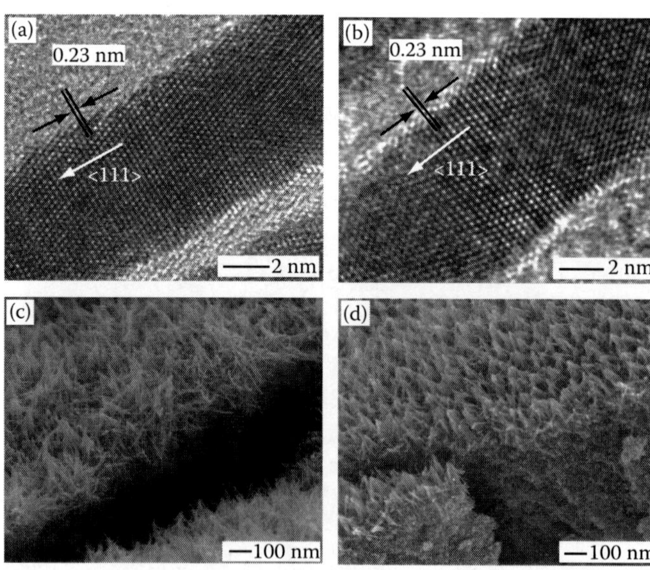

FIGURE 7.6 SEM and high-resolution TEM images of Pt nanowires on Pt (a, c) and W (b, d) gauzes. (Lee, E.P. et al., *ACS Nano*, 2, 2167–2173, 2008.)

nanostructure could also be prepared without support, where a multiarmed starlike NW was obtained (Sun et al. 2011a). It showed a very high catalytic durability toward methanol oxidation reaction (MOR) and ORR, due to the detachment of catalysts from the supports (e.g. carbon corrosion in Pt/C) is successfully eliminated for this kind of catalysts.

7.2.1.1.4 Core–Shell Configuration

Although single-crystalline 1-D Pt-based NWs possess high catalytic durability and better performance as compared with 0-D nanoparticles, the loading amount of expensive noble metal electrocatalyst is still very high due to the relative large diameters. To efficiently reduce the loading amount, recent developments in the field of fuel cell catalysts include the synthesis of new core–shell nanostructures by spontaneous deposition of Pt submonolayers on metal and other supports (e.g., C, Ni, Ag, or MoO) (Arico et al. 2005; Brankovic et al. 2001; Eichhorn et al. 2009; Stamenkovic et al. 2007; Tedsree et al. 2011) or dealloying from a bimetallic alloy precursor (e.g., Pt-Mo, Pt-Ag, Pt-Ni, Pt-Cu, or Pt-Cu-Co) to form a Pt-rich shell (Eichhorn et al. 2009; Peng et al. 2010; Strasser et al. 2007; Wang et al. 2011a). Apart from the economic benefits, this configuration is also attractive in terms of the highly enhanced reactivity (Eichhorn et al. 2009; Strasser et al. 2010; Tedsree et al. 2011). It has been demonstrated that platinum-rich shell exhibits compressive strain, which results in a shift of the electronic band structure of Pt and weakening chemisorption of oxygenated species, resulting in a very high reactivity toward ORR, as shown in Figure 7.7 (Strasser et al. 2010). Furthermore, this mechanism is not restricted to Pt, and it can be expanded to other expensive noble metals. For example, an Ag-Pd core–shell nanocatalyst with the shell containing between 1 and 10 layers of Pd atoms. The Pd shell contains terrace sites and is electronically promoted by the Ag core, leading to significantly enhanced catalytic properties (Tedsree et al. 2011). Further advances concern a better understanding of the surface chemistry in electrocatalyst nanoparticles and the effects of strong metal–support interactions that influence both the dispersion and electronic nature of Pt/Pd sites.

If we keep reducing the Pt loading amount in Pt-rich shell, subnanometer Pt clusters or even single atoms could possibly be achieved. Recent theoretical and experimental results demonstrated that subnanometer clusters have better catalytic activity and/or selectivity than nanometer-sized particles (Remediakis et al. 2005; Vajda et al. 2009). Of course, fabrication of practical and stable single-atom catalysts remains a significant challenge because, typically, single atoms are too mobile and easy to sinter under realistic reaction conditions. And, this is also the reason why a 3-nm Pt-based catalyst nanoparticle is usually preferred in low-temperature fuel cell catalysts. A very important finding on this topic was given by Zhang et al. (2007), who demonstrated that Pt electrocatalysts can be stabilized against dissolution under potential cycling regimes by modifying with Au clusters. The Au clusters are two to three monolayers thick and 2–3 nm in diameter, conferring activity by raising the Pt oxidation potential. There were insignificant changes in the activity and surface area of Au-modified Pt over the course of potential cycling between 0.6 and 1.1 V in over 30,000 cycles, in contrast to sizable losses observed with the pure Pt catalyst under the same conditions. Nevertheless, a synergy is achieved here by hybrid clusters with nanoparticles, which is a little bit similar to the shift of the electronic band structure

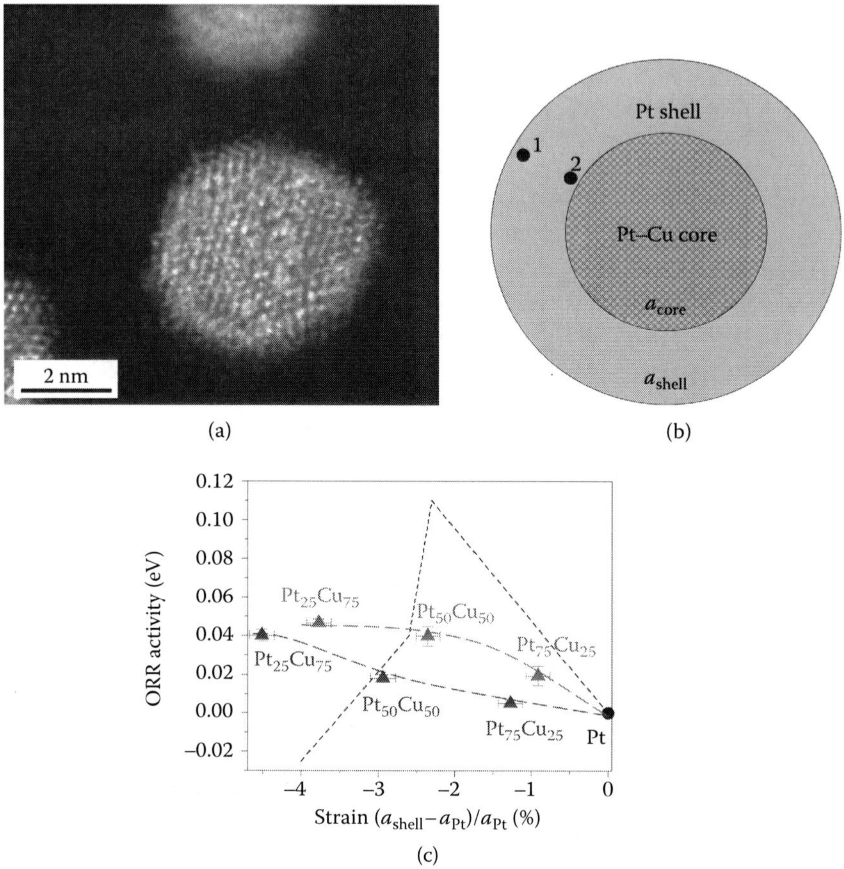

FIGURE 7.7 **(See color insert.)** (a) Subangstrom resolution image of a Pt-Cu dealloyed nanoparticle (~4 nm). (b) Scheme of a simple two-phase structural model for the dealloyed state of a bimetallic particle. (c) Experimental and predicted relationships between electrocatalytic oxygen reduction reaction (ORR) activity and lattice strain (annealing temperatures of 950°C [blue] and 800°C [red]). The dashed line is the predicted density functional theory (DFT) of the ORR activity. (Strasser, P. et al., *Nature Chemistry*, 2, 454–460, 2010.)

in the core–shell configuration mentioned in the preceding discussion in some degree. This means that if a support is thoroughly designed, the metal clusters or even single atoms could be anchored and stabilized by synergy with the support. For example, with FeO_x as oxide support, by a simple chemical coprecipitation method followed by heat treatment, Qiao et al. (2011) successfully synthesized a novel catalyst that consists of only isolated single Pt atoms uniformly dispersed on FeO_x nanocrystallites of high surface area. The Pt single-atom catalyst exhibits very high activity and stability for both CO oxidation and preferential oxidation of CO in H_2. The authors attributed this improvement to more vacant *d* orbitals of the single Pt atoms. The electron transfer from Pt atoms to the FeO_x surface is responsible both for the strong binding and stabilization of single Pt atoms and for providing positively charged Pt atoms, which

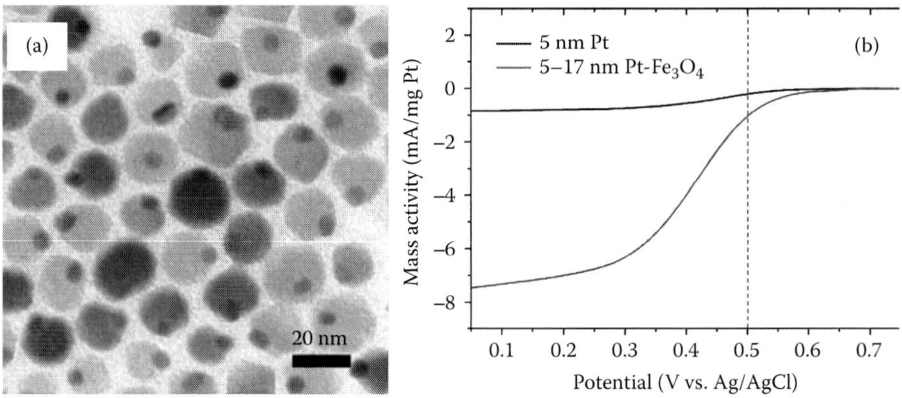

FIGURE 7.8 (a) TEM image of 5–12-nm Pt-Fe_3O_4 NPs and (b) oxygen reduction reaction mass activities for 5-nm Pt and 5–17-nm Pt-Fe_3O_4 NP catalysts measured in 0.5 M H_2SO_4 with the rotating disc electrode rotation speed at 1600 rpm and sweeping rate at 10 mV/s. (Sun, S.H. et al., *Nano Letters*, 9, 1493–1496, 2009b.)

ultimately account for the excellent catalytic activity of the Pt/FeO_x catalysts. For fuel cell applications, this synergy between Pt and FeO_x was also found in monodisperse dumbbell-like Pt-Fe_3O_4 nanoparticles (Sun et al. 2009b), where the nanoparticle size is tunable from 2 to 8 nm for Pt and from 6 to 20 nm for Fe_3O_4. This structure shows a 20-fold increase in mass activity toward ORR compared with the single-component Pt nanoparticles and the commercial 3-nm Pt particles, as shown in Figure 7.8. This synergy is not restricted to Pt and FeO_x, for example, Abruna et al. (2010b) demonstrated that the decoration with a small amount of Pt also enhanced the stability and electrocatalytic activity of Pd-Co@Pd nanoparticles toward ORR. By this approach, if a support could be found with the same synergy function and a good stability in acidic fuel cell environments, the Pt loading could be expected to be reduced to an extremely low level and the barrier of cost could be then resolved.

7.2.1.2 Catalyst Support in Low-Temperature Fuel Cells

We know that fuel cell performance is essentially governed by the electrocatalyst support materials as they strongly influence the electrocatalyst performance, durability, and efficiency. It is well known that carbonaceous support materials suffer from carbon corrosion (oxidation) especially at high potentials over time. Thus, alternative low-cost, high-performing, and noncorrosive electrocatalyst support materials are urgently required.

7.2.1.2.1 Carbon-Based Nanostructures

The use of carbon materials as catalyst supports for precious metals started from 1960s (Chang et al. 2010), but unlike the hot research efforts on electrocatalysts, only recently the study experienced a rapid increase on catalyst support, due to the continuous advancements in nanotechnologies and fuel cells. The main requirements of suitable supports for fuel cell catalysts are high surface area, good electrical conductivity, suitable porosity to allow good reactant flux, and acceptable stability (e.g., corrosion resistant) in acidic fuel cell environment. The high availability and low cost make CBs the most used support for low-temperature fuel cell catalysts. Conventionally, highly conductive CBs of turbostratic structures with high surface areas, such as Vulcan XC-72R (Cabot Corp., 250 m^2/g), Shawinigan (Chevron, 80 m^2/g), Black Pearl 2000 (BP2000, Cabot Corp., 1500 m^2/g), Ketjen Black (KB EC600JD & KB EC600J, Ketjen International, 1270 m^2/g and 800 m^2/g, respectively), and Denka Black (DB, Denka, 65 m^2/g), are currently used as low-temperature fuel cell electrocatalyst supports to ensure large electrochemical reaction surfaces. They are usually submitted to chemical activation to increase anchoring centers for catalysts (to increase metal loading and dispersion). The disadvantage of these materials is the presence of a high amount of micropores (<2 nm), where supply of a fuel may not occur smoothly and the activity of the catalyst may be limited. Moreover, the corrosion of CB by oxygen during the cycling processes of fuel cells strongly reduces the electrocatalytic activities of the catalysts (Antolini 2009; Hartl et al. 2011).

Recent studies have revealed that the physical properties of the carbon support can greatly affect the electrochemical properties of the fuel cell catalyst. It has been reported that carbon materials with both high surface area and good crystallinity cannot only provide a high dispersion of Pt nanoparticles but also facilitate electron transfer, resulting in better performance (Antolini 2009; Antolini et al. 2009; Hartl et al. 2011). On this basis, novel nonconventional carbon materials have attracted much interest as electrocatalyst support because of their good electrical and mechanical properties and their versatility in pore size and pore distribution tailoring. These materials present a different morphology than CBs both at the nanoscopic level in terms of their pore texture (e.g., mesoporous carbon) and at the macroscopic level in terms of their form (e.g., microsphere). The examples are supports produced from ordered mesoporous carbons (OMCs) (Chang et al. 2007; Xia et al. 2008), carbon aerogels (Moreno-Castilla et al. 2005), carbon nanohorns (Sano et al. 2006; Yoshitake et al. 2002), carbon nanocoils (Fuertes et al. 2007; Hyeon et al. 2003), carbon nanofibers (Hwang et al. 2007), CNTs (Wong et al. 2009), graphene, and graphene oxide (Sun et al. 2011b). Some of these works are summarized in Antolini's (2009) recent review. Among these novel nanostructures, because of the high surface area and proper porosity, the mesoporous carbons (MC, e.g., OMC and carbon aerogels), CNTs, and graphene are expected to be the most suitable for the preparation of catalytic particles and for efficient diffusion and transport of reactants and by-products.

The high surface area and high amount of mesopores of OMCs and carbon gels allow high metal dispersion and good reactant flux (Chang et al. 2007). So, catalysts supported on these carbons showed higher catalytic activity than the same catalysts supported on CB. It has to be remarked that the synthesis methods of OMCs and carbon aerogels are simple and not too expensive. However, suitable carbon supports must possess high mesoporosity in the pore size range of 20–40 nm for a high accessible surface area. Indeed, the Nafion binder solution, which is generally used in electrode preparation, is constituted by ionomers that do not enter or may occlude pores narrower than 20 nm (Yang et al. 2010), so that catalyst nanoparticles deposited in such pores are not in contact with the proton conductor and/or the fuel.

For this reason, the presence of mesopores with pore size <20 nm supports the gas flow, but decreases the active surface area of the catalyst. As a consequence, the electrochemical activity of these MCs could be lower than that of microporous carbons. On the basis of their high versatility in pore size and pore distribution tailoring, among the MCs, carbon gels seem more promising than OMCs.

Among the new carbon materials, CNTs are the most investigated as catalyst support for low-temperature fuel cells (Antolini 2009; Wong et al. 2009). They normally possess an outer diameter of 10–50 nm, an inside diameter of 3–15 nm (pore size), and a tube length of 10–50 μm. The high crystallinity of CNTs makes these materials highly conductive; the high specific surface area (SSA) and high amount of mesopores result in a high metal dispersion and a good reactant flux in tubular structures (Wong et al. 2009). Moreover, CNTs have a positive effect on Pt structure, resulting in a higher catalytic activity and a higher stability than CBs (Zhang et al. 2010a). A problem for the commercialization of CNTs is their higher cost compared with that of CBs. During synthesis of the catalyst using this support, Pt particles (2–5 nm size) present on the pore mouths of CNTs will take part in the chemical reaction. However, there is a great possibility for the existence of Pt particles inside the nanotube, depending on Pt particle size (Wong et al. 2009). These particles will take little part in the chemical reaction. The number of the Pt particles inside the tube will be more when the tube length of CNT increases. So, a decrease in the Pt active area and the electrochemical activity of the catalyst has to be expected.

Graphene, a single-layer graphite with closely packed conjugated hexagonal lattices, is recognized as the basic building block of all dimensional graphitic materials (Geim et al. 2007). Also, graphene nanoribbons can be prepared by longitudinal unzipping of CNTs (Tour et al. 2009). So, graphene not only possesses all advantages of CNTs as catalyst supports, for example, high electrical conductivity and huge SSA, but also avoids the drawback of catalyst particles existing inside the nanotube. Furthermore, the large size of graphene (layer diameter >1 μm) can also eliminate the possible health problems caused by CNT pollution. Thus, graphene has been exploited as the support of Pt catalyst. For example, Liu et al. (2009a) compared the SSA, electrocatalytic activity for ORR, and stability of the Pt nanoparticles supported on the functionalized graphene sheets (Pt-FGS) and CBs (*E-TEK*). Pt-FGS showed not only larger SSA and higher ORR activity but also excellent stability after 5000 cyclic voltammetry cycles. These improved properties were attributed to the smaller aggregation of Pt particles immobilized on graphene. In addition, Pt nanoparticles were supported on graphene nanoplatelets, and this catalyst also showed a good electrochemical durability (two to three times that of the Pt/CNT or *E-TEK* Pt/C) (Wang et al. 2010b,c, 2011c). Beside Pt nanoparticles, the graphene-supported Pt-on-Pd bimetallic nanodendrites (Wang et al. 2010a) (shown in Figure 7.9) and Pt-Ru nanoparticles (Dong et al. 2010) were also prepared for methanol and ethanol oxidations. Compared with CB, graphene strongly enhanced the oxidation efficiencies of both methanol and ethanol.

By comparing MCs, CNTs, and graphene, taking into account the cost of the materials, the complexity of the synthesis methods, and the versatility in pore size and pore

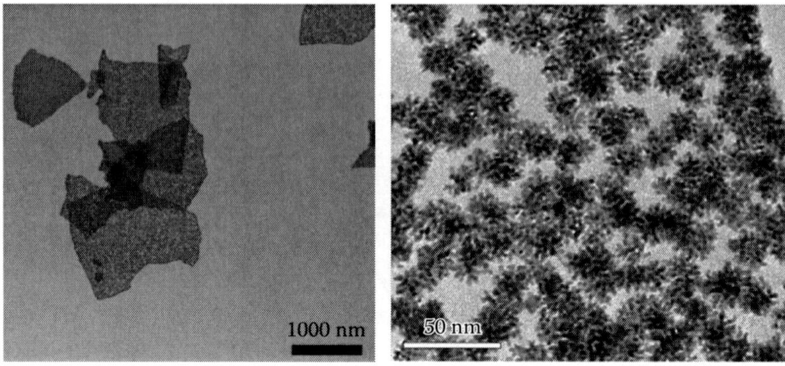

FIGURE 7.9 TEM image of the Pt-on-Pd bimetallic nanodendrites supported on graphene sheets. (Wang, E.W. et al., *ACS Nano*, 4, 547–555, 2010a.)

distribution tailoring, the MCs seem to have more changes to substitute CBs as fuel cell catalyst substrate (e.g., as microporous layer). On the other hand, CNTs or graphene, for their high electronic conductivity, due to their unique structure, and their high stability during long-term tests in acidic media, ascribed to the strong metal–carbon interactions, seem to be more suitable than MCs for use as a support for fuel cell catalysts. The stability in fuel cell conditions of MC-supported catalysts is similar to that of CB-supported catalysts, but can be increased by graphitization of the MCs.

Apart from pure carbon materials, nitrogen-doped carbon material is recognized as a good support for Pt catalysts. Data suggest that nitrogen functional groups introduced into a carbon support appear to influence at least three aspects of the catalyst/support system (Zhou et al. 2010): (1) modified nucleation and growth kinetics during catalyst nanoparticle deposition, which results in smaller catalyst particle size and increased catalyst particle dispersion; (2) increased support/catalyst chemical binding (or *tethering*), which results in enhanced durability; and (3) catalyst nanoparticle electronic structure modification, which enhances intrinsic catalytic activity. N-doped carbon-based nanostructures have much more effect in fuel cell applications; this includes the research on mesoporous N-doped carbon (Ma et al. 2009; Yang et al. 2011), N-doped CNTs (Chen et al. 2011; Dai et al. 2009), and N-doped graphene (Geng et al. 2011; Lin et al. 2010; Ramaprabhu et al. 2010; Song et al. 2010; Tang et al. 2009). For example, N-doped graphene can be synthesized by chemical vapor deposition (CVD) of methane in the presence of ammonia as reported by Dai et al. (Liu et al. 2010b). Compared with the commercialized Pt catalysts, this material showed higher electrocatalytic activity, longer-term stability, and improved tolerance to CO for ORR. Considering the low price and high electrocatalytic activity, N-doped graphene can be used as an efficient ORR catalyst for fuel cells. However, all these researches only performed the electrical measurement in KOH, their performance in PEMFCs and DMFCs is still unknown.

From the perspective of the replacement of CBs with MCs or CNTs as catalyst supports, or to directly use N-doped carbon materials as catalyst for ORR, further tests in fuel cells have to be performed to evaluate the electrochemical activity and the long-term stability of the catalysts supported on these new promising materials.

7.2.1.2.2 Noncarbonaceous Supports

An example of the alternative approach for avoiding the use of CBs is the fabrication of porous silicon catalyst support structures with a 5-μm pore diameter and a thickness of about 500 μm (Miu et al. 2010). These structures have high SSA, and they are of interest for miniature low-temperature fuel cells. A finely dispersed, uniform distribution of nanometer-scale catalyst particles deposited on the walls of the silicon pores creates, in contact with the ionomer, an efficient three-phase reaction zone capable of high-power generation in DMFCs (Manthiram et al. 2008).

Another group of noncarbonaceous support is tungsten carbide (WC) (Binder et al. 1969) and metal oxides, for example, ZrO_2, TiO_2, indium oxides, alumina, silica and tungsten oxide, ceria, and conducting polymer materials (Abruna et al. 2010a; Antolini 2009; Antolini et al. 2009; Miu et al. 2010). It has been demonstrated some of these materials are highly stable to oxidative decomposition compared to currently used CBs. Furthermore, the low catalytic ability of WC can effectively enhance the reactivity and durability of Pt catalysts (Hara et al. 2007; Liang et al. 2010a). However, the greater difficulty in the preparation limits its application.

7.2.1.3 Electrodes in SOFCs

Due to the fact that SOFCs operate at high temperatures, the particle size of ceramic electrocatalysts is more than one order of magnitude larger than Pt catalysts used in low-temperature fuel cells. Even like this, nanostructured electroceramic materials are increasingly used in intermediate-temperature SOFCs (IT-SOFCs) (Brett et al. 2008). For example, SOFC microstructured components fabricated by starting with nanosized particles possess different electrocatalytic and ion-conduction properties from the typical polycrystalline materials (Schoonman 2003). Nanosized yttria-stabilized zirconia (YSZ) and ceria-based (gadolliunium, samaria, and yttria-doped ceria) powders permit a reduction of the firing temperature during the membrane-forming step in the cell fabrication procedure because their sintering properties differ from those of polycrystalline powders (Ormerod 2003; Ruiz-Morales et al. 2010). Furthermore, by decreasing the particle size in the electroceramics, the quantum confinement effect causes an increase in the band gap and, thus, favors the occurrence of a purely ionic domain. For example, nanocrystalline ceria, which is characterized by mixed electronic ionic conduction properties, promotes the charge transfer reactions at the electrode–electrolyte interface (Brett et al. 2008; Suzuki et al. 2009).

Besides reduced particle size by using nanocrystalline materials, the SOFC electrode performance can also be improved by novel nanostructured design. It has been demonstrated that the electrode microstructure and the SOFC performance have a close correlation (Suzuki et al. 2009). The dual-scale porous SSC($Sm_{0.5}Sr_{0.5}CoO_3$)-GDC and NiO-GDC electrode for SOFCs can be synthesized with a foam-like thin film template (Liu et al. 2005). This porous structure contains large pores (0.8–1.5 μm) for rapid gas transport and small pores (approximate 2.5 and 35 nm) for fast electrochemical reactions (Figure 7.10). Highly porous and nanostructured SOFC cathodes consisting of nanograins of about 50 nm have also been fabricated using a combustion CVD process, and for which a very low interfacial polarization resistance (1.09 Ωcm^2 at 500°C) was achieved (Liu et al. 2004b). To increase the cathode/electrolyte interfacial area

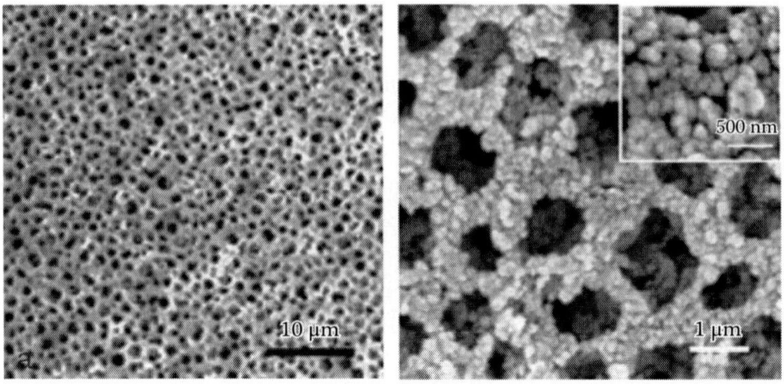

(a) (b)

FIGURE 7.10 (a, b) SEM images of a dual-scale porous NiO-GDC anode film fired at 800°C for 2 hours. The inset in (b) shows the nanostructure of the porous wall. (Liu, M.L. et al., *Advanced Materials*, 17, 487–491, 2005.)

density and to lower the polarization resistance present at the cathode/electrolyte interface, Yoon et al. (2009) introduced a binary vertically aligned nanocomposite interlayer of $(LSCO)_{0.5}(CGO)_{0.5}$ between the cathode and electrolyte to achieve a high-efficiency thin film SOFC. Cathode with very low polarization resistance was also prepared by cobaltite nanotubes (Bellino et al. 2007). Laberty-Robert et al. (2007) optimized electrical properties of GDC by creating nanoarchitectures interconnectivity established between the <10 nm crystallites in CGO aerogels for rate-critical applications. The high degree of network yields an electroceramic that responds as though it contains no grain boundaries and that exhibits long-range pathways for ionic diffusion.

7.2.2 Enhanced Ionic Conductivity and Structural Stability for Electrolyte

7.2.2.1 Electrolyte for PEMFCs and DMFCs

The electrolyte is an essential component of the fuel cell assembly. It is a quintessential nanomaterial in PEMFCs and DMFCs, with hydrophilic pores of around 10 nm in size through which protons are transported (Figure 7.11) (Buratto 2010). The most common polymer electrolyte membrane (PEM) material is the perfluorosulfonic polymer (e.g., Nafion), which is composed of a hydrophobic Teflon backbone and side chains terminated with hydrophilic sulfonic acid (SO_3H) groups, because of its excellent proton conductivity and electrochemical stability. Several decades of PEM research have resulted mostly in a phenomenological understanding of membrane performance. Nafion, despite its high cost and hydration requirement that restricts operation to 80°C, continues to be the most widely studied and employed membrane for PEMFCs (Devanathan 2008). Furthermore, even with the extensive experimental database on Nafion membrane structure and molecular transport, a deep understanding of the water network and proton transport processes is still evolving.

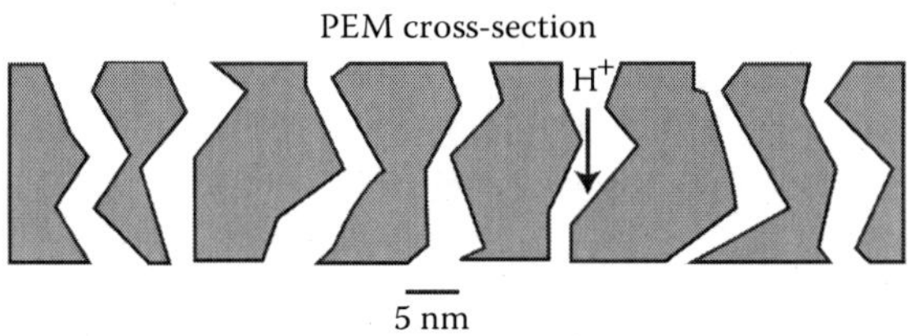

FIGURE 7.11 Schematic cross-section of a polymer electrolyte membrane (PEM). The proton-conducting channels are depicted in white. (Buratto, S.K., *Nature Nanotechnology*, 5, 176, 2010.)

Nafion suffers from several drawbacks such as methanol crossover and membrane dehydration. In a Nafion film, the hydrophilic pores form by phase separation of the side chains from the polymer backbone. The performance of the proton-conducting channels in these films is strongly dependent on the environmental conditions. At high temperature and low relative humidity, the proton-conducting channels shrink and the conductivity decreases markedly. This means that under high-temperature conditions—where the catalysts in the electrodes perform at their best—the membrane performs at its worst. Therefore, to optimize the performance of fuel cells, it is useful to design a membrane with high conductivity at high temperature and low humidity. The research is usually carried out in nanotechnologies by two ways: modifying the Nafion membrane and using alternative materials.

Various relationships between membrane nanostructure and transport characteristics, including conductivity, diffusion, permeation, and electroosmotic drag, have been observed (Arico et al. 2005; Paddison 2003). Fully or partially perfluorinated membranes can overcome some of the shortcomings of Nafion (Devanathan 2008). Modification of side chain and backbone length (Di Noto et al. 2010; Paddison 2003) or addition of inorganic fillers (Chan et al. 2011; Laberty-Robert et al. 2011; Spurgeon et al. 2011) and heteropoly acids (Herring et al. 2007; Ramani et al. 2005) may help retain water at low humidity. However, the effect of fillers on membrane morphology is not well understood.

Among the alternative materials to Nafion, one important finding was made by Moghaddam and colleagues (Buratto 2010; Moghaddam et al. 2010). They reported a new silicon-based proton exchange membrane with high proton conductivity over a wide range of relative humidity and temperature. Key to achieving these advantages is fabricating a silicon membrane with pores of diameters around 5–7 nm, adding a self-assembled molecular monolayer on the pore surface, and then capping the pores with a layer of porous silica. The silica layer reduces the diameter of the pores and ensures their hydration, resulting in a proton conductivity that is two to three orders of magnitude higher than that of Nafion at low humidity. A MEA constructed with this proton exchange membrane delivered an order of magnitude higher power density than that achieved previously with a dry hydrogen feed and an air-breathing cathode. Other research on alternative membranes includes materials based on aromatic backbone polymers (Park et al. 2010) and polystyrene sulfonic acid membranes (Jones et al. 2003).

However, most of the research products are only demonstrated in laboratory-scale tests; *real-life* tests of novel membranes during prolonged operation at elevated temperatures, low humidities, and oxidizing conditions are required to rigorously evaluate the membranes proposed as alternatives to Nafion. Furthermore, integrated experimental, theoretical, and computational studies that link proton transfer, hydrogen bonding, molecular transport, membrane morphology, and mechanical properties are needed for rational development of the next generation of PEMs.

7.2.2.2 Electrolyte for SOFCs

SOFCs use an ion-conducting ceramic material as an electrolyte. The most established technology is based on an YSZ electrolyte, which requires the cell to operate between 800°C and 1000°C. These high operating temperatures demand expensive materials for fuel cell interconnectors, cause thermal stresses, and require long start-up times and large energy inputs to heat the cell up to the operating temperature. Therefore, if SOFCs could be designed to give a reasonable power output at intermediate temperatures (IT, 400–700°C), tremendous benefits might result. However, in the attempt to reduce the operating temperature, the increasing electrolyte resistivity becomes one of the problems to overcome, since the ion-transport mechanism in ceramic electrolytes is a thermally activated process. However, two approaches on the development of nanomaterials for SOFC electrolyte can be followed: decreasing the electrolyte thickness to nanoscale and using alternative nanomaterials that have better ionic conductivity in the IT range than YSZ.

The thin electrolyte SOFCs usually have a configuration with electrode support, and the electrolytes are typically less than 25 μm thick (De Jonghe et al. 2003). For electrolyte less than ~100 nm thick, applications are mainly limited to micro-SOFC (μ-SOFC) (Takagi et al. 2011). Scaling-up remains a significant challenge because large-area membranes are susceptible to mechanical failure, while for optimum cell performance the nanostructured electrolyte must have a fully dense structure to minimize reactant crossover. Atomic layer deposition has been considered as a promising fabrication method (Prinz et al. 2011). Shim et al. (2007) prepared freestanding 60-nm YSZ films on amorphous Si_3N_4, and maximum power density of 270 mW/cm^2 was observed at 350°C. By depositing microfabricated metallic grids to function as mechanical supports, Tsuchiya et al. (2011) successfully deposited a scalable 54-nm-thick 8 mol% YSZ electrolyte film for μ-SOFCs, from which a power density of 155 mW/cm^2 was achieved at 510°C. Some details of a platinum-grid-supported 5 × 5 mm μ-SOFC is shown in Figure 7.12.

Regarding the alternative materials with high ionic conductivity, one way is focused on composition by materials optimization for both new and conventional electrolyte materials or by exploring new compounds that support fast oxide ion or proton conductivity. This topic was recently reviewed by Malavasi et al. (2010) in their publication. Another broad area attracting increasing attention is nanomaterials or *nanoionics* for the electrolyte. These systems are characterized by size effects, short diffusion lengths, a high density of interfaces (e.g., grain boundaries), and, in some cases, enhanced transport properties. More detailed discussions are found in

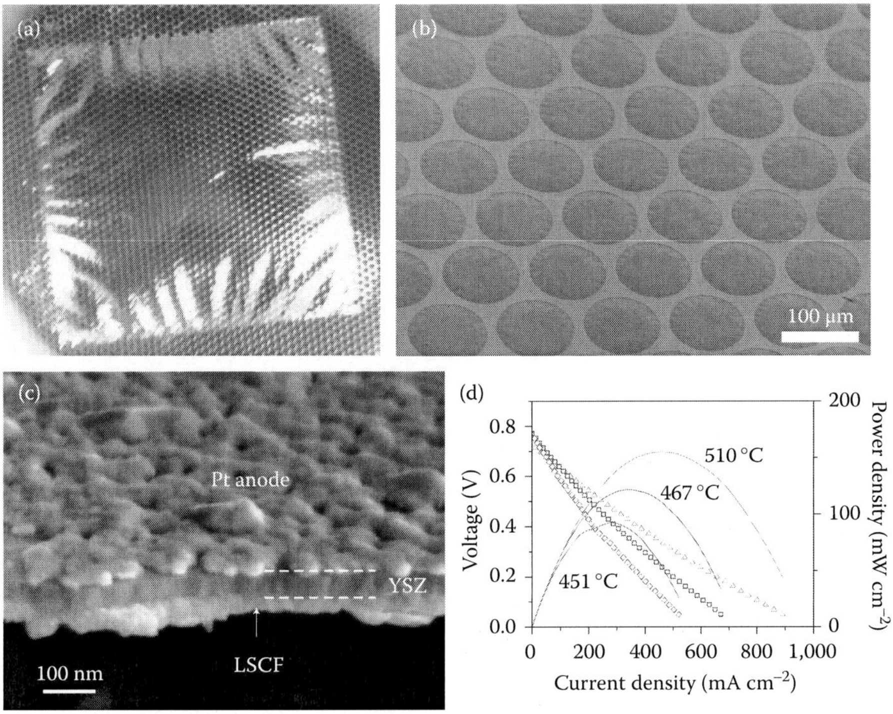

FIGURE 7.12 **(See color insert.)** Platinum-grid-supported 5 × 5 mm micro-solid oxide fuel cell (μ-SOFC): (a) Optical micrographs taken from the cathode side at 480°C; (b) SEM micrograph of slightly buckled freestanding membranes before testing; (c) cross-sectional micrograph of μSOFCs after testing; (d) current voltage sweep of platinum-grid-supported 5 × 5 mm μSOFC at three different temperatures. (Reprinted by permission from Macmillan Publishers Ltd. Tsuchiya [*Nature Nanotechnology*] [M. et al. 2011.], Copyright 2011.)

reviews by Maier (2005), Chadwick et al. (2006), and Arico et al. (2005), in which an important finding was given by Santamaria et al. (2008). They reported an eight-order-of-magnitude enhancement of YSZ conductivity in epitaxial heterostructures consisting of two-unit-cell-thick YSZ thin films sandwiched between two thicker layers of strontium titanate ($SrTiO_3$, STO). The origin of this huge enhancement in conductivity and associated decrease in activation energy (from 1.1 to 0.6 eV) is thought to lie in the YSZ/STO interfaces, in which the number of mobile ions increases, as does the volume through which they can move. This interface has been investigated by means of energy loss spectroscopy, showing that epitaxial growth between YSZ and STO leads to a modified interface with a highly disordered oxygen plane. This effect is a little similar to the synergy effect between alloyed catalysts and catalyst with support from low-temperature fuel cells (Qiao et al. 2011; Zhang et al. 2007). It is expected these kinds of novel processing strategies could lead to exciting new technologies based on nanostructured low-temperature oxide ion conductors and thus represent a potentially fruitful avenue of research.

7.3 MEMBRANE ELECTRODE ASSEMBLY

7.3.1 Novel MEAs for Low-Temperature Fuel Cells

Within the low-temperature fuel cells, the supported or unsupported catalyst is intimately mixed with the electrolyte ionomer (e.g., Nafion) to form a composite catalyst layer. The benefit of this approach is an enhancement of the interfacial region between catalyst particles and ionomer, extending the triple-phase boundary (or three-phase reaction zone). However, as discussed in MC supports, the large size range at 20–200 nm of the ionomer makes it very difficult to soak deeply in the small pores of the active layer. Thus, the reaction area is limited to an interface between catalyst particles distributed on the outer surface of carbon (or themselves in the unsupported case) agglomerates and the ionomer (as shown in Figure 7.13), resulting in a very low catalyst utilization with a typically value 20%–30%.

So, much research has been carried out to improve the catalyst utilization ratio, such as by reducing catalyst loading for a thinner active layer. One important contribution is by 3M Corporation for the Pt-coated nanostructured whisker supports, as shown in Figure 7.14 (Arico et al. 2005; Debe 2006; Garcia-Martinez 2010). Part of the MEA is a nanostructured thin film catalyst based on platinum-coated nanowhiskers. The approach uses highly oriented, high-aspect-ratio single-crystalline whiskers of an organic pigment material. This support permits suitable specific activity of the applied catalysts and aids processing and manufacturing. The electrocatalytic activities so obtained are comparable to catalyst–ionomer inks.

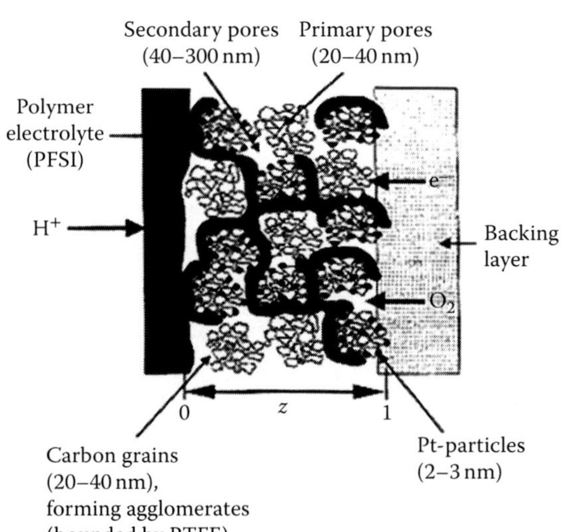

FIGURE 7.13 Schematic of a conventional composite (carbon-supported catalyst, ionomer) catalytic electrodes for PEMFCs ($z \sim 10$ μm). (Debe, M.K., *Nanostructured Thin Film Catalysts (NSTFC) for Next Generation PEM Fuel Cells*, University of Minnesota, Minneapolis, MN, 2006.)

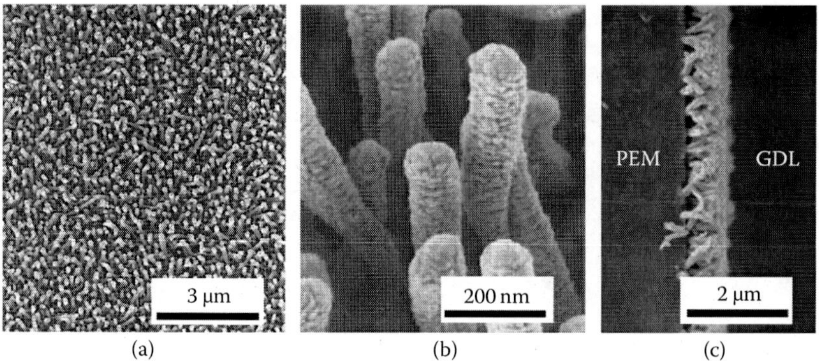

FIGURE 7.14 3M Corporation platinum (Pt)-coated nanostructured whisker supports (0.25 mg/cm^2), in plane view (a) and 45° view (b). The nanostructured film of the membrane electrode assembly (c) shows the Pt-coated nanowhiskers sandwiched between the polymer electrolyte membrane (PEM) and the gas diffusion layer (GDL). (Debe, M.K., *Nanostructured Thin Film Catalysts (NSTFC) for Next Generation PEM Fuel Cells*, University of Minnesota, Minneapolis, MN, 2006.)

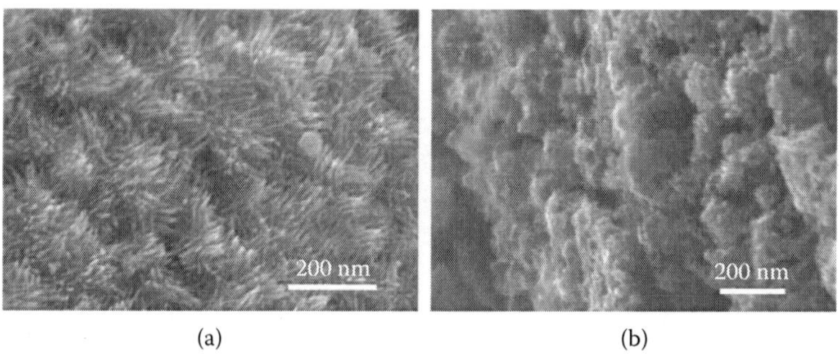

FIGURE 7.15 Gas diffusion electrode with a platinum-nanowire array grown on a gas diffusion layer surface, at plane view (a) and side view (b).

In our group, we developed gas diffusion electrodes (GDEs) with a thin Pt-NW catalyst layer (Du 2010), with a thickness about 0.5–1 μm. The single-crystalline Pt-NW array was directly grown on a gas diffusion layer (GDL) surface by using a reducing precursor of formic acid at room temperature in aqueous solution, without using any templates, organic solvents, or induced-growing catalysts. The catalyst synthesis process is the same as used by Sun et al. 2007 mentioned earlier, but by replacing the carbon support with GDL surface, the GDE can be fabricated by just one simple step without the catalyst-ionomer ink and coating process for fabricating the catalyst electrode. Figure 7.15 shows the SEM images of the Pt-NW-based GDE. The measurement in hydrogen/air single cell showed a higher power density than the state-of-the-art *E-TEK* GDE 120E-W. A good adhesion between the Pt NWs with support was also confirmed by the sonication treatment. A higher performance and better durability were also obtained by testing a DMFC cathode (Du et al. 2011b). This high performance and good durability of the Pt-NW electrodes, in addition to the simple and cheap manufacture process, may offer a large potential in practical applications. However, as a completely novel approach, it has been demonstrated that a new procedure was necessary to fabricate MEAs from these GDEs in fuel cells, such as changing the electrolyte ionomer loading amount (Du et al. 2011a). More work on process optimization needs to be done before it can be put into practical applications.

7.3.2 A Nonelectrolyte-Separator SOFCs

As we mentioned in Section 7.1, fuel cells are constructed with three functional components: anode, electrolyte, and cathode. In SOFCs, to decrease the high operating temperatures, many efforts have also been made by reducing the thickness of the electrolyte and nanoscale membranes have previously been fabricated, as discussed in Section 7.2.2.1. Even like this, the interfaces between the electrolyte and the electrodes (anode and cathode) produce major polarization losses. The electrolyte acts as a bottleneck that limits the improvement of fuel cell performance and thus delays its commercialization. Of course, if the electrolyte can be eliminated, all these problems can then be ignored. Zhu et al. (2011a,b,d) reported the nonelectrolyte-separator fuel cell (NEFC), creating a fuel cell from a single homogenous layer.

The layer is made from a mixture of ionically conducting materials of $(Na/K)_2CO_3$–$Ce_{0.8}Sm_{0.2}O_{2-\delta}$ (NKSDC) (Li et al. 2010a, 2011) and nanoparticles of a LiNiZn(CuFe)-based oxide and has both ionic and semiconducting properties. Moreover, when hydrogen and air are supplied to either side of the layer, the composite can act as a catalyst for both the oxidation of hydrogen and the reduction of oxygen. On one side, hydrogen is broken down into protons and electrons, a function similar to that of the anode of a typical fuel cell; whereas on the other, electrons are received through an external circuit and oxygen from the air is split into negative oxygen ions, just like a fuel cell's cathode. Water is then thought to be generated through the direct combination of protons and oxygen ions on the surface of the particles.

A schematic of the device working principles was given by the team, as shown in Figure 7.16. It works quite similar to dye solar cells (Oregan et al. 1991). A p–n junction may be formed within the LiNiCuZn(Fe) materials to keep an effective charge separation, where doping of NiO and ZnO shows p- and n-type conductivity, respectively. NKSDC and LiNiCuZn(Fe) form percolating paths among different phases. It can thus guarantee both ionic (H^+ and O^{2-}) and electronic (n and p) conduction in a continuous network throughout the component to support current transferring paths and continuous outputs; the electrons and holes do not pass through the component(s) but move to the corresponding current collectors because of the p–n junction and its operating principle, the same as in the solar cell. The exact mechanism of the underlying processes is still unclear, but the team showed that the one-layer fuel cells could convert hydrogen and air into electricity and water producing a power output of more than 680 mW/cm^2 at 550°C (Zhu 2003; Zhu et al. 2011a,c,d).

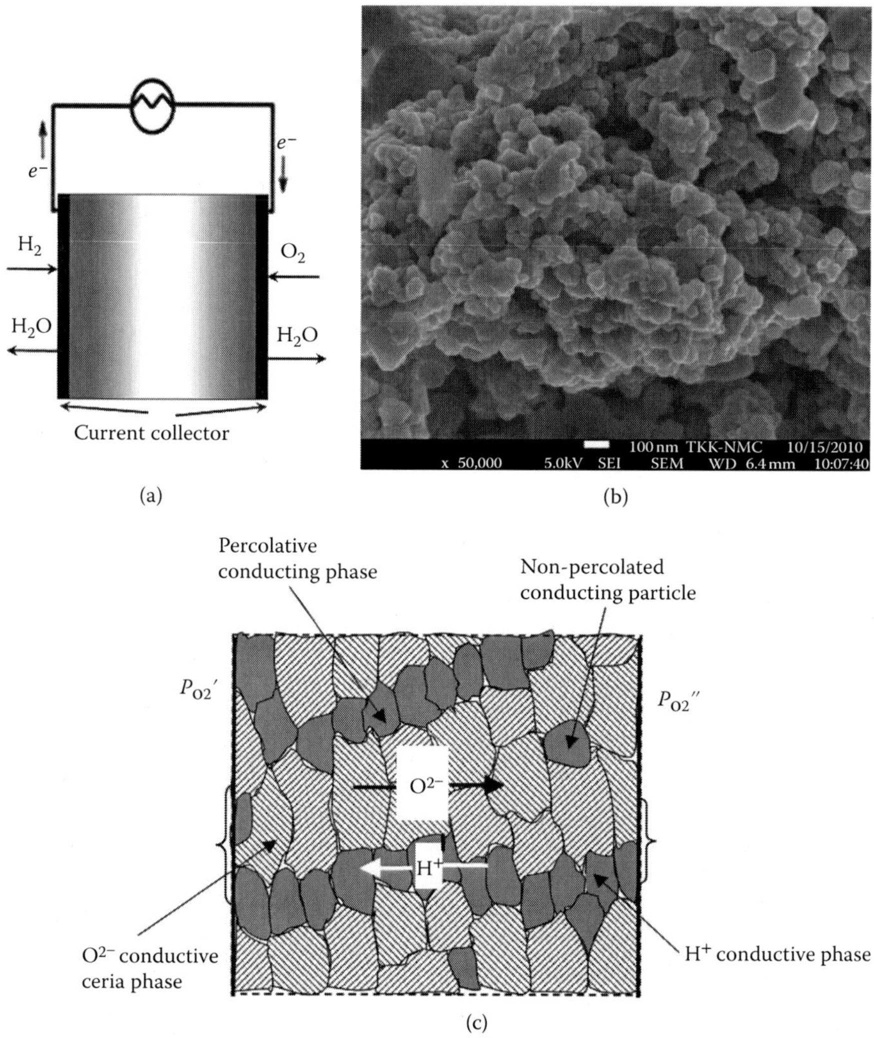

FIGURE 7.16 (a) Schematic of the configuration for a single-component nonelectrolyte-separator fuel cell device; (b) SEM image of the cross section of the LiNiCuZn (Fe) oxide–SDC single-component device; and (c) the working principles behind the fuel cell. (Zhu, B., *Journal of Power Sources*, 114, 1–9, 2003; Zhu, B. et al., *Electrochemistry Communications*, 13, 225–227, 2011a; Zhu, B. et al., *Journal of Power Sources*, 196, 6362–6365, 2011c.)

7.4 CURRENT LIMITATIONS IN NANOMATERIALS FOR FUEL CELLS

Notwithstanding the importance of nanomaterials in the progress of fuel cell development, there are still some issues needed to be solved before these nanomaterials can be really put to practical applications, and the performance of practical fuel cells remains limited by scale-up, stack housing design, gas manifold, and sealing. Some of them can be overcome by the methods mentioned in the preceding discussion.

7.4.1 Fabrication of MEAs and Fuel Cells

Although progress has been achieved in the direct electrochemical oxidation of alcohol and hydrocarbon fuels, fuel cells are still mostly fed by hydrogen (Arico et al. 2005). The achievements with electrocatalysts in low-temperature fuel cells are mostly limited to the performance measured by ex situ electrochemistry measurement, for example, cyclic voltammetry, rotating disk electrode, and half-cell measurement in lab test (Du 2012; Jaouen et al. 2011). The tests in practical fuel cells usually suffer from the very low quantity of the products obtained due to the large difficulty on scale-up, caused by the high cost and the limitation of the fabrication method. For example, in the preparation of electrocatalysts by template electrodeposition methods (Alia et al. 2010; Ponrouch et al. 2010) or organic solvent methods (Eichhorn et al. 2009; Lee et al. 2008), large quantities of organic solvents and special precursors are used, as well as the sacrificed template and the complex process, leading to very high costs; the large difficulty in fabricating nanoscale electrolyte membranes in IT-SOFC limits the applications only in μ-SOFC in lab research (Takagi et al. 2011; Tsuchiya et al. 2011). Another major challenge is the MEA fabrication using these novel nanomaterials. They usually possess quite different features when compared to the conventional Pt or Pt/C nanoparticle electrocatalysts, offering these nanomaterials some special advantages, for example, high reactivity and stability, but also unusual shapes, structures, or other features, such as the nonisotropic NWs resulting in a thick catalytic layer with loose structures (Cademartiri et al. 2009; Choi et al. 2003; Du 2012); the intrinsic feature of the preparation methods for core–shell structure limit their feasibility to be used for making supported electrocatalysts (Eichhorn et al. 2009; Strasser et al. 2010), thus a high catalyst loading amount is necessary in MEA fabrication. Innovative solutions for common problems are urgently needed in practical applications.

Indeed, novel approaches for MEA fabrication could be a solution to tackle these important issues. For example, simple procedure and configurations, low preparation cost, and easy scale-up process could make them promising techniques. However, 3M electrodes can still not eliminate the drawbacks from the spherical nanocatalysts (e.g., easily aggregated to each other and Ostwald ripening of the small nanoparticles). While Pt-NW-based GDE and NEFC are completely new approaches, the optimization of the process for commercial applications still needs a very long time (Du et al. 2011a), and their long-term durabilities in *real operating* conditions are still unknown.

7.4.2 Disadvantages of Nanomaterials

7.4.2.1 Poor Durability in Fuel Cells

For nanomaterials, apart from the benefits brought by large SSA and high surface reactivity, the high surface energy also leads to drawbacks including low thermodynamic stability and high surface reactions. The low thermodynamic stability makes nanomaterials form aggregates in preparation and during operation (Du 2012), and the high surface reaction results in low stability of the whole MEA system, such as the dissolution and Oswald ripening of the electrocatalyst (Yu et al. 2007a), and the risk of secondary reactions involving electrolyte decomposition between the electrodes and the electrolyte caused by the large SSA of the nanocatalysts (Guo et al. 2008).

7.4.2.2 Toxicity from Nanomaterials

When we mention fuel cell technology, we usually emphasize their advantages as a clean technology, for example, environmental-friendly by reducing carbon emissions. Unfortunately, very little research has been focused on health and safety risks caused by nanoparticles in fuel cell systems or on their threats to the environment. However, in the fabrication and operation, these nanomaterials can be released to the environment and thus induce toxic problems to human body, the fauna, and flora. Among the nanomaterials used in fuel cell systems, many have been shown to induce toxicity problems, in particular metal nanocatalysts (Valiyaveettil et al. 2011), catalyst supports (e.g., CNT and carbon fiber) (Carriere et al. 2008; Porter et al. 2011) in low-temperature fuel cells, ceria, or zirconia-based nanoparticles (Lanone et al. 2009; Zhang et al. 2010b, 2011b) in SOFCs. Nanoparticles are small enough to penetrate human and animal cell membranes and defenses, yet they are large enough to cause trouble by interfering with normal cell processes. For example, the mean life span of nematodes was significantly decreased by 12% even at the exposure level of 1 nmol ($p < .01$) to nanoceria (8.5 nm) (Zhang et al. 2011b). Pt nanoparticles can induce hatching delays as well as a concentration-dependent drop in heart rate, touch response, and axis curvatures (Valiyaveettil et al. 2011). In the nanomaterial production and processing, the operator may be exposed to these nanoparticles by skin contact or breathing. In operation, these nanoparticles could detach from the support and then be released to the environment along with output water or gases.

Although detailed mechanisms are still not very well understood and studied, it has been shown that the toxicity of nanomaterials is determined by their size, shape, structure, composition, concentration, origin, and many other factors (Reijnders 2008), such as ceria nanoparticles synthesized in biocompatible media that could also be used as antioxidant in the treatment of medical disorders (Karakoti et al. 2008). Thus, the toxicity of nanomaterials to human and environment can be greatly eliminated by some approaches in production and processing as well as in operations. For example, in the fabrication of integrated Pt-NW electrodes, catalytic electrodes were directly achieved on GDL surfaces by simply one reduction step from precursor in aqueous solutions (Du 2010), whereby there is no risk to produce *free* nanoparticles like Pt nanocatalyst, CB, or catalyst ink, which are commonly used in MEA manufacturing process today and might be environmentally and health hazardous based on previous studies on asbestos and chrysolite (Xia et al. 2003). So, by the direct growth approach, the potential environmental and health issues that might arise from such nanoparticles produced in labs can be avoided.

7.5 CONCLUSIONS

Moving from bulk materials to the nanoscale can significantly change electrode and electrolyte properties and consequently their performance and durability in fuel cells. On one hand, nanomaterials show favorable and promising properties such as enhanced kinetics and activity, which are simply a consequence of a reduction in size, for example, when nanoparticulate electrodes or electrocatalysts lead to higher electrode/electrolyte contact areas and hence higher rates of electrode reaction; in other cases the effects may be more subtle, involving internally nanostructured materials or nanostructures with particular morphologies, for example, nanotubes. On the other hand, nanomaterials also bring disadvantages such as low thermodynamic stability, high surface reaction, as well as possible toxicity issues. The remaining challenges include (1) a better understanding of various nanosize effects and then developing new theories, (2) investigating fine details regarding the surface features of nanomaterials, (3) designing optimized hybrid nano/microstructures and surface modifications, (4) searching for new synthetic routes and new material systems, and (5) moving nanomaterials from lab to fuel cells in practical applications. Solving these challenges will require researchers from a wide range of disciplines, including materials chemistry and surface science, as both are necessary to elucidate the role and effect of nanomaterials. The development of novel nanomaterials will play an important role in improving fuel cell performance and durability and breakthroughs are urgently required to bring this green and sustainable energy device into real practical applications.

ACKNOWLEDGMENT

This work was supported by the Fellowships from Advantage West Midlands (AWM) Science City Research Alliance (SCRA) awarded to Shangfeng Du.

REFERENCES

Abruna HD, Wang DL, Subban CV, Wang HS, Rus E, DiSalvo FJ. (2010a) Highly stable and CO-tolerant Pt/Ti(0.7)W(0.3)O(2) electrocatalyst for proton-exchange membrane fuel cells. *Journal of the American Chemical Society* 132:10218–10220. DOI: 10.1021/ja102931d.

Abruna HD, Wang DL, Xin HL, Yu YC, Wang HS, Rus E, Muller DA. (2010b) Pt-decorated PdCo@Pd/C core-shell nanoparticles with enhanced stability and electrocatalytic activity for the oxygen reduction reaction. *Journal of the American Chemical Society* 132:17664–17666. DOI: 10.1021/ja107874u.

Alia SM, Zhang G, Kisailus D, Li DS, Gu S, Jensen K, Yan YS. (2010) Porous platinum nanotubes for oxygen reduction and methanol oxidation reactions. *Advanced Functional Materials* 20:3742–3746. DOI: 10.1002/adfm.201001035.

Antolini E. (2009) Carbon supports for low-temperature fuel cell catalysts. *Applied Catalysis B-Environmental* 88:1–24. DOI: 10.1016/j.apcatb.2008.09.030.

Antolini E, Gonzalez ER. (2009) Polymer supports for low-temperature fuel cell catalysts. *Applied Catalysis A-General* 365:1–19. DOI: 10.1016/j.apcata.2009.05.045.

Arico AS, Bruce P, Scrosati B, Tarascon JM, Van Schalkwijk W. (2005) Nanostructured materials for advanced energy conversion and storage devices. *Nature Materials* 4:366–377. DOI: 10.1038/Nmat1368.

Barton SC, Gallaway J, Atanassov P. (2004) Enzymatic biofuel cells for implantable and microscale devices. *Chemical Reviews* 104:4867–4886. DOI: 10.1021/cr020719k.

Baxter J, Bian ZX, Chen G, Danielson D, Dresselhaus MS, Fedorov AG, Fisher TS, Jones CW, Maginn E, Kortshagen U, Manthiram A, Nozik A, Rolison DR, Sands T, Shi L, Sholl D, Wu YY. (2009) Nanoscale design to enable the revolution in renewable energy. *Energy & Environmental Science* 2:559–588. DOI: 10.1039/b821698c.

Bellino MG, Sacanell JG, Lamas DG, Leyva AG, de Reca NEW. (2007) High-performance solid-oxide fuel cell cathodes based on cobaltite nanotubes. *Journal of the American Chemical Society* 129:3066–3067. DOI: 10.1021/ja068115b.

Berger DJ. (1999) Fuel cells and precious-metal catalysts. *Science* 286:49.

Binder H, Kohling A, Kuhn W, Lindner W, Sandsted, G. (1969) Tungsten carbide electrodes for fuel cells with acid electrolyte. *Nature* 224:1299–1300.

Bohm H. (1970) New non-noble metal anode catalysts for acid fuel cells. *Nature* 227:483–484.

Brandon NP, Skinner S, Steele BCH. (2003) Recent advances in materials for fuel cells. *Annual Review of Materials Research* 33:183–213. DOI: 10.1146/annurev.matsci.33.022802.094122.

Brankovic SR, Wang JX, Adzic RR. (2001) Pt submonolayers on Ru nanoparticles: A novel low Pt loading, high CO tolerance fuel cell electrocatalyst. *Electrochemical and Solid State Letters* 4:A217–A220.

Brett DJL, Atkinson A, Brandon NP, Skinner SJ. (2008) Intermediate temperature solid oxide fuel cells. *Chemical Society Reviews* 37:1568–1578. DOI: 10.1039/b612060c.

Brodd RJ, Winter M. (2004) What are batteries, fuel cells, and supercapacitors? *Chemical Reviews* 104:4245–4269. DOI: 10.1021/cr020730k.

Buratto SK. (2010) Fuel cells: Engineering the next generation. *Nature Nanotechnology* 5:176. DOI: 10.1038/nnano.2010.39.

Cademartiri L, Ozin GA. (2009) Ultrathin nanowires: A materials chemistry perspective. *Advanced Materials* 21:1013–1020. DOI: 10.1002/adma.200801836.

Carriere M, Simon-Deckers A, Gouget B, Mayne-L'Hermite M, Herlin-Boime N, Reynaud C. (2008) In vitro investigation of oxide nanoparticle and carbon nanotube toxicity and intracellular accumulation in A549 human pneumocytes. *Toxicology* 253:137–146. DOI: 10.1016/j.tox.2008.09.007.

Chadwick AV, Savin SLP. (2006) Structure and dynamics in nanoionic materials. *Solid State Ionics* 177:3001–3008. DOI: 10.1016/j.ssi.2006.07.046.

Chan WK, Haverkate LA, Borghols WJH, Wagemaker M, Picken SJ, van Eck ERH, Kentgens APM, Johnson MR, Kearley GJ, Mulder FM. (2011) Direct view on nanoionic proton mobility. *Advanced Functional Materials* 21:1364–1374. DOI: 10.1002/adfm.201001933.

Chang H, Joo SH, Pak C. (2007) Synthesis and characterization of mesoporous carbon for fuel cell applications. *Journal of Materials Chemistry* 17:3078–3088. DOI: 10.1039/b700389g.

Chang H, Pak C, Kang S, Choi YS. (2010) Nanomaterials and structures for the fourth innovation of polymer electrolyte fuel cell. *Journal of Materials Research* 25:2063–2071. DOI: 10.1557/Jmr.2010.0280.

Chen YG, Feng LY, Yan YY, Wang LJ. (2011) Nitrogen-doped carbon nanotubes as efficient and durable metal-free cathodic catalysts for oxygen reduction in microbial fuel cells. *Energy & Environmental Science* 4:1892–1899. DOI: 10.1039/c1ee01153g.

Chen AC, Holt-Hindle P. (2010) Platinum-based nanostructured materials: Synthesis, properties, and applications. *Chemical Reviews* 110:3767–3804. DOI: 10.1021/cr9003902.

Chen Z, Waje M, Li W, Yan Y. (2007) Supportless Pt and PtPd nanotubes as electrocatalysts for oxygen-reduction reactions. *Angewandte Chemie International Edition* 46:4060–4063. DOI: 10.1002/anie.200700894.

Choi SM, Kim JH, Jung JY, Yoon EY, Kim WB. (2008) Pt nanowires prepared via a polymer template method: Its promise toward high Pt-loaded electrocatalysts for methanol oxidation. *Electrochimica Acta* 53:5804–5811. DOI: 10.1016/j.electacta.2008.03.041.

Choi WC, Woo SI. (2003) Bimetallic Pt-Ru nanowire network for anode material in a direct-methanol fuel cell. *Journal of Power Sources* 124:420–425. DOI: 10.1016/S0378-7753(03)00812-7.

Dai LM, Gong KP, Du F, Xia ZH, Durstock M. (2009) Nitrogen-doped carbon nanotube arrays with high electrocatalytic activity for oxygen reduction. *Science* 323:760–764. DOI: 10.1126/science.1168049.

De Jonghe LC, Jacobson CP, Visco SJ. (2003) Supported electrolyte thin film synthesis of solid oxide fuel cells. *Annual Review of Materials Research* 33:169–182. DOI: 10.1146/annurev.matsci.33.041202.103842.

Debe MK. (2006) *Nanostructured Thin Film Catalysts (NSTFC) for Next Generation PEM Fuel Cells*, 2006 Northern Nano Workshop, University of Minnesota, Minneapolis, MN.

Devanathan R. (2008) Recent developments in proton exchange membranes for fuel cells. *Energy & Environmental Science* 1:101–119. DOI: 10.1039/b808149m.

Di Noto V, Negro E, Sanchez JY, Iojoiu C. (2010) Structure-relaxation interplay of a new nanostructured membrane based on tetraethylammonium trifluoromethanesulfonate ionic liquid and neutralized nafion 117 for high-temperature fuel cells. *Journal of the American Chemical Society* 132:2183–2195. DOI: 10.1021/ja906975z.

Dodelet JP, Lefevre M, Proietti E, Jaouen F. (2009) Iron-based catalysts with improved oxygen reduction activity in polymer electrolyte fuel cells. *Science* 324:71–74. DOI: 10.1126/science.1170051.

Dong LF, Gari RRS, Li Z, Craig MM, Hou SF. (2010) Graphene-supported platinum and platinum-ruthenium nanoparticles with high electrocatalytic activity for methanol and ethanol oxidation. *Carbon* 48:781–787. DOI: 10.1016/j.carbon.2009.10.027.

Du SF. (2010) A facile route for polymer electrolyte membrane fuel cell electrodes with in situ grown Pt nanowires. *Journal of Power Sources* 195:289–292.

Du SF. (2012) Pt-based nanowires as electrocatalysts in proton exchange fuel cells. *International Journal of Low-Carbon Technologies* 7:44–54.

Du SF, Majewski A. (2011b) Integrated electrodes with Pt nanowires in direct methanol fuel cells. *In the Proceedings of the European Fuel Cell Forum 2011: A0707.*

Du SF, Millington B, Pollet BG. (2011a) The effect of Nafion ionomer loading coated on gas diffusion electrodes with in-situ grown Pt nanowires and their durability in proton exchange membrane fuel cells. *International Journal of Hydrogen Energy* 36:4386–4393. DOI: 10.1016/j.ijhydene.2011.01.014.

Eichhorn B, Liu Z, Hu JE, Wang Q, Gaskell K, Frenkel AI, Jackson GS. (2009) PtMo alloy and MoO(x)@Pt core-shell nanoparticles as highly CO-tolerant electrocatalysts. *Journal of the American Chemical Society* 131:6924–6925. DOI: 10.1021/ja901303d.

Fang B, Chaudhari NK, Kim MS, Kim JH, Yu JS. (2009a) Homogeneous deposition of platinum nanoparticles on carbon black for proton exchange membrane fuel cell. *Journal of the American Chemical Society* 131:15330–15338. DOI: Doi 10.1021/Ja905749e.

Fang JY, Zhang J. (2009b) A general strategy for preparation of Pt 3d-transition metal (Co, Fe, Ni) nanocubes. *Journal of the American Chemical Society* 131:18543–18547. DOI: 10.1021/ja908245r.

Fang JY, Zhang J, Yang HZ, Zou SZ. (2010) Synthesis and oxygen reduction activity of shape-controlled Pt(3)Ni nanopolyhedra. *Nano Letters* 10:638–644. DOI: 10.1021/nl903717z.

Fuertes AB, Sevilla M, Lota G. (2007) Saccharide-based graphitic carbon nanocoils as supports for PtRu nanoparticles for methanol electrooxidation. *Journal of Power Sources* 171:546–551. DOI: 10.1016/j.jpowsour.2007.05.096.

Garbarino S, Ponrouch A, Pronovost S, Gaudet J, Guay D. (2009) Synthesis and characterization of preferentially oriented (1 0 0) Pt nanowires. *Electrochemistry Communications* 11:1924–1927. DOI: 10.1016/j.elecom.2009.08.017.

Garcia-Martinez J. (Ed.) (2010) *Nanotechnology for the Energy Challenge*, Wiley-VCH, Weinheim, Germany.

Geim AK, Novoselov KS. (2007) The rise of graphene. *Nature Materials* 6:183–191.

Geng DS, Chen Y, Chen YG, Li YL, Li RY, Sun XL, Ye SY, Knights S. (2011) High oxygen-reduction activity and durability of nitrogen-doped graphene. *Energy & Environmental Science* 4:760–764. DOI: 10.1039/c0ee00326c.

Gong XZ, Yang Y, Huang SM. (2011) Mn3O4 catalyzed growth of polycrystalline Pt nanoparticles and single crystalline Pt nanorods with high index facets. *Chemical Communications* 47:1009–1011. DOI: 10.1039/C0cc03656k.

Gorzny ML, Walton AS, Evans SD. (2010) Synthesis of high-surface-area platinum nanotubes using a viral template. *Advanced Functional Materials* 20:1295–1300. DOI: 10.1002/adfm.200902196.

Guo Y-G, Hu J-S, Wan L-J. (2008) Nanostructured materials for electrochemical energy conversion and storage devices. *Advanced Materials* 20:2878–2887. DOI: 10.1002/adma.200800627.

Hara Y, Minami N, Matsumoto H, Itagaki H. (2007) New synthesis of tungsten carbide particles and the synergistic effect with Pt metal as a hydrogen oxidation catalyst for fuel cell applications. *Applied Catalysis A-General* 332:289–296. DOI: 10.1016/j.apcata.2007.08.030.

Harnisch F, Schröder U. (2010) From MFC to MXC: Chemical and biological cathodes and their potential for microbial bioelectrochemical systems. *Chemical Society Reviews* 39:4433–4448. DOI: 10.1039/c003068f.

Hartl K, Hanzlik M, Arenz M. (2011) IL-TEM investigations on the degradation mechanism of Pt/C electrocatalysts with different carbon supports. *Energy & Environmental Science* 4:234–238. DOI: 10.1039/c0ee00248h.

Herrero E, Sanchez-Sanchez CM, Solla-Gullon J, Vidal-Iglesias FJ, Aldaz A, Montiel V. (2010) Imaging structure sensitive catalysis on different shape-controlled platinum nanoparticles. *Journal of the American Chemical Society* 132:5622–5624. DOI: 10.1021/ja100922h.

Herring AM, Meng FQ, Aieta NV, Dec SF, Horan JL, Williamson D, Frey MH, Pham P, Turner JA, Yandrasits MA, Hamrock SJ. (2007) Structural and transport effects of doping perfluorosulfonic acid polymers with the heteropoly acids, H3PW12O40 or H4SiW12O40. *Electrochimica Acta* 53:1372–1378. DOI: 10.1016/j.electacta.2007.06.047.

Hwang KC, Hsin YL, Yeh CT. (2007) Poly(vinylpyrrolidone)-modified graphite carbon nanofibers as promising supports for PtRu catalysts in direct methanol fuel cells. *Journal of the American Chemical Society* 129:9999–10010. DOI: 10.1021/ja072367a.

Hyeon T, Han S, Sung YE, Park KW, Kim YW. (2003) High-performance direct methanol fuel cell electrodes using solid-phase-synthesized carbon nanocoils. *Angewandte Chemie-International Edition* 42:4352–4356. DOI: 10.1002/anie.200250856.

Jaouen F, Proietti E, Lefevre M, Chenitz R, Dodelet JP, Wu G, Chung HT, Johnston CM, Zelenay P. (2011) Recent advances in non-precious metal catalysis for oxygen-reduction reaction in polymer electrolyte fuel cells. *Energy & Environmental Science* 4:114–130. DOI: 10.1039/c0ee00011f.

Jones DJ, Roziere J. (2003) Non-fluorinated polymer materials for proton exchange membrane fuel cells. *Annual Review of Materials Research* 33:503–555. DOI: 10.1146/annurev.matsci.33.022702.154657.

Karakoti AS, Monteiro-Riviere NA, Aggarwal R, Davis JP, Narayan RJ, Self WT, McGinnis J, Seal S. (2008) Nanoceria as antioxidant: Synthesis and biomedical applications. *JOM* 60:33–37.

Kim JM, Joh HI, Jo SM, Ahn DJ, Ha HY, Hong SA, Kim SK. (2010) Preparation and characterization of Pt nanowire by electrospinning method for methanol oxidation. *Electrochimica Acta* 55:4827–4835. DOI: 10.1016/j.electacta.2010.03.036.

Kim YS, Kim HJ, Kim WB. (2009b) Composited hybrid electrocatalysts of Pt-based nanoparticles and nanowires for low temperature polymer electrolyte fuel cells. *Electrochemistry Communications* 11:1026–1029. DOI: 10.1016/j.elecom.2009.03.003.

Kim HJ, Kim YS, Seo MH, Choi SM, Kim WB. (2009a) Pt and PtRh nanowire electrocatalysts for cyclohexane-fueled polymer electrolyte membrane fuel cell. *Electrochemistry Communications* 11:446–449. DOI: 10.1016/j.elecom.2008.12.027.

Kim YS, Nam SH, Shim HS, Ahn HJ, Anand M, Kim WB. (2008) Electrospun bimetallic nanowires of PtRh and PtRu with compositional variation for methanol electrooxidation. *Electrochemistry Communications* 10:1016–1019. DOI: 10.1016/j.elecom.2008.05.003.

Kinoshita K. (1990) Particle size effects for oxygen reduction on highly dispersed platinum in acid electrolytes. *Journal of the Electrochemical Society* 137:845–848. DOI: 10.1149/1.2086566.

Koenigsmann C, Wong SS. (2011) One-dimensional noble metal electrocatalysts: A promising structural paradigm for direct methanol fuel cells. *Energy & Environmental Science* 4:1161–1176. DOI: 10.1039/c0ee00197j.

Laberty-Robert C, Long JW, Pettigrew KA, Stroud RM, Rolison DR. (2007) Ionic nanowires at 600°C: Using nanoarchitecture to optimize electrical transport in nanocrystalline gadolinium-doped ceria. *Advanced Materials* 19:1734–1739. DOI: 10.1002/adma.200601840.

Laberty-Robert C, Valle K, Pereira F, Sanchez C. (2011) Design and properties of functional hybrid organic-inorganic membranes for fuel cells. *Chemical Society Reviews* 40:961–1005. DOI: 10.1039/c0cs00144a.

Lanone S, Rogerieux F, Geys J, Dupont A, Maillot-Marechal E, Boczkowski J, Lacroix G, Hoet P. (2009) Comparative toxicity of 24 manufactured nanoparticles in human alveolar epithelial and macrophage cell lines. *Particle and Fibre Toxicology* 6:14(1–12).

Lee EP, Peng ZM, Cate DM, Yang H, Campbell CT, Xia Y. (2007) Growing Pt nanowires as a densely packed array on metal gauze. *Journal of the American Chemical Society* 129:10634–10635. DOI: 10.1021/Ja074312e.

Lee EP, Peng ZM, Chen W, Chen SW, Yang H, Xia YN. (2008) Electrocatalytic properties of Pt nanowires supported on Pt and W gauzes. *ACS Nano* 2:2167–2173. DOI: 10.1021/Nn800458p.

Lefèvre M, Proietti E, Jaouen F, Dodelet J-P. (2009) Iron-based catalysts with improved oxygen reduction activity in polymer electrolyte fuel cells. *Science* 324:71–74. DOI: 10.1126/science.1170051.

Li SH, Wang XD, Ma Y, Kashyout AH, Zhu B, Muhammed M. (2011) Ceria-based nanocomposite with simultaneous proton and oxygen ion conductivity for low-temperature solid oxide fuel cells. *Journal of Power Sources* 196:2754–2758. DOI: 10.1016/j.jpowsour.2010.11.033.

Liang CH, Ding L, Li CA, Pang M, Su DS, Li WZ, Wang YM. (2010a) Nanostructured WC(x)/CNTs as highly efficient support of electrocatalysts with low Pt loading for oxygen reduction reaction. *Energy & Environmental Science* 3:1121–1127. DOI: 10.1039/c001423k.

Liang ZX, Shi JY, Liao SJ, Zeng JH. (2010b) Noble metal nanowires incorporated Nafion (R) membranes for reduction of methanol crossover in direct methanol fuel cells. *International Journal of Hydrogen Energy* 35:9182–9185. DOI: 10.1016/j.ijhydene.2010.06.054.

Liang ZX, Zhao TS. (2007) New DMFC anode structure consisting of platinum nanowires deposited into a Nafion membrane. *Journal of Physical Chemistry C* 111:8128–8134. DOI: 10.1021/Jp0711747.

Lin YH, Shao YY, Zhang S, Engelhard MH, Li GS, Shao GC, Wang Y, Liu J, Aksay IA. (2010) Nitrogen-doped graphene and its electrochemical applications. *Journal of Materials Chemistry* 20:7491–7496. DOI: 10.1039/c0jm00782j.

Liu F, Lee JY, Zhou WJ. (2004a) Template preparation of multisegment PtNi nanorods as methanol electro-oxidation catalysts with adjustable bimetallic pair sites. *Journal of Physical Chemistry B* 108:17959–17963. DOI: 10.1021/Jp0472360.

Liu J, Kou R, Shao YY, Wang DH, Engelhard MH, Kwak JH, Wang J, Viswanathan VV, Wang CM, Lin YH, Wang Y, Aksay IA. (2009a) Enhanced activity and stability of Pt catalysts on functionalized graphene sheets for electrocatalytic oxygen reduction. *Electrochemistry Communications* 11:954–957. DOI: 10.1016/j.elecom.2009.02.033.

Liu ML, Liu Y, Zha SW. (2004b) Novel nanostructured electrodes for solid oxide fuel cells fabricated by combustion chemical vapor deposition (CVD). *Advanced Materials* 16:256–260. DOI: 10.1002/adma.200305767.

Liu LF, Pippel E, Scholz R, Gosele U. (2009b) Nanoporous Pt-co alloy nanowires: Fabrication, characterization, and electrocatalytic properties. *Nano Letters* 9:4352–4358. DOI: 10.1021/Nl902619q.

Liu Y, Qu LT, Baek JB, Dai LM. (2010b) Nitrogen-doped graphene as efficient metal-free electrocatalyst for oxygen reduction in fuel cells. *ACS Nano* 4:1321–1326. DOI: 10.1021/nn901850u.

Liu ML, Zhang YL, Zha SW. (2005) Dual-scale porous electrodes for solid oxide fuel cells from polymer foams. *Advanced Materials* 17:487–491. DOI: 10.1002/adma.200400466.

Liu QH, Zhu B. (2010a) Theoretical description of superionic conductivities in samaria doped ceria based nanocomposites. *Applied Physics Letters* 97:183115 (1–3).

Lux KW, Rodriguez KJ. (2006) Template synthesis of arrays of nano fuel cells. *Nano Letters* 6:288–295. DOI: 10.1021/Nl052150j.

Ma YW, Jiang SJ, Jian GQ, Tao HS, Yu LS, Wang XB, Wang XZ, Zhu JM, Hu Z, Chen Y. (2009) CN(x) nanofibers converted from polypyrrole nanowires as platinum support for methanol oxidation. *Energy & Environmental Science* 2:224–229. DOI: 10.1039/b807213m.

Ma Y, Wang X, Li S, Toprak MS, Zhu B, Muhammed M. (2010) Samarium-doped ceria nanowires: Novel synthesis and application in low-temperature solid oxide fuel cells. *Advanced Materials* 22:1640–1644. DOI: 10.1002/adma.200903402.

Maier J. (2005) Nanoionics: Ion transport and electrochemical storage in confined systems. *Nature Materials* 4:805–815. DOI: 10.1038/Nmat1513.

Malavasi L, Fisher CAJ, Islam MS. (2010) Oxide-ion and proton conducting electrolyte materials for clean energy applications: Structural and mechanistic features. *Chemical Society Reviews* 39:4370–4387. DOI: 10.1039/b915141a.

Manthiram A, Murugan AV, Sarkar A, Muraliganth T. (2008) Nanostructured electrode materials for electrochemical energy storage and conversion. *Energy & Environmental Science* 1:621–638. DOI: 10.1039/b811802g.

Minteer SD, Moehlenbrock MJ. (2008) Extended lifetime biofuel cells. *Chemical Society Reviews* 37:1188–1196. DOI: 10.1039/b708013c.

Miu M, Kleps I, Danila M, Ignat T, Simion M, Bragaru A, Dinescu A. (2010) Electrocatalytic activity of platinum nanoparticles supported on nanosilicon. *Fuel Cells* 10:259–269. DOI: 10.1002/fuce.200900202.

Moghaddam S, Pengwang E, Jiang YB, Garcia AR, Burnett DJ, Brinker CJ, Masel RI, Shannon MA. (2010) An inorganic-organic proton exchange membrane for fuel cells with a controlled nanoscale pore structure. *Nature Nanotechnology* 5:230–236. DOI: 10.1038/Nnano.2010.13.

Moreno-Castilla C, Maldonado-Hodar FJ. (2005) Carbon aerogels for catalysis applications: An overview. *Carbon* 43:455–465. DOI: 10.1016/j.carbon.2004.10.022.

Mukerjee S, Srinivasan S, Soriaga MP, Mcbreen J. (1995) Role of structural and electronic-properties of Pt and Pt alloys on electrocatalysis of oxygen reduction: An in-situ XANES and EXAFS investigation. *Journal of the Electrochemical Society* 142:1409–1422.

Murray CB, Kang YJ. (2010) Synthesis and electrocatalytic properties of cubic Mn-Pt nanocrystals (nanocubes). *Journal of the American Chemical Society* 132:7568–7569. DOI: 10.1021/ja100705J.

Oregan B, Gratzel M. (1991) A low-cost, high-efficiency solar-cell based on dye-sensitized colloidal TiO_2 films. *Nature* 353:737–740.

Ormerod RM. (2003) Solid oxide fuel cells. *Chemical Society Reviews* 32:17–28. DOI: 10.1039/b105764m.

Paddison SJ. (2003) Proton conduction mechanisms at low degrees of hydration in sulfonic acid-based polymer electrolyte membranes. *Annual Review of Materials Research* 33:289–319. DOI: 10.1146/annurev.matsei.33.022702.155102.

Park IS, Choi JH, Sung YE. (2008) Synthesis of 3 nm Pt nanowire using MCM-41 and electrocatalytic activity in methanol electro-oxidation. *Electrochemical and Solid State Letters* 11:B71–B75. DOI: 10.1149/1.2888220.

Park MJ, Kim SY, Kim S. (2010) Enhanced proton transport in nanostructured polymer electrolyte/ionic liquid membranes under water-free conditions. *Nature Communications* 1:88. DOI: 10.1038/ncomms1086.

Peng ZM, You HJ, Yang H. (2010) An electrochemical approach to PTAG alloy nanostructures rich in Pt at the surface. *Advanced Functional Materials* 20:3734–3741. DOI: 10.1002/adfm.201001194.

Ponrouch A, Garbarino S, Pronovost S, Taberna PL, Simon P, Guay D. (2010) Electrodeposition of arrays of Ru, Pt, and PtRu alloy 1D metallic nanostructures. *Journal of the Electrochemical Society* 157:K59–K65. DOI: 10.1149/1.3276500.

Porter AE, Nerl HC, Cheng C, Goode AE, Bergin SD, Lich B, Gass M. (2011) Imaging methods for determining uptake and toxicity of carbon nanotubes in vitro and in vivo. *Nanomedicine* 6:849–865. DOI: 10.2217/Nnm.11.87.

Prinz FB, Chao CC, Hsu CM, Cui Y. (2011) Improved solid oxide fuel cell performance with nanostructured electrolytes. *ACS Nano* 5:5692–5696. DOI: 10.1021/Nn201354p.

Proietti E, Jaouen F, Lefèvre M, Larouche N, Tian J, Herranz J, Dodelet J-P. (2011) Iron-based cathode catalyst with enhanced power density in polymer electrolyte membrane fuel cells. *Nature Communications* 2:416. Available at http://www.nature.com/ncomms/journal/v2/n8/suppinfo/10.1038-ncomms1427-unlocked-60x70_S1.html.

Qiao B, Wang A, Yang X, Allard LF, Jiang Z, Cui Y, Liu J, Li J, Zhang T. (2011) Single-atom catalysis of CO oxidation using Pt1/FeOx. *Nature Chemistry* 3:634–641. Available at http://www.nature.com/nchem/journal/v3/n8/abs/nchem.1095.html-supplementary-information.

Reijnders L. (2008) Hazard reduction in nanotechnology. *Journal of Industrial Ecology* 12:297–306. DOI: 10.1111/j.1530-9290.2008.00049.x.

Ramani V, Kunz HR, Fenton JM. (2005) Stabilized composite membranes and membrane electrode assemblies for elevated temperature/low relative humidity PEFC operation. *Journal of Power Sources* 152:182–188. DOI: 10.1016/j.jpowsour.2005.03.135.

Ramaprabhu S, Jafri RI, Rajalakshmi N. (2010) Nitrogen doped graphene nanoplatelets as catalyst support for oxygen reduction reaction in proton exchange membrane fuel cell. *Journal of Materials Chemistry* 20:7114–7117. DOI: 10.1039/c0jm00467g.

Remediakis IN, Lopez N, Nørskov JK. (2005) CO oxidation on rutile-supported au nanoparticles. *Angewandte Chemie International Edition* 44:1824–1826. DOI: 10.1002/anie.200461699.

Ruiz-Morales JC, Marrero-Lopez D, Galvez-Sanchez M, Canales-Vazquez J, Savaniu C, Savvin SN. (2010) Engineering of materials for solid oxide fuel cells and other energy and environmental applications. *Energy & Environmental Science* 3:1670–1681. DOI: 10.1039/c0ee00166j.

Sano N, Ukita S. (2006) One-step synthesis of Pt-supported carbon nanohorns for fuel cell electrode by arc plasma in liquid nitrogen. *Materials Chemistry and Physics* 99:447–450. DOI: 10.1016/j.matchemphys.2005.11.019.

Santamaria J, Garcia-Barriocanal J, Rivera-Calzada A, Varela M, Sefrioui Z, Iborra E, Leon C, Pennycook SJ. (2008) Colossal ionic conductivity at interfaces of epitaxial ZrO2: Y2O3/SrTiO3 heterostructures. *Science* 321:676–680. DOI: 10.1126/science.1156393.

Schmidt TJ, Noeske M, Gasteiger HA, Behm RJ, Britz P, Brijoux W, Bönnemann H. (1997) Electrocatalytic activity of PtRu alloy colloids for CO and CO/H2 electrooxidation: Stripping voltammetry and rotating disk measurements. *Langmuir* 13:2591–2595. DOI: 10.1021/la962068r.

Schoonman J. (2003) Nanoionics. *Solid State Ionics* 157:319–326.
Seal S, Kuchibhatla SVNT, Karakoti AS, Bera D. (2007) One dimensional nanostructured materials. *Progress in Materials Science* 52:699–913. DOI: 10.1016/j.pmatsci.2006.08.001.
Shao YY, Yin GP, Gao YZ. (2007) Understanding and approaches for the durability issues of Pt-based catalysts for PEM fuel cell. *Journal of Power Sources* 171:558–566. DOI: 10.1016/j.jpowsour.2007.07.004.
Shim JH, Chao CC, Huang H, Prinz FB. (2007) Atomic layer deposition of yttria-stabilized zirconia for solid oxide fuel cells. *Chemistry of Materials* 19:3850–3854. DOI: 10.1021/Cm070913t.
Song WG, Zhang LS, Liang XQ, Wu ZY. (2010) Identification of the nitrogen species on N-doped graphene layers and Pt/NG composite catalyst for direct methanol fuel cell. *Physical Chemistry Chemical Physics* 12:12055–12059. DOI: 10.1039/c0cp00789g.
Spurgeon JM, Walter MG, Zhou JF, Kohl PA, Lewis NS. (2011) Electrical conductivity, ionic conductivity, optical absorption, and gas separation properties of ionically conductive polymer membranes embedded with Si microwire arrays. *Energy & Environmental Science* 4:1772–1780. DOI: 10.1039/c1ee01028j.
Stamenkovic VR, Fowler B, Mun BS, Wang GF, Ross PN, Lucas CA, Markovic NM. (2007) Improved oxygen reduction activity on Pt3Ni(111) via increased surface site availability. *Science* 315:493–497. DOI: 10.1126/science.1135941.
Strasser P, Koh S, Anniyev T, Greeley J, More K, Yu CF, Liu ZC, Kaya S, Nordlund D, Ogasawara H, Toney MF, Nilsson A. (2010) Lattice-strain control of the activity in dealloyed core-shell fuel cell catalysts. *Nature Chemistry* 2:454–460. DOI: 10.1038/Nchem.623.
Strasser P, Srivastava R, Mani P, Hahn N. (2007) Efficient oxygen reduction fuel cell electrocatalysis on voltammetrically dealloyed Pt-Cu-Co nanoparticles. *Angewandte Chemie-International Edition* 46:8988–8991. DOI: 10.1002/anie.200703331.
Subhramannia M, Pillai VK. (2008) Shape-dependent electrocatalytic activity of platinum nanostructures. *Journal of Materials Chemistry* 18:5858–5870.
Sun SH, Jaouen F, Dodelet JP. (2008a) Controlled growth of Pt nanowires on carbon nanospheres and their enhanced performance as electrocatalysts in PEM fuel cells. *Advanced Materials* 20:3900–3904. DOI: 10.1002/adma.200800491.
Sun SG, Xu HY, Tang SH, Guo JS, Li HQ, Cao L, Zhou B, Xin Q, Sun GQ. (2006) Synthesis of PtRu nanowires and their catalytic activity in the anode of direct methanol fuel cells. *Chinese Journal of Catalysis* 27:932–936.
Sun SH, Wang C, Daimon H. (2009b) Dumbbell-like Pt-Fe(3)O(4) nanoparticles and their enhanced catalysis for oxygen reduction reaction. *Nano Letters* 9:1493–1496. DOI: 10.1021/nl8034724.
Sun YQ, Wu QO, Shi GQ. (2011b) Graphene based new energy materials. *Energy & Environmental Science* 4:1113–1132. DOI: 10.1039/c0ee00683a.
Sun SH, Yang DQ, Villers D, Zhang GX, Sacher E, Dodelet JP. (2008b) Template- and surfactant-free room temperature synthesis of self-assembled 3D Pt nanoflowers from single-crystal nanowires. *Advanced Materials* 20:571–574. DOI: 10.1002/adma.200701408.
Sun SH, Yang D, Zhang G, Sacher E, Dodelet JP. (2007) Synthesis and characterization of platinum nanowire-carbon nanotube heterostructures. *Chemistry of Materials* 19:6376–6378. DOI: 10.1021/Cm7022949.
Sun SH, Zhang GX, Geng DS, Chen YG, Banis MN, Li RY, Cai M, Sun XL. (2010) Direct growth of single-crystal Pt nanowires on Sn@CNT nanocable: 3D electrodes for highly active electrocatalysts. *Chemistry-A European Journal* 16:829–835. DOI: 10.1002/chem.200902320.
Sun SH, Zhang GX, Geng DS, Chen YG, Li RY, Cai M, Sun XL. (2011a) A highly durable platinum nanocatalyst for proton exchange membrane fuel cells: Multiarmed starlike nanowire single crystal. *Angewandte Chemie-International Edition* 50:422–426. DOI: 10.1002/anie.201004631.

Sun S, Zhang G, Zhong Y, Liu H, Li R, Zhou X, Sun X. (2009a) Ultrathin single crystal Pt nanowires grown on N-doped carbon nanotubes. *Chemical Communications* December 7:7048–7050.

Suzuki T, Hasan Z, Funahashi Y, Yamaguchi T, Fujishiro Y, Awano M. (2009) Impact of anode microstructure on solid oxide fuel cells. *Science* 325:852–855. DOI: 10.1126/science.1176404.

Takagi Y, Lai B-K, Kerman K, Ramanathan S. (2011) Low temperature thin film solid oxide fuel cells with nanoporous ruthenium anodes for direct methane operation. *Energy & Environmental Science* 4:3473–3478.

Tang LH, Wang Y, Li YM, Feng HB, Lu J, Li JH. (2009) Preparation, structure, and electrochemical properties of reduced graphene sheet films. *Advanced Functional Materials* 19:2782–2789. DOI: 10.1002/adfm.200900377.

Tedsree K, Li T, Jones S, Chan CWA, Yu KMK, Bagot PAJ, Marquis EA, Smith GDW, Tsang SCE. (2011) Hydrogen production from formic acid decomposition at room temperature using a Ag-Pd core-shell nanocatalyst. *Nature Nanotechnology* 6:302–307. DOI: 10.1038/Nnano.2011.42.

Tian M, Kumar N, Xu SY, Wang JG, Kurtz JS, Chan MHW. (2005) Suppression of superconductivity in zinc nanowires by bulk superconductors. *Physical Review Letters* 95:076802. DOI: 10.1103/PhysRevLett.95.076802.

Tian N, Zhou Z-Y, Sun S-G, Ding Y, Wang ZL. (2007) Synthesis of tetrahexahedral platinum nanocrystals with high-index facets and high electro-oxidation activity. *Science* 316:732–735. DOI: 10.1126/science.1140484.

Tour JM, Kosynkin DK, D. V, Higginbotham AL, Sinitskii A, Lomeda JR, Dimiev A, Price BK. (2009) Longitudinal unzipping of carbon nanotubes to form graphene nanoribbons. *Nature* 458:872–876. DOI: 10.1038/nature07872.

Tsuchiya M, Lai BK, Ramanathan S. (2011) Scalable nanostructured membranes for solid-oxide fuel cells. *Nature Nanotechnology* 6:282–286. DOI: 10.1038/Nnano.2011.43.

Vajda S, Pellin MJ, Greeley JP, Marshall CL, Curtiss LA, Ballentine GA, Elam JW, Catillon-Mucherie S, Redfern PC, Mehmood F, Zapol P. (2009) Subnanometre platinum clusters as highly active and selective catalysts for the oxidative dehydrogenation of propane. *Nature Materials* 8:213–216. Available at http://www.nature.com/nmat/journal/v8/n3/suppinfo/nmat2384_S1.html.

Valiyaveettil S, Asharani PV, Yi LW, Gong ZY. (2011) Comparison of the toxicity of silver, gold and platinum nanoparticles in developing zebrafish embryos. *Nanotoxicology* 5:43–54. DOI: 10.3109/17435390.2010.489207.

Wang C, Chi MF, Wang GF, van der Vliet D, Li DG, More K, Wang HH, Schlueter JA, Markovic NM, Stamenkovic VR. (2011a) Correlation between surface chemistry and electrocatalytic properties of monodisperse Pt(x)Ni(1-x) nanoparticles. *Advanced Functional Materials* 21:147–152. DOI: 10.1002/adfm.201001138.

Wang C, Daimon H, Onodera T, Koda T, Sun S. (2008) A general approach to the size- and shape-controlled synthesis of platinum nanoparticles and their catalytic reduction of oxygen. *Angewandte Chemie International Edition* 47:3588–3591. DOI: 10.1002/anie.200800073.

Wang EK, Guo SJ. (2011b) Noble metal nanomaterials: Controllable synthesis and application in fuel cells and analytical sensors. *Nano Today* 6:240–264. DOI: 10.1016/j.nantod.2011.04.007.

Wang EW, Guo SJ, Dong SJ. (2010a) Three-dimensional Pt-on-Pd bimetallic nanodendrites supported on graphene nanosheet: Facile synthesis and used as an advanced nanoelectrocatalyst for methanol oxidation. *ACS Nano* 4:547–555. DOI: 10.1021/nn9014483.

Wang Y, Kou R, Shao YY, Mei DH, Nie ZM, Wang DH, Wang CM, Viswanathan VV, Park S, Aksay IA, Lin YH, Liu J. (2011c) Stabilization of electrocatalytic metal nanoparticles at metal-metal oxide-graphene triple junction points. *Journal of the American Chemical Society* 133:2541–2547. DOI: 10.1021/ja107719u.

Wang JQ, Liu S, Zeng J, Ou JF, Li ZP, Liu XH, Yang SR. (2010b) "Green" electrochemical synthesis of Pt/graphene sheet nanocomposite film and its electrocatalytic property. *Journal of Power Sources* 195:4628–4633. DOI: 10.1016/j.jpowsour.2010.02.024.

Wang Y, Shao YY, Zhang S, Wang CM, Nie ZM, Liu J, Lin YH. (2010c) Highly durable graphene nanoplatelets supported Pt nanocatalysts for oxygen reduction. *Journal of Power Sources* 195:4600–4605. DOI: 10.1016/j.jpowsour.2010.02.044.

Whittingham MS, Savinell RF, Zawodzinski T. (2004) Introduction: Batteries and fuel cells. *Chemical Reviews* 104:4243–4244. DOI: 10.1021/cr020705e.

Williams MC. (2001) Status and promise of fuel cell technology. *Fuel Cells* 1:87–91. DOI: 10.1002/1615-6854(200107)1:2<87::aid-fuce87>3.0.co;2-r.

Wong SS, Peng XH, Chen JY, Misewich JA. (2009) Carbon nanotube-nanocrystal heterostructures. *Chemical Society Reviews* 38:1076–1098. DOI: 10.1039/b811424m.

Xia YN, Lim B, Jiang MJ, Camargo PHC, Cho EC, Tao J, Lu XM, Zhu YM. (2009) Pd-Pt bimetallic nanodendrites with high activity for oxygen reduction. *Science* 324:1302–1305. DOI: 10.1126/science.1170377.

Xia BY, Wang JN, Wang XX, Niu JJ, Sheng ZM, Hu MR, Yu QC. (2008) Synthesis and application of graphitic carbon with high surface area. *Advanced Functional Materials* 18:1790–1798. DOI: 10.1002/adfm.200701263.

Xia YN, Yang PD, Sun YG, Wu YY, Mayers B, Gates B, Yin YD, Kim F, Yan YQ. (2003) One-dimensional nanostructures: Synthesis, characterization, and applications. *Advanced Materials* 15:353–389.

Yan YS, Wang C, Waje M, Wang X, Tang JM, Haddon RC. (2004) Proton exchange membrane fuel cells with carbon nanotube based electrodes. *Nano Letters* 4:345–348. DOI: 10.1021/nl034952p.

Yang W, Fellinger TP, Antonietti M. (2011) Efficient metal-free oxygen reduction in alkaline medium on high-surface-area mesoporous nitrogen-doped carbons made from ionic liquids and nucleobases. *Journal of the American Chemical Society* 133:206–209. DOI: 10.1021/ja108039j.

Yang H, Kang YY, Ren MJ, Yuan T, Qiao YJ, Zoa ZQ. (2010) Effect of Nafion aggregation in the anode catalytic layer on the performance of a direct formic acid fuel cell. *Journal of Power Sources* 195:2649–2652. DOI: 10.1016/j.jpowsour.2009.11.025.

Yeager E. (1961) Fuel cells: They produce more electricity per pound of fuel than any other nonnuclear method of power production. *Science* 134:1178–1786.

Yoon J, Cho S, Kim JH, Lee J, Bi ZX, Serquis A, Zhang XH, Manthiram A, Wang HY. (2009) Vertically aligned nanocomposite thin films as a cathode/electrolyte interface layer for thin-film solid-oxide fuel cells. *Advanced Functional Materials* 19:3868–3873. DOI: 10.1002/adfm.200901338.

Yoshitake T, Shimakawa Y, Kuroshima S, Kimura H, Ichihashi T, Kubo Y, Kasuya D, Takahashi K, Kokai F, Yudasaka M, Iijima S. (2002) Preparation of fine platinum catalyst supported on single-wall carbon nanohorns for fuel cell application. *Physica B-Condensed Matter* 323:124–126.

You H, Komanicky V, Iddir H, Chang KC, Menzel A, Karapetrov G, Hennessy D, Zapol P. (2009) Shape-dependent activity of platinum array catalyst. *Journal of the American Chemical Society* 131:5732–5733. DOI: 10.1021/ja900459w.

Yu X, Ye S. (2007a) Recent advances in activity and durability enhancement of Pt/C catalytic cathode in PEMFC: Part II: Degradation mechanism and durability enhancement of carbon supported platinum catalyst. *Journal of Power Sources* 172:145–154. DOI: 10.1016/j.jpowsour.2007.07.048.

Yu XW, Ye SY. (2007b) Recent advances in activity and durability enhancement of Pt/C catalytic cathode in PEMFC—Part I. Physico-chemical and electronic interaction between Pt and carbon support, and activity enhancement of Pt/C catalyst. *Journal of Power Sources* 172:133–144. DOI: 10.1016/j.jpowsour.2007.07.049.

Zelenay P, Wu G, More KL, Johnston CM. (2011) High-performance electrocatalysts for oxygen reduction derived from polyaniline, iron, and cobalt. *Science* 332:443–447. DOI: 10.1126/science.1200832.

Zhang ZY, Li MJ, Wu ZL, Li WZ. (2011a) Ultra-thin PtFe-nanowires as durable electrocatalysts for fuel cells. *Nanotechnology* 22:015602. DOI: 10.1088/0957-4484/22/1/015602.

Zhang XY, Lu W, Da JY, Wang HT, Zhao DY, Webley PA. (2009) Porous platinum nanowire arrays for direct ethanol fuel cell applications. *Chemical Communications* January 8:195–197. DOI: 10.1039/B813830c.

Zhang ZY, Ma YH, Kuang LL, He X, Bai W, Ding YY, Zhao YL, Chai ZF. (2010b) Effects of rare earth oxide nanoparticles on root elongation of plants. *Chemosphere* 78:273–279. DOI: 10.1016/j.chemosphere.2009.10.050.

Zhang J, Sasaki K, Sutter E, Adzic RR. (2007) Stabilization of platinum oxygen-reduction electrocatalysts using gold clusters. *Science* 315:220–222. DOI: 10.1126/science.1134569.

Zhang WM, Sherrell P, Minett AI, Razal JM, Chen J. (2010a) Carbon nanotube architectures as catalyst supports for proton exchange membrane fuel cells. *Energy & Environmental Science* 3:1286–1293. DOI: 10.1039/c0ee00139b.

Zhang ZY, Zhang HZ, H. F, He XA, Zhang P, Li YY, Ma YH, Kuang YS, Zhao YL, Chai ZF. (2011b) Nano-CeO(2) exhibits adverse effects at environmental relevant concentrations. *Environmental Science & Technology* 45:3725–3730. DOI: 10.1021/Es103309n.

Zhao F, Slade RCT, Varcoe JR. (2009) Techniques for the study and development of microbial fuel cells: An electrochemical perspective. *Chemical Society Reviews* 38:1926–1939. DOI: 10.1039/b819866g.

Zhao GY, Xu CL, Guo DJ, Li H, Li HL. (2006) Template preparation of Pt-Ru and Pt nanowire array electrodes on a Ti/Si substrate for methanol electro-oxidation. *Journal of Power Sources* 162:492–496. DOI: 10.1016/j.jpowsour.2006.06.082.

Zhao GY, Xu CL, Guo DJ, Li H, Li HL. (2007) Template preparation of Pt nanowire array electrode on Ti/Si substrate for methanol electro-oxidation. *Applied Surface Science* 253:3242–3246. DOI: 10.1016/j.apsusc.2006.07.015.

Zhong Y, Xu CL, Kong LB, Li HL. (2008) Synthesis and high catalytic properties of mesoporous Pt nanowire array by novel conjunct template method. *Applied Surface Science* 255:3388–3393. DOI: 10.1016/j.apsusc.2008.09.056.

Zhou YK, Neyerlin K, Olson TS, Pylypenko S, Bult J, Dinh HN, Gennett T, Shao ZP, O'Hayre R. (2010) Enhancement of Pt and Pt-alloy fuel cell catalyst activity and durability via nitrogen-modified carbon supports. *Energy & Environmental Science* 3:1437–1446. DOI: 10.1039/c003710a.

Zhu B. (2003) Functional ceria–salt-composite materials for advanced ITSOFC applications. *Journal of Power Sources* 114:1–9. Available at http://dx.doi.org/10.1016/S0378-7753(02)00592-X.

Zhu B, Ma Y, Wang XD, Raza R, Qin HY, Fan LD. (2011a) A fuel cell with a single component functioning simultaneously as the electrodes and electrolyte. *Electrochemistry Communications* 13:225–227. DOI: 10.1016/J.Elecom.2010.12.019.

Zhu B, Raza R, Abbas G, Singh M. (2011b) An electrolyte-free fuel cell constructed from one homogenous layer with mixed conductivity. *Advanced Functional Materials* 21:2465–2469. DOI: 10.1002/adfm.201002471.

Zhu B, Raza R, Qin HY, Fan LD. (2011c) Single-component and three-component fuel cells. *Journal of Power Sources* 196:6362–6365. DOI: 10.1016/J.Jpowsour.2011.03.078.

Zhu B, Raza R, Qin HY, Liu QH, Fan LD. (2011d) Fuel cells based on electrolyte and non-electrolyte separators. *Energy & Environmental Science* 4:2986–2992. DOI: 10.1039/c1ee01202a.

8 Nanostructured Metal Oxide Catalysts

Vicente Cortés Corberán, Vicente Rives, Natalia V. Mezentseva, Vladislav A. Sadykov, and Eduardo Martínez-Tamayo

CONTENTS

8.1 INTRODUCTION

8.1.1 Catalysis and Nanotechnology

Catalysis and related technologies have a substantial impact in a broad range of processes and goods production, leading to an improvement in many areas of human activity, such as food production, health care, clothing, efficient energy use, new materials, transport, environmental protection, and waste reduction. In fact, improvement of the life standards in developed countries is closely related to the use of catalysts. Catalysis has a major impact on the chemical and related industries, forming the basis of about 90% of chemical manufacturing processes and more than 20% of the world's industrial production (IEA 2013). Current catalyst market is around 10,000 million euros per year, and the value of products dependent on process catalysts, including petroleum products, chemicals, fragrances, pharmaceuticals, synthetic rubber and plastics, and many others, is said to be around 300,000–380,000 million euros per year (Yoneyama et al. 2010).

The widespread use of catalysts is based on their ability to accelerate a chemical reaction without being consumed itself. Most of the chemical processes are implicitly complex, and catalysts allow controlling their direction to the desired products, by increasing their rate of formation and productivity among the others. This makes catalytic processes to be environmentally superior to noncatalytic ones because they typically consume less raw materials, energy, and time. Thus, efficiency is the main objective pursued by the use of catalysis technology, which has become the key for sustainability.

Nanotechnology is not new for catalysis research. Actually, the relevant length scale in heterogeneous catalysis has been known by researchers to be that of a few nanometers, or even smaller, for many years. Nanoparticles of metals, oxides, and sulfides have been developed and used as catalysts for hydrocarbon conversion, partial oxidation, and combustion reactions since the 1920s. Thus, they probably stand for the oldest commercial application of nanotechnology and it is foreseen as one of its main application lines in the future. However, the rise of the interest of this industrial area in nanotechnology results from the recent developments in the general understanding of the distinct behavior of ensembles of molecules or particles when their size is in this length range and in instrumentation and techniques that allow manipulating and observing matter, molecules, and atoms at that scale. Therefore, catalysis is expected to take advantage from the fast advance in the knowledge of nanoscience and nanotechnology. It should be underlined that catalysis is the only application of nanotechnology that depends on chemical properties, whereas only physical (electrical, electronic, magnetic, mechanical, and optical) properties are involved in other applications.

8.1.2 Metal Oxide Catalysts: Relevance and Specific Features

Metal oxides are present in the composition of most of the industrial heterogeneous catalysts, either as catalysts themselves or as supports (e.g., for metals, sulfides, or other oxides). Metal oxide catalysts are one of the main catalyst types and play a key role in refinery processes, production of petrochemicals, intermediates, and fine chemicals. Their estimated worldwide market is over 3,000 million euros, and their economical impact is even much higher due to the relevant number of chemicals produced using these catalysts (Cortés Corberán and Conesa 2006).

Oxide catalysts are intrinsically multifunctional. Besides the surface functionalities common to all heterogeneous catalysts (adsorption, electron transfer, and Lewis acidity/basicity), they exhibit one more, namely, interaction with hydrogen (or water), which generates Brønsted acidity/basicity through formation of hydroxyl groups.

Nevertheless, their most specific feature is that, unlike metal catalysts, the atoms in their structure (those of oxygen) may participate as reactants (and be consumed) in the catalyzed reaction. This oxygen loss results in an inactive *reduced* site. To recover its catalytic activity, the site must be reoxidized by replacing the lost oxygen with oxygen coming from another crystallographic site of the oxide (i.e., through the bulk) and ultimately by oxygen coming from the reaction medium (i.e., on the surface). This is known as the Mars–van Krevelen mechanism or redox cycle (Mars and van Krevelen 1954). This brings along two important consequences: (1) Oxygen mobility on, and from bulk to, the oxide surface plays a key role in the overall kinetics of the redox processes and (2) the selectivity of the oxidation reaction depends on the easiness with which oxygen is released from the surface, which makes the binding energy of the metal-to-oxygen (M-O) bond to be the key factor controlling the selectivity.

8.1.3 Metal Oxide Catalysts and the Nanoscale

The ideal method for preparing a catalyst should be to design the active center by determining the nature and distribution of its constituting atoms, similarly as, in principle, it can be hopefully done with homogeneous catalysts (e.g., in metal complexes). But local properties of individual active centers in heterogeneous (solid) catalysts depend partially on collective properties at the crystal or surface scale. This boosts the interest on solid particles at the nanoscale (<100 nm), caused by their chemical and physical behavior, remarkably different from those of the bulk. In the case of catalysts, the decrease of the particle dimension to this scale brings several effects of relevance in their catalytic performance. First of all, it increases the surface-to-volume ratio, which translates into a higher specific surface area (SSA), and a higher number of active centers per unit of mass, which translates itself into a higher specific activity. But it also brings other catalytically relevant effects, such as different distribution of exposed crystalline planes, edges, corners, and defects; redistribution of electronic density in the particle and its surroundings; increased number of interphase perimeter atoms when supported on other materials (usually oxidic supports); and, specifically in oxides, an increase of coordinatively unsaturated (CUS) metal cations. These atoms show oxidation states and M-O binding energy values that are unusual for the bulk oxide.

It should be stressed that the situation with metal oxide catalysts in redox processes is unique, not only among the applications of oxides but also among those of nanotechnology. At the working state, the oxygen atoms of the oxide catalyst are being continuously replaced, while the structure and nature of the oxide remain. This implies that, in addition to surface properties, also those of the bulk are relevant for the overall catalytic performance. As a result, the nanoscale is relevant not only concerning the size of the primary particle (nanosized) but also concerning the size of the crystalline domains inside complex materials (nanocrystalline) with larger particle sizes. These nanostructured phases differ from the bulk phases not only by the small size of crystalline grains (usually ranged from 5 to 20 nm) but also by the significant content of atoms located in so-called noncoherent interfaces or grain boundaries.

8.1.4 Scope of the Chapter

Redox applications of metal oxide catalysts are very broad, ranging from total oxidation (combustion to carbon dioxide) to selective oxidation reactions (where the desired product is any but carbon oxides). They show some specific features, the most relevant being that, because of the reaction mechanism, the working state of the catalyst is under dynamic equilibrium with the reaction medium. On the other hand, when oxygen is used as oxidant (as in most of the gas phase-catalyzed reactions), its molecules can be thermally activated in the gas phase, adding new reactant species (radicals) to the reaction medium. This may add additional complexity to the kinetics and mechanism (Sinev et al. 2000) of the reaction and may be relevant when confined spaces exist or are formed inside the oxide catalysts, for instance, inside the pores of microporous and mesoporous oxide catalysts. The development of metal oxide catalysts must consider these specific features, and the effect of the nanoscale in the oxide catalysts will add to the overall complexity of these catalytic processes.

This chapter reviews the application of nanostructured oxide catalysts in a variety of oxidation reactions such as combustion, selective oxidation of light alkanes, methane and oxygenates conversion into syngas, and so on. It should be mentioned that some metal oxide nanocatalysts (e.g., titania and related oxides or composites) are very successful in photocatalytic processes, but, as these are described in Chapter 9, will not be deeply discussed here.

The first step of the application of nanotechnology is to develop oxide catalysts with a size below 100 nm and a size distribution as narrow as possible. Section 8.2 reviews concisely the methods used to prepare nanosized oxides in a controlled way. It covers also specific methods used to prepare nanotitania with different morphologies.

As mentioned in Section 8.1.3, *nanosize* may refer either to the primary particle dimension or to the crystalline domain size in oxide composites. In the first case, the nanoparticle itself is a single crystalline domain, and the size effects can be attributed to the change of properties of this only crystallite and its external surface. However, in the second case, the multiplicity of nanocrystalline domains within a single particle brings along the additional influence of internal interfaces between these domains on the phenomena taking place in the bulk (e.g., oxygen transfer).

For these reasons both cases will be considered separately. So, Section 8.3 analyzes the size effects at the nanoscale in nanoparticle oxide catalysts providing several examples of their use for combustion and selective oxidation of hydrocarbons. Section 8.4 reviews the size effects of nanodomains in nanocomposite oxide catalysts, aiming at elucidating the effect of their interfaces in their use, both as oxide catalysts and as oxide support for metals, in various reactions for synthesis gas production. Finally, Section 8.5 reviews the additional effect of nanostructure of internal void spaces in oxides with ordered porous systems.

8.2 METHODS FOR PREPARING NANOSIZED OXIDE CATALYSTS

8.2.1 Conventional Methods and Their Limitations

The main target when preparing catalytic materials is, basically, to obtain solids showing the desired activity and the highest selectivity and with the largest SSA. To meet these objectives, the preparative methods should provide an effective control on both their chemical composition and their microstructure (Thomas and Thomas 1996). Moreover, from an applied point of view, these preparative methods should be transferable to industry, that is, they must be efficient, versatile, cheap, and scalable to be used at an industrial scale.

Procedures developed to prepare bulk solids can be broadly classified in two main groups, namely, synthesis in the solid state (dry route) and synthesis in solution (wet route). Conventionally, the synthesis in the solid state has been carried out by means of the so-called ceramic method. In this synthetic procedure, the reagents are dosed, mixed, and milled until a "homogeneous" mixture is obtained. This mixture is calcined and milled, the process being repeated until completion of the reaction. Therefore, this process implies the application of high temperatures (to facilitate ionic mobility and diffusion) during long periods of time. A somewhat related process is the so-called mechanochemical synthesis; in this method, the reagents are treated in high-energy mills during long periods of time. Here thermal energy is substituted by mechanical energy, but the objective is the same, namely, to increase the mobility of the ions, fundamentally those located in the surface. In both cases, the basic reaction can be written as follows:

$$x\,A(s) + y\,B(s) \rightarrow A_xB_y(s)$$

Although the ceramic method is used thoroughly in the preparation of materials with applications in different fields, catalysis among them, the derived disadvantages of the use of this procedure are well known. Its main drawbacks are as follows:

Completion of the reaction is hardly detected, and the process has to be halted to perform a suitable analysis (XRD, IR-Raman, element chemical analysis, etc.) to detect the presence of reagents or their complete consumption.

There is scarce control of the chemical homogeneity and of the microstructure of the final material.

Due to sintering upon application of high reaction temperatures, materials with a large particle size are obtained.

Phases with low limit of thermal stability (thermodynamic control) cannot be prepared.

There is possible segregation of phases, with eventual losses by evaporation of volatile compounds, especially in complex systems.

It is very expensive from the point of view of energy consumption and reaction time.

The mechanochemical route shares some of the disadvantages of the ceramic method, although, in general, the particle size of the obtained products can be very small, of the order of some nanometers (Schäfer 1971; Rao and Gopalakrishnan 1997).

8.2.2 Nonconventional Methods: Use of Precursors

Researchers and technologists have assumed during the past decades the necessity of designing alternative procedures to overcome the problems associated with the ceramic method. One should aim to design synthesis procedures to prepare materials with predetermined composition, structure, and properties. C.N.R. Rao pointed out, "The ideal condition for carrying out a solid state reaction in order to obtain a homogeneous product in the shortest time at the lowest possible temperatures is to ensure homogeneous mixing of the reactants on an atomic scale. This, however, cannot be achieved by the ceramic method or its modifications. The only way to achieve this is to prepare a single phase (a chemical compound) in which the reactants are present in the required stoichiometry" (Rao and Gopalakrishnan 1987).

In reactions among solids, the reagent particles have an average size ranging between a few and several micrometers. These distances are enormous if compared with the average length of a chemical bond, 3–5 Å. For this reason, the atoms of the reagents must diffuse through very long pathways to approach each other, encountering steric hindrance due to "bottlenecks" as well as due to electrostatic fields, until reaching the adequate distance for forming bonds in the final product. The magnitude of these difficulties can be expressed by means of the diffusion coefficients, whose values are usually very small, except in the case of small sized species that can move rather easily through an open network, as in the case of ionic conductors. To overcome these obstacles, it is possible to design different strategies, but all of them have the same objective: "to minimize the distances that the species of the reactants must travel to form the chemical bonds present in the reaction products" (West 1984).

All of these strategies are based on the use of *precursors*. A precursor is a compound, or phase, that possesses all the necessary chemical elements, in the appropriate ratio, to form the pursued material, usually after an appropriate thermal treatment. A useful classification of these systems is as *ordered precursors* and *disordered precursors*.

The ordered precursors are chemical compounds with perfectly defined stoichiometry and crystal structure. These compounds must contain the necessary elements, in the required proportion, to form the pursued products. Moreover, the counterions existing in the precursor must comply with certain conditions in relation to their thermal stability, namely, (1) they should produce gaseous species or easily removable vapors during their pyrolysis and (2) they should not form volatile species nor refractory compounds with any element forming part of the desired product. It is worthy to note that the atmosphere in which the reaction takes place can be a source of some of the elements existing in the desired product, but they have a very high mobility (Alconchel et al. 1999; Oyama et al. 1999; Reyes de la Torre et al. 2004). Some examples are given in the following:

$$(NH_4)_2\, MoO_4\, (\text{air atm.}) \rightarrow MoO_3$$

$$(NH_4)_2 MoO_4 (NH_3 atm.) \rightarrow Mo_2N$$

$$(NH_4)_2 MoO_4 (CH_4 atm.) \rightarrow Mo_2C$$

The atoms in compounds with well-defined crystal structures and stoichiometries, such as the ordered precursors (OPs), are occupying predetermined positions in the network, so that their spatial distribution is homogeneous and ordered. In the network, these atoms are separated only by a few angstroms, so the diffusion distances are much smaller than in the case of the conventional ceramic procedure. Furthermore, during the decomposition, the network collapses, facilitating the mobility of the species (Rao et al. 1986). Some classical examples are given in the following equations, where in each pair of reactions, the first one represents the ceramic procedure, while the second one stands for the precursor method:

$$BaCO_3(s) + TiO_2(s) \rightarrow BaTiO_3(s) + CO_2(g)$$

$$BaTiO(C_2O_4)_2(s) + O_2(g) \rightarrow BaTiO_3(s) + 2\ CO_2(g)$$

$$La_2O_3(s) + Co_2O_3(s) \rightarrow LaCoO_3(s)$$

$$LaCo(CN)_6 \cdot 6H_2O \rightarrow LaCoO_3(s) + 6\ H_2O(g) + 6\ CO_2(g) + 3\ N_2(g)$$

OPs are the *ideal* precursors: they contain all the elements required, in the appropriate proportions, together with other chemical species that are evolved as gases during thermal decomposition. Unfortunately, these kinds of precursors only can be found in very few occasions, and most materials cannot be prepared using this strategy through lack of appropriate precursors.

An alternative to these ideal precursors (stoichiometric and ordered) are the so-called *disordered* precursors. These are usually amorphous phases that contain the necessary elements in the appropriate ratio (stoichiometry) to yield the pursued final materials. These elements might be accompanied by other chemical species that should already comply with the same conditions (easiness to form gaseous or volatile products, not forming volatile neither refractory products with the elements forming the final products) applicable to the ordered ones.

The amorphous character of these phases implies disorder in the spatial distribution of their elements. This disorder causes chemical inhomogeneity at a local scale. The chemical composition can vary depending on the considered net volume, although only very slightly. This disadvantage with regard to the OPs is balanced by the other following favorable factors:

The stoichiometry of the precursors can be tightly controlled in a broader range and be adjusted more easily to that of the final material, even for complex, multicomponent, or doped materials.

The amorphous networks are thermodynamically less stable than the ordered ones and, thus, they are more sensitive to thermal decomposition.

Because of the differences between both types of systems, the ways to prepare them are different. The *ordered* ones can be prepared by crystallization from solutions under controlled conditions (nature of solvent, temperature, pH, and concentration); well crystallized particles are obtained under adequate conditions. The size of these particles can be modulated by controlling the nucleation and particle growth rates: when the solubilities of the reaction products are very low, these rates are usually very high and particles with small size are formed by sudden precipitation reactions (Patzke et al. 2011).

Procedures to obtain *disordered precursors* are varied, although in all of them the objective is to assure that the network of the precursor contains all the elements required, in the appropriate ratio, to prepare the final material. Although these networks are usually amorphous, crystalline precursors with very low chemical homogeneity can be sometimes isolated, especially in complex systems, when all cations existing in the medium should coprecipitate to form the precursor compound.

8.2.3 Synthesis of Disordered Precursors

The more widely methods used to prepare disordered precursors are the following:

Sol-gel: Solvolysis of certain compounds leads to the formation of monomeric and oligomeric species. These species can react among them through condensation reactions that result in the formation of extended networks (Livage 1994). Hopefully, a single network extended to the whole system is formed:

$$M(OR)_n + n \bullet H_2O \rightarrow M(OH)_n + n \bullet ROH \quad \text{solvolysis}$$

$$M(OH) + M(OH) \rightarrow M\text{-}O\text{-}M + H_2O \quad \text{condensation}$$

$$M(OH) + M(OR) \rightarrow M\text{-}O\text{-}M + ROH \quad \text{condensation}$$

The whole process is known as *gelation*: solution → solvolysis → condensation → oligomers → micelles → gel.

To isolate the precursors from the gel, the solvent must be removed and, depending on the conditions of the drying process, solids with different nature can be obtained (Dutoit et al. 1996): a gel from conventional drying, or an aerogel/xerogel by supercritical drying. Phases isolated by this procedure are always amorphous, and although they can constitute the final material, the pursued product is usually obtained after thermal treatment of the precursor.

Spray-drying: Salt solutions with the appropriate stoichiometry of the required elements are transformed into aerosols by injection in a carrier gas, through an appropriate nozzle. When these aerosols are mixed with a hot gas in a drying column, the sudden evaporation of the solvent causes the precipitation of the solute. Generally, the precipitates obtained are amorphous or poorly crystalline. Ideally, these solids maintain the same chemical homogeneity existent in the original solution (Xu et al. 1990).

Freeze-drying (FD): Solutions with the required composition are dropwise added to a cryogenic bath where the droplets freeze instantly. Later on, the frozen solvent is removed by sublimation, and the solute is isolated as an amorphous solid that contains all the components existing in the original solution. The ions are randomly dispersed in the isolated solid, maintaining the chemical homogeneity of the solution. Consequently, the precursors give rise to the final material upon calcination (González et al. 1997).

The solids isolated by these procedures (sol-gel, spray-drying, and FD) are thermodynamically unstable with regard to the crystalline phases with the same compositions. This instability is inherent to their amorphous structure and it is due to the presence of local fluctuations of their chemical homogeneity: although the distribution of the species in the initial solution is homogeneous at a "macro" scale, they are randomly distributed at a local scale. This intrinsic instability allows decomposition of the precursors under relatively mild conditions, that is, low temperatures and short calcinations times. Under these mild conditions, the particle size cannot grow excessively, and nanostructured materials, with particle size smaller than 100 nm are usually obtained.

These preparative methods are widely used for the synthesis of massive materials. However, there are other synthetic routes that also lead to small particle size materials, either from precursors or directly. Some of these procedures are: precipitation/coprecipitation, micelle formation, solvothermal procedures, reactive plasmas or flames, chemical vapor decomposition or physical vapor decomposition.

All these methods are used for the preparation of solid phases in the laboratory and in the industry. Nevertheless, each method presents its own advantages and disadvantages, which mostly depend on the precise properties of the materials to be prepared.

Concerning binary materials (SiO_2, Al_2O_3, ZrO_2, etc.), the operative factors (i.e., productivity, price, and access to the equipment) usually determine the election of the preparative method. However, each synthetic procedure has its own limitations in the case of complex materials, such as $Zr_{(1-x)}Y_xO_{(2-x)}$, $LaMnO_3$, and $YBa_2Cu_3O_{(7-x)}$.

On considering the chemical reactions taking place during formation of the precursors (hydrolysis in the sol-gel pathway, or precipitation in the isolation of solids by coprecipitation), both thermodynamical and kinetics parameters must be analyzed: the first ones because the equilibrium constant (K) describes the extension of the reaction when the system has reached the equilibrium and the kinetics ones because the rate constant (V) indicates the rate with which the system reaches the equilibrium. Thus, for any complex process such as follows:

$$
\begin{array}{ll}
A_1 + B_1 \rightarrow C_1 & K_1, V_1 \\
A_2 + B_2 \rightarrow C_2 & K_2, V_2 \\
\ldots & \\
A_n + B_n \rightarrow C_n & K_n, V_n
\end{array}
$$

where formation of the reaction products implies a variety of parallel reactions, the control of the chemical homogeneity of the phases formed is extremely difficult, except in the unlikely case in which all the equilibrium constants (K_i) are very high and the rate constants of all the reactions involved are similar and preferably high ($V_1 = V_2 = \ldots = V_n$). For this reason, the complex materials prepared by these procedures usually lack homogeneous composition, although the compositional fluctuations can be minimized by means of subsequent thermal treatments that facilitate the ionic diffusion in the solid (Schwarz et al. 1995).

As advantageous aspects, one can emphasize the high productivity of the spray-drying and the FD methods, besides the great versatility of the latter. Also, in both cases, the required equipment is broadly implemented in certain industrial sectors: food, drugs, detergents, ceramics, and so on.

With regard to economic aspects, while the cost of the reagents is usually high in the case of the sol-gel method, the reagents needed in the other methods are common and relatively cheap.

Finally, one should mention that scarcely productive synthetic routes requiring expensive reagents and equipment are less used at an industrial scale. They are exclusively used to prepare certain materials with high added value, such as synthetic diamonds, ceramics with very high mechanical modules for the aerospace industry, and biomaterials.

8.2.4 Nanosized Titania as a Case Study

We can look at the specific methods developed to prepare nanosized titania in a variety of morphologies. It is well known that quantum effects develop below a given particle size, strongly affecting the catalytic performance of some materials, such as titania. Consequently, methods have been developed to tailor the morphology and particle size of titania to be used in photocatalytic reactions or other light-related devices. One-dimensional architectures play outstanding roles in functional devices due to their dimensionality and quantum confinement phenomena. Preparation of these systems can be attained by conventional methods as those described in Section 8.2.3 (e.g., hydrolysis, hydrothermal, and sol-gel), but alternative, specific, methods have been recently developed.

The use of *ionic liquids* (acting as a sort of template) has led to formation of nanosized (5 nm) anatase, active in methyl orange photodecomposition at room temperature (Zhai et al. 2007), from [BMIM][PF_6] and tetra n-butyl titanate; however, titania prepared by reacting titanium tetra-isopropoxide with water and 1-(3-hydroxypropyl)-3-methylimidazolium-bis(trifluoro-methylsulfonyl) amide was less active, probably because of the lack of a small amount of rutile together with the major anatase component.

Anatase (5–6 nm) prepared by heating titanyl nitrate and oxalyldihydrazide in a *combustion-assisted procedure* also showed an enhanced activity for methyl orange photodegradation (Deshpande et al. 2011).

Gas flame combustion has been largely used to prepare powdery products (Li et al. 2010); for instance, Akurati et al. (2007) used titanium tetraisopropoxide diluted in nitrogen, with methane as a fuel and oxygen as an oxidant; the specific properties (size and morphology) of the powders prepared depended on the relative concentrations of reactants and the flame geometry, but they were more active for methyl orange photodegradation than commercial titania P25 from Degussa.

Ultrasonic irradiation has been used simultaneously in the flame synthesis procedure (Lee and Byeon 2009), leading to powders whose crystallite size changed upon calcination depending on the maximum ultrasound amplitude during synthesis. The quality of the titania nanoparticles has been improved by using premixed flames on a rotating surface (Memarzadeh et al. 2011). Ultrasound applied without other simultaneous procedure has been used to prepare nanosized titania with a particle size ranging between 5 and 8 nm, depending on the specific reaction conditions (Ghows and Enterazi 2010; González-Reyes et al. 2010), a change in the band gap being observed with the particle size. Hollow titania microspheres, active for photodegradation of active dyes, have been prepared by Wang et al. (2008) using reverse microemulsions.

Procedures starting from metallic titanium have been also reported. Rutile nanowire arrays (up to 1000 nm length) have been prepared (Zhou et al. 2011) by reaction of HCl vapor with a Ti foil (Yang et al. 2011) under different experimental conditions (time, temperature, and acid concentration); the nanowires are formed from highly organized and intertwined nanofibers, made of tetragonal rutile nanocrystals. A nanorod array (1500 nm length, 20–50 nm diameter) was formed by soaking a titanium foil in concentrated H_2O_2 at 80°C for 72 hours (Mu et al. 2011); its performance for determination of organics in aqueous solutions was very high. Oxidation of titanium foils with H_2O_2 containing HF (Ng et al. 2011) leads to anatase nanoparticles (1–11 nm).

Anodization of titanium leads to formation of titania nanotubes or titania foils. The properties of the former depended on the nature of the salts existing in the electrolytic solution (Aw et al. 2011; Liu et al. 2011; Xin et al. 2011; Zheng et al. 2011) or the presence of added organic compounds (Liu et al. 2008; Park et al. 2008; Su and Zhou 2008; Yoriya et al. 2011; Kamalov et al. 2012). The nature of the electrolyte also controls the final properties of titania films (Diamanti et al. 2011; de Souza Sikora et al. 2011).

In some cases, commercial titania has been surface-modified to improve its properties by developing new structures. A general approach consists of generating alkali titanate nanofibers by hydrothermal reaction in alkaline solutions, exchanging alkali cations with protons forming H-titanates, and then thermal dehydration or hydrothermal treatment to produce TiO_2 nanofibers or nanotubes (Armstrong et al. 2005a,b; Yu and Xu 2007; Costa and Prado 2009; Mu et al. 2010). For instance, hydrothermal treatment of commercial Degussa P25 for 24 hours in a strongly basic (NaOH 10 M) medium at 150°C or 180°C led to formation of nanotubes or nanowires, respectively (Guo et al. 2011); a similar approach was followed by Fen et al. (2011) to prepare titania nanotubes.

Bilayer inverse opal TiO_2 electrodes have been prepared from commercial titania nanoparticles (ca. 15 nm) mixed with in situ prepared polystyrene (Shin and Moon 2011); the former filled the cavities of the polystyrene colloidal crystals and burning off the polystyrene at 500°C left behind the inverse opal titania layer.

Chemical vapor synthesis at atmospheric pressure, using $TiCl_4$ as a precursor, led to titania nanoparticles (Rahiminezhad-Soltani et al. 2011) whose crystallite size was in the 25–87 nm range.

The main drawback in the use of titania as a photocatalyst is that it may not "use" visible light, but ultraviolet wavelengths represents only 3%–4% of solar radiation reaching the earth surface. Attempts have been made to extend the light absorption to the visible range by the simultaneous use of colored transition metal cations. Uyguner and Bekbolet (2007) studied the photocatalytic oxidation of humic acids by a titania suspension containing Cr(VI) or Mn(II), following the reaction by decolorization or removal of aromatic moieties; the presence of chromium was largely detrimental, and manganese decreased the rate constant by 15%, a fact attributed to a competitive complexation and oxidation-reduction reactions in the medium (for chromium) or formation of a weak complex with humic acid, blocking the active sites on titania, in the presence of manganese. Anpo and Takeuchi (2001) demonstrated that doping of titania with transition metal cations develop new energy levels in the band gap of titania, depending on the precise nature of the doping cation and its concentration, but does not shift the absorption edge of titania to the visible region; however, *ion implantation* actually shifts the absorption edge toward the visible range, depending on both the concentration and nature (V > Cr > Mn > Fe > Ni) of the implanting cation. The difference arises from the formation of aggregates of metal-containing species upon doping, but isomorphical metal/Ti substitution upon implantation, as concluded from extended x-ray absorption fine structure spectroscopy (EXAFS) studies. However, Herrmann (2012) has recently refuted the suitability of these systems; in the particular case of doping with Cr(III) cations, the detrimental effect is attributed to an increase in electron-hole recombination at the Cr(III) ion sites, which play the role of acceptor centers, which, once filled, attract electrons. The use of decolorization as a test reaction to check the goodness of these systems has been also criticized (Vautier et al. 2001), as it is not a catalytic reaction, but a limited photo-assisted reaction with a limited electron transfer from the photo-excited dye to titania.

Titania doped with gold nanoparticles has shown, however, an improvement of 50% and 100%, respectively, for removal of 4-chlorophenol and methyl tert-butyl ether (Orlov et al. 2006).

In the past decade, a different approach has been used by several teams, namely, oxide-nitride substitution in the titania crystalline lattice (Asashi et al. 2001; In et al. 2006; Liu et al. 2009a), as such a substitution red-shifts the absorption spectrum of titania. However, this approach has been also criticized by Herrmann (2012), stating that is should be proved that nitrogen is really in the nitride state, isomorphically substituting oxide anions in the lattice, and that under oxidizing reaction conditions, nitride is not expulsed from the anionic sublattice, as oxidation from the N(-III) state is favorable in the sense that it involves a sharp decrease in the ionic radius.

8.3 NANOPARTICLE OXIDE CATALYSTS

8.3.1 Introduction

Catalytic applications of metal oxides for oxidative transformation of hydrocarbons and organic compounds can be roughly divided into total and selective oxidation. Total oxidation consists of destruction of the organic molecular structure and its transformation into carbon dioxide, either to improve the efficiency of energy conversion in combustion or to fully destroy harmful or pollutant compounds (which sometimes is termed as *mineralization*). The rates of deep oxidation reactions depend on the surface metal-oxygen bond strength: the weaker the bond, the higher the rate. Transition metals single oxides in high oxidation states, such as cobalt and manganese, are active for these reactions, although binary or multiple oxides can exhibit a higher activity if their microstructure favors oxygen mobility and release, as in perovskites.

On the contrary, selective oxidation reactions are aimed to preserve the whole (or most) molecular structure and to add new chemical functionalities to it by oxygen insertion or by partial hydrogen removal (oxidative dehydrogenation [ODH]). In this case, the first reaction step is the activation of the hydrocarbon, and oxygen release capacity must be moderate. This higher complexity may explain why single oxides able to catalyze selective oxidation are so scarce and that most catalysts are multicomponent, or at least binary, oxides. In this section, representative examples of nanoparticulated simple or multimetal oxide catalysts useful for both total oxidation (combustion) and selective oxidation of hydrocarbons are presented.

Preparation of oxide catalyst particles with nanosize demands to limit crystal growth and to avoid sintering. This involves low calcination temperatures (T_C) and, hence, the use of *soft* synthesis methods, that is, those looking for minimizing the diffusion paths in the synthesis of the solid, by the use of precursors. One should note that this use of *low* calcination temperatures provides an additional advantage: it allows the preparation of metastable phases and avoids sublimation of components, which allows obtaining unusual catalyst compositions unreachable by other methods. This leads to novel catalytic performances, which may add to the new features brought by the nanoscale. It will be shown that as a consequence some oxides that are typical catalysts for combustion become efficient for selective oxidation reactions too.

8.3.2 Nanosized Oxide Catalysts for Combustion

Perovskite-type oxides are probably the most widely investigated type among the mixed metal oxides, due to the interest of their broad variety of physical properties (ferroelectricity, piezoelectricity, pyroelectricity, magnetic, and electro-optical effects) and their high activity and operational stability in various catalytic reactions. They have the general formula ABO_3 (cation A with a larger size than B) and their diversity of properties derives from two facts: around 90% of the nonsynthetic metals can be stable in this type of oxide structure and multicomponent perovskites can be synthesized by partial substitution of cations in positions A and B, which allows to accommodate a certain degree of oxygen nonstoichiometry (Tejuca et al. 1989).

It is commonly accepted that the activity of perovskites in deep oxidation reactions depends on the nature of the transition metal in the B sublattice and on the electronic state of this metal, which in turn depends on substitutional cations in the A and B sublattices. Thus, $LaBO_3$ (B = Co, Mn, Ni, Fe) and perovskites with Ca, Sr, Ba, or Ce substitution for lanthanum are typical catalysts for alkane combustion (Arai et al. 1986).

Early research on nanoperovskites looked for increasing SSA, while completing crystallization at moderate temperatures (Johnson et al. 1976). With this goal, González et al. (1997) selected $NdCoO_3$ as a model compound difficult to prepare with high SSA because (1) in the homologous $LaMO_3$ series, cobalt gave the lowest areas; (2) in the $ACoO_3$ (A = rare earth metal) series, Nd leads to lower areas, but its activity in combustion was among the highest found; and (3) unsubstituted perovskites show lower SSA than partially substituted ones. They compared two different strategies: *OP* (samples), by synthesis of a stoichiometric bimetallic complex with DTPA (Figure 8.1), and *disordered precursor*, prepared by FD (samples) of nitrate solutions. The perovskite structure already formed by calcining at 400°C and was completed at temperatures as low as 600°C for OP samples or 700°C for the FD ones. Homogeneous nanosized particle distribution were obtained at $T_C \leq 700$°C (Table 8.1). The obtained surface areas, although moderate, were the highest reported until then for this perovskite. Only recently, Chen et al. (2007) have reported higher surface areas for this perovskite composition. By introduction of NaCl in the conventional combustion synthesis process, followed by calcination at 500°C, they obtained well-dispersed perovskite nanoparticles with an average particle size close to 9 nm and increased SSA of the solid product, from 1.7 up to 43 m^2/g. Surface characterization techniques with different depth range, that is, electron dispersive analysis of X-rays (EDX) and X-ray photoelectron spectroscopy (XPS), show that OPs produced fully homogeneous atomic compositions, while the FD method produced solids with compositional gradients and a surface enrichment of Co, which decreases with increasing T_C (Table 8.1). Their specific activity for isobutene oxidation was very different, even considering the differences in SSA: the surface-specific rates varied more than 60% among the samples, a difference that cannot be due to the surface Co content (compare XPS values in Table 8.1). The highest activity of sample FD-700 was ascribed to its oxygen stoichiometric excess ($\delta > 0$). As shown later for many other cases, the differences in oxidation catalysis performance of nanooxides is closely related to their oxygen mobility and nonstoichiometry (vacancies, M-O bond energy, and coordinative insaturation).

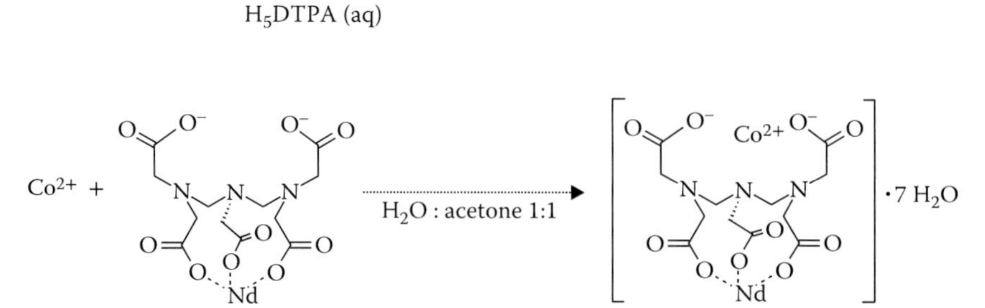

FIGURE 8.1 Synthesis of heteronuclear complex with DTPA (diethylentriaminpentaacetic acid) as ordered precursor for nano-$NdCoO_3$ preparation.

TABLE 8.1
Influence of the Synthesis Method on the Physicochemical Properties of Nano-$NdCoO_{3+\delta}$ and Kinetic Parameters of Catalytic Oxidation of Isobutene on Them at 275°C

Sample[a]	Co:Nd Atom Ratio		BET Area (m^2/g)	Particle Size (nm)	Specific Rate (mmol/h.g cat)	Surface-Specific Rate (μmol/h.m^2)	δ[b]
	EDX	XPS					
OP-600	0.96	0.96	10.8	40	4.03	373	≈0
OP-700	0.98	0.99	6.7	70	2.02	301	−0.16
FD-700	1.65	2.46	12.3	30–50	6.09	495	+0.56
FD-900	1.30	1.57	1.6	600	0.55	304	−0.07

Source: Adapted from González, A. et al., *Catal. Today*, 33, 361–369, 1997.
[a] FD, freeze-drying; OP, ordered precursor. The number indicates the calcination temperature in °C.
[b] Nonstoichiometry of oxygen.

A representative example of unusual catalyst compositions using low T_C is that of potassium-substituted perovskites, because potassium sublimates at high temperature. Nanosized perovskites, $Ln_{1-x}K_xMnO_{3+x}$ (Ln = La, Nd), with substitution degrees up to 25% ($0 < x < 0.25$), were prepared by FD of spray-drying of nitrates (Johnson et al. 1976) and FD of acetates (Lee et al. 1999) and were tested for ethane oxidation (Lee et al. 1997, 1999). The particles obtained with $x \leq 0.15$ were monophasic according to XRD, with a homogeneous morphology and narrow particle size distribution (11–16 nm and 30–50 nm for T_C = 600°C and 900°C, respectively). SSAs were high, between 10 and 25 m^2/g. The unsubstituted $LaMnO_{3+x}$ and $NdMnO_{3+x}$ samples showed a high combustion activity. However, at a variance of the alkaline earths or cerium substitution for lanthanum, which modifies the activity but not the selectivity, potassium addition had a marked effect on the selectivity: when its substitution degree (x) increased, selectivity to ethene for a given conversion also increased, and reached values of 15%–20% at 10%–20% conversion. Even with these low values, this showed the possibility to produce ethene from ethane at temperatures as low as 300–375°C. This effect was much more marked in the Nd-containing samples and was parallel to the decrease of the oxygen nonstoichiometry (δ) from +0.15 to −0.04 (Lee et al. 1999) and of the reaction order in oxygen (Lee et al. 1997).

When comparing the catalytic activity of transition metals simple oxides active for catalytic combustion of different hydrocarbons, Co_3O_4 shows always the best performance. It is also active for CO oxidation at low temperatures (Xie et al. 2009). This has boosted the research on nano-Co_3O_4. Solsona et al. (2007) studied combustion of propane at low temperature on nanocrystalline Co_3O_4 (SSA = 160 m^2/g) prepared by solid state reaction between cobalt nitrate and ammonium hydroxycarbonate at T_C = 300°C. Its activity was exceptionally high, starting at 100°C and total conversion being reached at 200°C. In comparison, on a conventional Pd/alumina catalyst, known to be one of the most active ones for deep oxidation of hydrocarbons, alkane conversion started at 250°C and reached 100% only at 400°C. In fact, it is noteworthy that nanocrystalline Co_3O_4 is significantly more active than a series of conventional noble metal catalysts. Unsupported nanocrystalline cobalt oxide catalysts were more active than any cobalt oxide catalyst supported on alumina, due to the dispersion of cobalt on the surface and formation of inactive Co–O–Al species. The catalytic activity decreased with an increase in the crystallite size (Solsona et al. 2008). Liu et al. (2009b) prepared similar nanocrystalline Co_3O_4-based catalysts by means of an innovative soft reactive grinding of citric acid with the Co(II) basic carbonate precursors (-GC), citrate sol-gel (-SG), and coprecipitation methods (-CC), followed by calcination at 300°C or 350°C. The particle size of Co_3O_4-GC (ca. 13 nm) was much smaller as compared to those of Co_3O_4-CC (21 nm) and Co_3O_4-SG (23 nm). XRD patterns and Raman spectra revealed a lattice distortion induced by grinding in -GC samples, which also showed the highest SSA, approximately 120 m^2/g. These catalysts show exceptionally high specific rate for propane combustion, which is attributed to the beneficial formation of highly strained cobalt spinel nanocrystals responsible for the favorable development of a significantly large amount of O^- species on the surface.

8.3.3 Nanosized Oxide Catalysts for ODH

Most intermediates and monomers in the petrochemical industry are currently produced from olefins and aromatics. In the future, the shift in hydrocarbon sources from oil to natural gas and the implementation of Fischer–Tropsch and biomass transformation processes will bring an increasing amount of alkanes. The challenge is, hence, to convert light alkanes directly into unsaturated oxygenates or, at least, to olefins, to use them by applying currently available technologies.

Catalytic dehydrogenation is the most direct way to get olefins from alkanes. However, the reaction being endothermic, it faces conversion limitations by equilibrium, limitations that become more and more important as the hydrocarbon chain length decreases, becoming specially critical for ethane dehydrogenation: at 1 bar, to reach an equilibrium conversion of 50% requires a temperature of 750°C. This leads to energy-intensive processes and fast catalyst deactivation (Cavani and Trifirò 1995). These drawbacks are avoided by the exothermic oxidative route. But in this case, selectivity drops very fast with increasing conversion, leading to low yields and productivity of olefins (Cavani and Trifirò 1992), the main by-product is carbon dioxide, whose emission is strongly undesired, due to its known greenhouse effect. Therefore, the key issue for the oxidative activation routes, that is, ODH and selective oxidation, is to improve catalyst selectivity and productivity up to a level that makes them economically feasible and competitive with currently used technologies.

In spite of its inherent interest, little attention has been paid in the literature on the use of nanosized oxide catalysts for ODH. The most studied catalytic systems for ODH of light alkanes operate at temperatures above 300°C and are based on vanadium (Mamedov and Cortés Corberán 1995) or molybdenum (Madeira et al. 2004) oxides. Free vanadium pentoxide is a poor catalyst for selective oxidation, but vanadium oxides become highly selective when they are well dispersed onto a support oxide. It has been found that the most selective active centers are vanadium species in isolated tetrahedral coordination (Valenzuela and Cortés Corberán 2000). This has probably precluded the research of nanovanadia catalysts, but boosted that of alternative ways to disperse vanadium oxides (see Section 8.5).

Simple molybdenum oxide MoO_3 is a poorly active and moderately selective catalyst too. But combined with other metal elements in molybdates it is highly selective and efficient. Thus, bismuth molybdates are the reference catalysts for selective oxidation/ammoxidation of light olefins, and cobalt and nickel molybdates are being studied as catalysts for ODH of alkanes.

The catalytic performance of nanosized cobalt and nickel molybdates for ODH of propane (Valenzuela et al. 2001a; Vie et al. 2004) has been compared to that of their homologues of similar SSA prepared by the conventional citrate method. The FD method has been applied to obtain isomorphically substituted $Co_{1-x}Ni_xMoO_4$ in the whole $0 \leq x \leq 1$ range (Vie et al. 2004). Calcination of the amorphous precursors at T_C between 300°C and 700°C, obtained from mixed solutions of Ni or Co nitrates and ammonium heptamolybdate, produced monophasic molybdates, with particle sizes <50 nm and SSA of 20–40 m^2/g. Formation of the molybdate phase was already completed at T_C = 300°C or 400°C, respectively, for the Co-Mo and Mo-Ni systems. Meaningful differences appear depending on the nature of the cation (Ni or Co). XRD shows that only β-$CoMoO_4$ (the phase usually stable at high temperatures) is formed for $T_C \geq$ 300°C in Co-Mo oxides, while in Ni-Mo oxides only β-$NiMoO_4$ is formed at $T_C \leq$ 500°C, the α phase also forming above this T_C, accompanied by some MoO_3 up to $T_C \leq$ 700°C. EDX studies of FD samples show that the surface is enriched with Co or Ni, but the surface M/Mo (M = Co, Ni) ratio decreases by increasing the calcination temperature (Table 8.2).

All samples were tested for ODH of propane at 400–525°C with a propane: oxygen: helium feed of 2:4.6:93.4 mol ratio and a residence time of 70 g h/mol of C_3H_8. The apparent activation energy for ODH of propane is the same (ca. 20 kcal/mol), within experimental error, whichever the molybdate tested, indicating the same reaction mechanism. However, the surface-specific rate for Co molybdates is lower for the nanosized sample than that for the citrate sample, while the reverse is true for the Ni molybdates. Nevertheless, the most relevant finding

TABLE 8.2
Surface Properties of the Molybdate Catalysts and Apparent Activation Energy (EA) of Propane ODH over Them

Catalyst	T_C (°C)	EDX M/Mo Atomic Ratio	SSA (m^2/g)	EA (kcal/mol)
$CoMoO_4$ (cp)	550	0.86[a]	10	17 ± 1
$CoMoO_4$ (FD)	300	2.0	30	20 ± 3
$CoMoO_4$ (FD)	400	1.5	18	22 ± 5
$NiMoO_4$ (FD)	400	2.33	39	19 ± 1
$NiMoO_4$ (FD)	500	1.26	42	20 ± 2
$NiMoO_4$ (cp)	550	1.16[a]	46	

[a] Measured by XPS.

is that, in both systems, nanosized molybdates show a much higher selectivity to olefin at isoconversion (20–25 percentage points), which means a twofold increase in the case of Ni molybdate (Figure 8.2). This means a 30%–40% increase of the olefin yield per pass for the nanosized molybdates compared to their conventional counterparts.

Molybdenum atoms are considered the active centers in ODH in these molybdate catalysts. This makes even more striking the effects of the nanosize on the selectivity in ODH reported recently for oxides conventionally behaving as total oxidation catalysts, such as NiO or Co_3O_4. He et al. (2006) reported that nanosized NiO (particle size 9.9 nm), prepared by a modified sol-gel method, is effective for ODH of propane at unusually low temperatures: at just 275°C, conversion reaches 27%, with 23% olefin selectivity, a temperature markedly lower than that used for V, Mo-based catalysts. The particle size decreases (to about 7 nm) upon addition of Ti (Ti/Ni = 0.1) or Zr (Zr/Ni = 0.12). This gave rise to a slight decrease of conversion, but selectivity improved markedly, resulting in an improvement of the overall catalytic performance: 11.5% yield (with 48% propene selectivity) and 10.3% yield are obtained on Ti-Ni-O and Zr-Ni-O, respectively, at the same reaction temperature. The Ti-Ni-O catalyst was active even at lower temperatures, giving 22.7% conversion with 40.5% selectivity (using gas hourly space velocity (GHSV) = 9 L/h/g and feed C_3H_8:O_2:N_2 = 1.1:1:4) at 250°C. The effects of titanium content and calcination temperature T_C were studied in the Ti-Ni-O system (Wu et al. 2006). The mixed oxides obtained by calcination at T_C = 400°C exhibit nanosized particles (4–7 nm) and high SSAs (150–210 m^2/g), in the whole range of TiO_2 content explored (≤50 wt.%). Among these, the highest yield was found for 9.1 wt.% TiO_2. However, when the same catalyst composition was prepared by calcination at T_C = 600°C, the particle size was >100 nm, and the catalyst became almost inactive: at 300°C it converted 0.3% propane versus 28.4% conversion by the sample calcined at 400°C. This clearly evidences that the

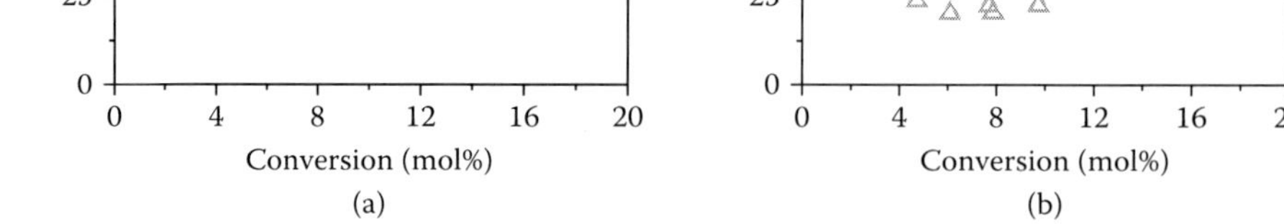

FIGURE 8.2 Effect of the preparation method on the oxidative dehydrogenation of propane over molybdates: variation of olefin selectivity as a function of propane conversion. (a) Nanosized Co-Mo-O prepared by freeze-drying (FD) calcined at T_C = 300°C (circles) or 400°C (squares) or 500°C and prepared by coprecipitation of citrates (triangles). (b) Nanosized Ni-Mo-O prepared by FD calcined at T_C = 400°C (circles) or 500°C (squares) and prepared by coprecipitation of citrates (triangles). (Adapted from Vie, D. et al., *Chem. Mater.*, 16, 1697–1703, 2004.)

calcination temperature is a key parameter for obtaining nanosized oxides and their outstanding catalytic performance. Indeed, the SSA is much lower (11 vs. 156 m^2/g when calcined at $T_C = 600°C$), but not enough to explain this sharp drop, which implies that the nanosize also influences the chemical state of the surface atoms. The improvement of olefin selectivity brought by titanium addition to NiO is due to the decreased reducibility, and the activity decreases with the TiO_2 content because of the decreased oxygen adsorption.

Nevertheless, propane ODH can be catalyzed by nano-oxides even at lower temperatures. Davies et al. (2006) reported that nanocrystalline Co_3O_4, with SSA 159 m^2/g, is able to activate propane, yielding propene, under ambient conditions. The oxide was obtained by calcining at 300°C for 2 hours in static air, the precursor obtained by intimate mixing of cobalt nitrate and ammonium hydrogen carbonate by grinding in a mortar for 0.5 hours, a method also used by Solsona et al. (2007). Under the reaction conditions tested (1% propane in air, contact time = 250 g_{cat} h/mol C_3), propene was obtained with near 100% selectivity between 25°C and 80°C. In comparison, bulk Co_3O_4 catalyst was inactive below 120°C. This exceptional performance of the nanocrystalline oxide catalyst seems to be related to a surprising low-intensity reduction feature observed in reduction thermograms at 90–100°C, which is absent for the bulk oxide. Propane conversion remained constant above 120°C with prolonged time on stream, but propene selectivity was <100%; below 120°C, the nanocrystalline oxide deactivated totally after a period of time that depended on the contact time employed, and no further propane conversion was observed. However, its full activity was recovered after reoxidation at just 180°C. Formation of CO_2 during temperature programmed desorption in oxygen indicates that deactivation may be most probably due to species adsorbed on the surface, which are unable to desorb at the low reaction temperatures used. It is worth to note that even if propane conversion is almost negligible (<1% at the contact times above indicated), these results are promising, as they allow to envisage a cyclic operation process, operated at quite moderate temperatures.

Nanosized Co_3O_4 catalysts have been also used for selective oxidation of other hydrocarbons, such as oxidation of cyclohexane to cyclohexanol and cyclohexanone. Zhou et al. (2005) synthesized Co_3O_4 nanocrystals with particle sizes of 30–50 nm, using cobalt nitrate as a precursor and Cetyl trimethyl ammonium bromide (CTAB) and in situ-generated triethylamine salt as a stabilizer to prevent agglomeration of the nanocrystals during the process. These catalysts showed obviously higher activities as compared to Co_3O_4 prepared by the conventional methods, Co_3O_4/Al_2O_3, or even homogeneous cobalt catalyst under comparable reaction conditions. The 89.1% selectivity to cyclohexanol and cyclohexanone at 7.6% conversion of cyclohexane was reached over 50 nm sized Co_3O_4 nanocrystals at 100°C for 6 hours.

8.3.4 Nanoxide Catalysts for Using CO_2 as Oxidant

The main problem in the ODH of light alkanes is the overoxidation of the olefinic product, much more reactive (oxidizable) than the original alkane. To overcome this problem, one alternative is the use of an oxygen source with a lower oxidizing power than molecular O_2. Carbon dioxide is attractive for this purpose, as it is a renewable resource, easily available (especially in petrochemical environments), inexpensive (even with a negative cost, after implementation of the Kyoto protocol), it avoids the risks of flammability of the reacting mixtures, and its subproduct, CO, can be easily used with currently used technologies, which could possibly lead to processes with a 100% carbon utilization (Cortés Corberán 2005). One might argue the drawback of its very low concentration in air, but industrial and power plants generate gaseous streams with much higher concentrations, ranging from 8% for oil refinery and petrochemical plant fired heaters up to >80% in coal–oxygen combustion (Thambimuthu et al. 2002).

Catalysts efficient for ODH of alkanes can be roughly grouped into three types: ceria-based oxides, transition metal (Mn, Fe, and Cr) oxides, and gallium oxide catalysts (Krylov et al. 1995; Wang and Zhu 2004).

Valenzuela et al. (1998, 2000) reported the ODH of ethane with CO_2 for the first time on pure ceria and calcium-doped ceria showing that the reaction with CO_2 over ceria-based catalysts is a heterogeneous catalytic reaction. Bulk ceria is effective for ODH of ethane with CO_2, with ethene selectivities of 58%–65% in the temperature range 680–750°C. They proposed that the redox couple Ce^{+4}/Ce^{+3} can activate CO_2 to produce active oxygen species for the reaction. No carbon formation is observed on the catalyst surface, and no hydrogen was observed in the gas products, but CO formation rates were always higher than the ethene ones. Therefore, besides catalytic ODH of ethane by CO_2 ($C_2H_6 + CO_2 \rightarrow C_2H_4 + CO + H_2O$), unselective catalyzed oxidation reaction ($C_2H_6 + 5\ CO_2 \rightarrow 7\ CO + 3\ H_2O$) must be considered to explain such results. Doping ceria with 10 mol% of CaO (catalyst 10CaCe) further improved the olefin selectivity, up to 88%–94% with 60%–75% CO_2 use efficiency, by inhibiting the unselective reaction, in the same temperature range. To improve activity and ethene productivity, nanosized catalysts with the same compositions (hereinafter denoted as CeO2-FD and 10CaCe-FD, respectively) were prepared by the FD method (Valenzuela et al. 2001b). After calcination at merely 300°C, both samples were monophasic and nanostructured (formed by agglomeration of spherical particles of ca. 7 nm diameter) with very high SSA (117 m^2/g for CeO2-FD and 91 m^2/g for 10CaCe-FD). These catalysts were much more efficient than their "bulk" counterparts: over undoped ceria, yield of ethene increases with the reaction temperature up to 30.3% (selectivity 71.4%) at 750°C; Ca-doped ceria (10CaCe-FD) was less active, but much more selective, reaching a maximum yield of 22.6% (with a selectivity of 90.5%). However, this catalyst deactivated quite rapidly at temperatures above 650°C under reaction conditions: tests conducted after operation at 750°C showed a marked decrease of CO formation, while ethene formation remained similar, indicating the inhibition of the unselective oxidation reaction. As a consequence, the selectivity was much improved. Further tests of this sample showed no change of activity within experimental error (less than 10%), indicating a stabilization of the catalyst in the new state, reached upon reaction at high temperatures. Interestingly, the apparent activation energy of the reaction varied from 37 ± 3 to 52 ± 5 kcal/mol after this modification, a value similar to that observed for the bulk 10CaCe catalyst, which seems to confirm the new state of the catalyst surface. This suggests that, in addition to its role as a selective oxidant, CO_2 may play a role in the formation of highly efficient phases and/or surface states.

The positive role of the Ca dopant seems to be related to its ability in increasing oxygen mobility in the crystalline framework, but it favors sintering. The interaction with Zr modifies the oxygen storage capacity of ceria, thus modifying its properties for catalytic oxidation reactions, and also increases its thermal stability. This led to study nanostructured ceria-zirconia catalysts with a variable composition, $Ce_xZr_{1-x}O_2$ ($0 \leq x \leq 1$), prepared by the FD method as catalysts for this reaction (Navas et al. 2007). Particle sizes of 20–40 nm were obtained for all compositions at low $T_C = 400°C$, and while Zr-rich compositions stabilized the tetragonal phase, the increase in the Ce content stabilizes cubic phase, characteristic of CeO_2. When the calcination temperature T_C was increased up to 1000°C, the CeO_2 particle size increases up to 150–300 nm, but incorporation of Zr atoms into the ceria network increases the thermal stability of the (nano)particles: sizes of all Zr-containing samples were <100 nm. The crystalline phases depended on composition: a higher Zr content favored the formation of the tetragonal and monoclinic phases. Nanosized ZrO_2 was highly selective, but little active, for ODH of ethane with CO_2. However, the best catalytic performance corresponds to compositions with high zirconia contents (in the range 30–40 atom% Ce), where the nanosize stabilizes the tetragonal phase, reaching a maximum of 24% ethene yield with 90% selectivity at 750°C on these catalysts. This could be due to the fact the monoclinic structure of ZrO_2 brings about stronger surface adsorption sites for CO_2 than the tetragonal structure (Bachiller-Baeza et al. 1998). A variety of ZrO_2-based mixed oxides, namely, TiO_2-ZrO_2, MnO_2-ZrO_2, CeO_2-ZrO_2, K_2O/TiO_2-ZrO_2, B_2O_3/TiO_2-ZrO_2, and CeO_2-ZrO_2/SBA-15, have been investigated by the group of S.E. Park for the closely related reaction of ethylbenzene dehydrogenation to styrene with CO_2 (Reddy et al. 2008). Its possible utilization for ODH of alkanes has not been investigated yet and possibly deserves much attention.

Chromium oxides have been also widely studied for this reaction. Deng et al. (2003) have used nanosized chromia catalysts, prepared by sol-gel coupling with azeotropic distillation method and compared them with bulk Cr_2O_3. Using a concentrated feed (C_2H_4:CO_2:Ar = 3:9:3 mL/min, 0.2 g catalyst, 550–700°C), ethane conversions up to 77% with selectivity ranging between 70% and 90% were obtained on nano-Cr_2O_3, compared to conversion ≤11% on bulk Cr_2O_3. The performance of the nano-Cr_2O_3 was improved by preparing Cr_2O_3-ZrO_2 nanocomposites by the same method (Deng et al. 2007). Under the same reaction conditions, unmodified nano-Cr_2O_3-ZrO_2 (particle size 7–12 nm, SSA 147 m^2/g) gives 29% ethene yield, with 78% selectivity, even at 550°C. To improve its performance, Co, Ni, Mn, and Fe were added to the nanocomposite. Only zirconia phases were detected by XRD. Interestingly, the incorporation of Ni, totally deleterious to the ODH activity, promoted the transformation to the monoclinic phase, while addition of the other cations decreased the surface area, but led to stabilization of the tetragonal phase of ZrO_2. The nanocomposites were formed by single crystalline particles, with a narrow size distribution in the range 5–12 nm, and SSAs of 125–172 m^2/g. Incorporation of Co, Mn and, specially, Fe, decreases the conversion, but greatly improves the selectivity, thus giving higher yields: 50% ethene (with 93% selectivity) for the Fe5-Cr10/Zr composite (5% Fe, 10% Cr) at 650°C. It should be noted that these values are approaching those of economic feasibility of the process.

8.4 NANOCOMPOSITE OXIDE CATALYSTS

8.4.1 Introduction

Nanocrystalline fluorite-like and perovskite-like oxides and their nanocomposites, showing high surface/lattice oxygen mobility and reactivity, are among the most fascinating and promising systems for such advanced applications as automotive catalysis (Shelef 2002), transformation of hydrocarbons or oxygenates into syngas by partial oxidation or steam/dry/autothermal reforming (Pavlova et al. 2005, 2007), anode and cathode materials for solid oxide fuel cells (SOFCs) (Atkinson et al. 2004; Wincewicz and Cooper 2005; Sadykov et al. 2008a), and so on. Their tailor-made design for any application requires elucidation of the relationships between their composition, real structure/microstructure, and oxygen mobility and reactivity. A specific feature of nanocrystalline/nanocomposite materials is that their functional characteristics, namely, surface/lattice oxygen mobility and reactivity, ability to activate reagents (O_2, mild oxidants-H_2O and CO_2, and fuels), stability against sintering, and coking, are determined in a great extent by interfaces–domain boundaries in oxide nanocomposites, metal–support interaction (including epitaxy, decoration, and incorporation of metal species into the support), and so on.

In this section, results of research (Efremov et al. 2008, Sadykov et al. 2005, 2006a,b, 2007a–e, 2008a–c, 2009a–e, 2010a–d, 2011a–e, 2012) aimed at elucidating these effects of interfaces for the following systems are presented:

1. Pt-supported nanocrystalline $Ln_x(Ce_{0.5}Zr_{0.5})_{1-x}O_{2-\delta}$ (Ln = La^{3+}, Gd^{3+}, $Pr^{3+/4+}$, $x = 0.05$–0.3) solid solutions prepared by the Pechini route
2. Nanocomposite catalysts composed of Ni/YSZ and fluorite (doped ceria and ceria-zirconia) or LnPrMnCrO perovskite oxide promoted by noble metals
3. Nanocomposites composed of $La_{0.8}Sr_{0.2}MnO_3$ (LSM) or $La_{0.8}Sr_{0.2}Fe_{1-x}Ni_xO_{3-\delta}$ ($x = 0.1$–0.4) ($LSFN_x$) perovskites (P) and $Sc_{0.1}Ce_{0.01}Zr_{0.89}O_{2-x}$ (ScCeSZ) or $Ce_{0.9}Gd_{0.1}O_{2-\delta}$ (GDC) fluorites as cathode materials for SOFC

Details of synthesis procedures and techniques of characterization of the real structure/surface properties/oxygen mobility and reactivity of the samples are described elsewhere (Sadykov et al. 2005, 2006a,b, 2007a–e, 2008a–c, 2009a–e, 2010a–d, 2011a–e, 2012). Nanocrystalline oxides and nanocomposites composed of perovskite ($La_{0.8}Pr_{0.2}Mn_{0.2}Cr_{0.8}O_3$) or fluorites ($Ln_x(Ce_{0.5}Zr_{0.5})_{1-x}O_{2-\delta}$) oxides in combination with NiO and YSZ ($Y_{0.08}Zr_{0.92}O_{2-\delta}$) were synthesized using different modifications of the polyester ethylene glycol-citric acid (Pechini) method. Pt, Pd, and Ru were supported by incipient wetness impregnation, followed by drying and calcination at 800°C. Cathode nanocomposites were prepared through ultrasonic dispersion of the mixture of perovskite (LSM, LSFN, LSFC, etc.) and electrolyte (GDC, ScCeSZ, etc.) powders in isopropanol with addition of polyvinyl butyral followed by drying, pressing pellets or supporting on electrolytes, and sintering in air up to 1300°C (Sadykov et al. 2010c, 2012). Catalytic properties were evaluated using both diluted and concentrated feeds in lab-scale and pilot-scale setups equipped with gas sensors and gas chromatographic analysis (Sadykov et al. 2009a).

8.4.2 Structural Features

8.4.2.1 Microstructure

The Pechini method was shown to provide single-phase composition of nanocrystalline complex oxides with the perovskite and fluorite structures not easily achieved by other traditional methods (e.g., precipitation) (Sadykov et al. 2005, 2006a,b, 2007a–e, 2010c; Kharlamova et al. 2008a). They usually consist of stacked nanodomains with typical sizes in the 5–30-nm range depending on their composition and sintering temperature. Typical TEM images (Figures 8.3 and 8.4) illustrate this feature for both fluorite-like oxides and their cathode/anode nanocomposites. The dense ceramics formed by sintering of perovskite-fluorite nanocomposites even at high (up to 1200°C) temperatures retain the nanodomain structure due to mutual hampering of

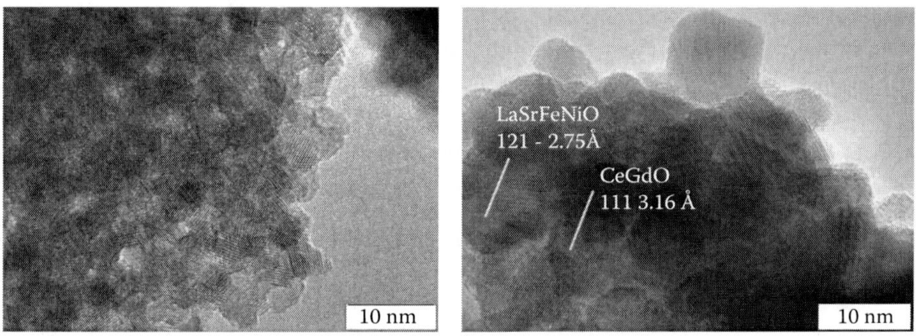

(a) (b)

FIGURE 8.3 High-resolution TEM image of Pt-supported particles of $Pr_{0.3}(Ce_{0.5}Zr_{0.5})_{0.7}O_2$ samples (a) and LSFN-GDC nanocomposite sintered at 700°C (b).

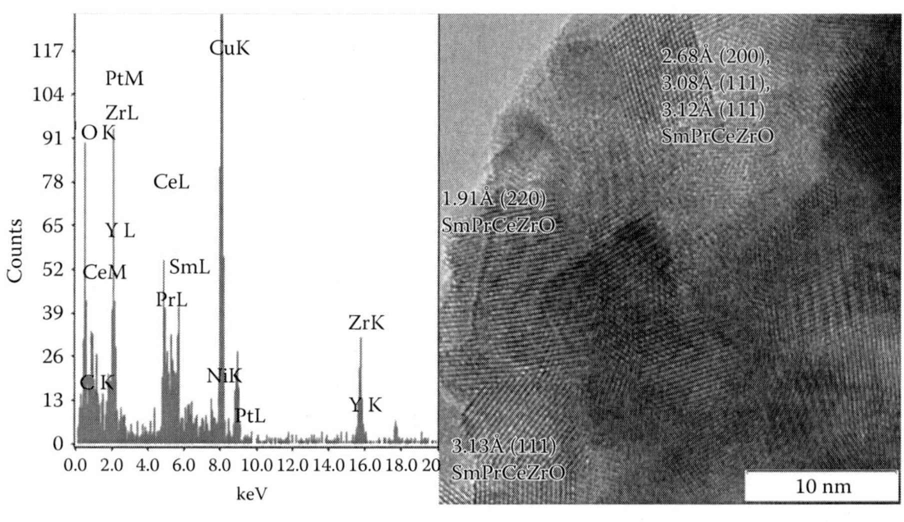

FIGURE 8.4 High-resolution TEM image and electron dispersive analysis of X-rays data for 1% Pt-80% SmPrCeZr-10% NiO-10% $Zr_{0.92}Y_{0.08}O_2$ nanocomposite.

sintering (Sadykov et al. 2010c, 2012). As follows from EDX data, pronounced redistribution of elements between perovskite and fluorite domains takes place: transition metal cations are incorporated into fluorite domains (mainly, surface layers), while Sc, Ce, and Zr cations migrate into perovskite domains, mostly without forming new phases. Domain boundaries appear to be rather coherent, demonstrating a good epitaxy between the perovskite and fluorite phases. Hence, perovskite–fluorite interfaces with the chemical composition and structural and transport properties different from those of perovskite and fluorite domains are provided by these nanocomposites.

Pronounced interaction between nanocrystalline fluorite and perovskite oxides with NiO/YSZ cermet components was revealed for anode nanocomposites as well (Sadykov et al. 2008b,c, 2009d, 2010a). X-ray diffraction patterns of composites prepared through one-pot Pechini route correspond to the mixture of NiO, YSZ, and promoter oxide phases, so new phases are not formed. The diffraction maxima of the promoter oxide phases are broad, due to their high dispersion (X-ray crystallite size ~10 nm). Reflections of NiO phase are narrower due to larger (>100 nm) X-ray crystallite sizes.

In anode nanocomposites with a low content of promoters (a big NiO/YSZ content), large (100–150 nm) NiO particles are decorated with epitaxially intergrown small particles of YSZ and an oxide promoter. In nanocomposites with a small content of NiO and YSZ, EDX revealed the presence of Ni cations and Pt group metals even in the regions where their separate particles are not observed, thus proving their incorporation into the surface layers of perovskite or fluorite particles. After contacting with a reducing reaction feed, metal (nano) particles (Ni or Ni-based alloys) remain to be in the epitaxy registry with oxide support and/or are decorated by oxidic fragments. Hence, developed metal–oxide interfaces are retained in anode nanocomposites even under real reaction conditions.

8.4.2.2 Defect Structure

For doped ceria-zirconia solutions, domain sizes decline with the dopant content with simultaneous decrease of the microstrains density $\Delta d/d$ (Figure 8.5). Decline of domain size with doping level is caused by dopant segregation at domain boundaries,

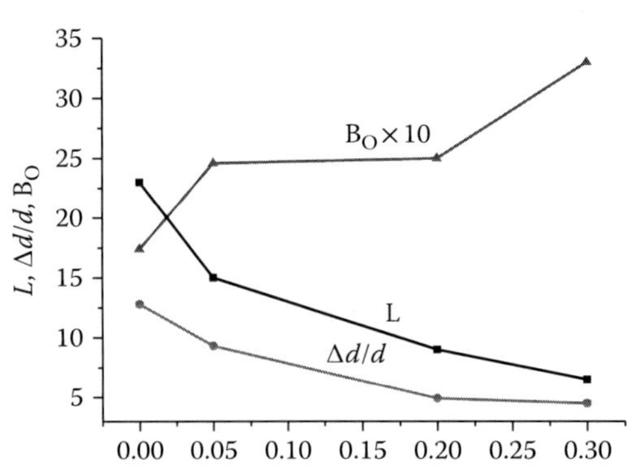

FIGURE 8.5 Domain size L (nm), $\Delta d/d$ 10^3 and B_O for Pr-doped samples based on neutron diffraction studies.

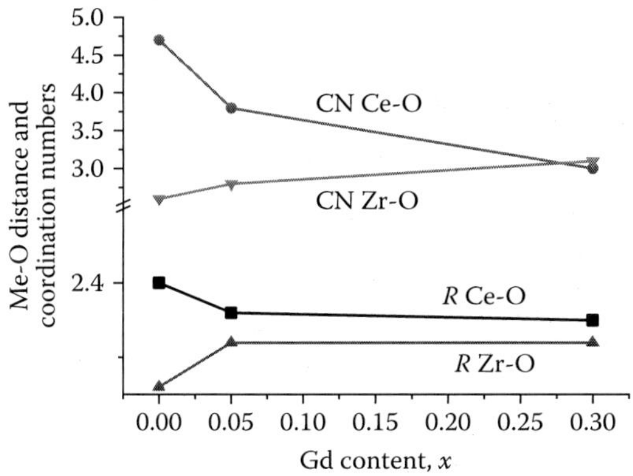

FIGURE 8.6 Variation of long Me-O distances (*R*) and respective coordination numbers (CN) for GdCeZrO samples.

thus hampering their sintering. From neutronographic studies on the nanoscale heterogeneity of $Ce_{0.5}Zr_{0.5}O_2$ oxide composed of zirconia-enriched domains in ceria-enriched matrix, Mamontov et al. (2000, 2003) observed a decrease in the microstrain density with doping, which can be explained by relieving the lattice strains due to replacing small Zr^{4+} cations by large Ln^{3+} cations. However, the isotropic temperature parameter for the oxygen atoms B_O increases with the Pr content due to a complex rearrangement of the oxygen sublattice (Figure 8.5). Indeed, according to EXAFS data (Sadykov et al. 2006a, 2007a,b), doping of ceria-zirconia solid solution provides a specific rearrangement of the coordination polyhedra of different cations. Thus, for a Gd-doped series, an increase in the Gd content is accompanied by an increase of the Zr-O integral coordination numbers (CN) from 6.7 to 8.0, while distances increase from 2.10–2.26 Å to 2.15–2.32 Å. On the contrary, despite the unit cell expansion, Ce-O distances contract with doping (from 2.20–2.40 Å to 2.18–2.35 Å), while CN decrease from 2.8 + 4.7 to 2.6 + 3.0, respectively (Figure 8.6). A similar trend—decrease of coordination numbers and distances with the doping level—was revealed for the Gd-O coordination sphere making it more dense and symmetric. Magnetic susceptibility data revealed clustering of Gd^{3+} cations at a high doping level (Sadykov et al. 2007c), suggesting that some ordering of the Gd distribution in the lattice occurs, causing such a local structure rearrangement. This rearrangement removes Ce^{3+}cations (as well as associated "free" anion vacancies) existing in ceria-zirconia samples only at a low doping level. Similarly, magnetic susceptibility measurements revealed the appearance of Pr^{4+} cations at a high Pr content, their mole fraction reaching up to 30 rel.% Pr at $x = 0.3$. Hence, clustering of doping cations at a high doping level can be a general trend and a driving force for the structure rearrangement.

8.4.3 Surface Features

According to XPS and secondary ions mass spectrometry (SIMS) data (Sadykov et al. 2007a–c), the surface of doped ceria-zirconia oxides is enriched by larger Pr, Ce, and La cations and is depleted by smaller Zr, Gd cations. This suggests that domain boundaries are as well enriched by larger cations.

Three different states of Pt (BE 71, 72, 73, and 75 eV) corresponding to species in the 0, 2+, and 4+ states (Sadykov et al. 2007b, 2011b) have been observed for oxidized Pt-supported doped ceria-zirconia samples. For samples with a low (~0.5 wt.%) Pt loading, the Pt° state was not detected. The Pt/(Pr + Ce + Zr) surface concentrations ratio increases from ~0.2% to ~0.45% and to ~1% when the Pt loading was increased from 0.5 to 1.6 and to 4.9 wt.%, respectively. Such a low surface concentration of Pt species suggests incorporation of Pt cations into the surface/subsurface layers of fluorite-like oxides and its strong interaction with the support, probably decoration of Pt clusters by support oxidic species (Sadykov et al. 2011b).

As revealed by fourier transform infrared spectroscopy of adsorbed CO (Sadykov et al. 2006b, 2007b,c, 2009b), CUS isolated Ce^{4+}/Zr^{4+} cations (carbonyl bands at 2160–2180 cm^{-1}) and clustered Me^{3+} cations (bands at 2110–2125 cm^{-1}) exist on the surface of doped ceria-zirconia oxides. As a typical example, Figure 8.7 shows the dependence of the normalized intensities of the Me^{4+}-CO and Me^{3+}-CO bands on the Pr content for Pr-doped ceria-zirconia samples. The decrease of the intensity of the Me^{4+}-CO band with the Pr content correlates with the decrease in the density of microstrains, while the subsequent increase can be explained by the increase of the density of domain boundaries (see Figure 8.6). Hence, Lewis acid sites—CUS Me^{4+} cations—seem to be located at outlets of extended defects including domain boundaries.

FTIRS of CO adsorbed on Pt-supported doped ceria-zirconia samples revealed several Pt species, namely, Pt°, Pt^{+}, and Pt^{2+} (Sadykov et al. 2006b, 2007b,c, 2009b). The intensities of the high wavenumbers carbonyl bands corresponding to CO complexes with isolated Pt^{2+} cations (Figure 8.8) are strongly affected by the type and content of doping cation. The intensity of the Pt^{2+}-CO band for Pr-doped samples correlates with that of the Me^{4+}-CO band, thus suggesting that extended defects (domain boundaries) play some role in stabilization of Pt^{2+} species on the surface. The highest intensity of the Pt^{2+}-CO band for the La-doped samples is explained by the positive role of the largest (and strongly basic) La cations in stabilizing the cationic Pt species. The weakening of this band at the highest La content can be explained by Pt incorporation into domain boundaries or subsurface layers.

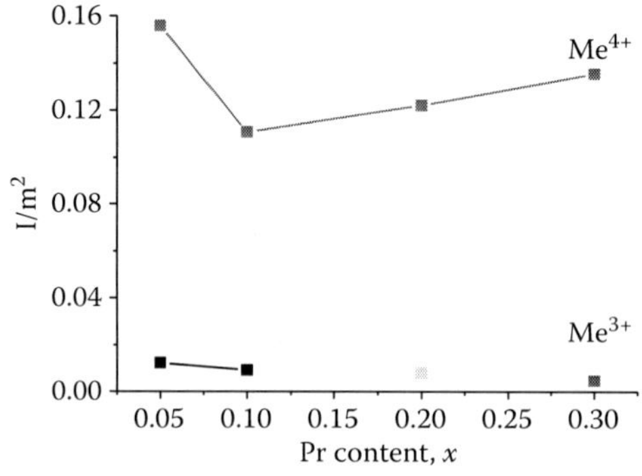

FIGURE 8.7 Normalized intensity of Me^{4+} and Me^{3+} carbonyl bands for Pr-doped ceria-zirconia samples.

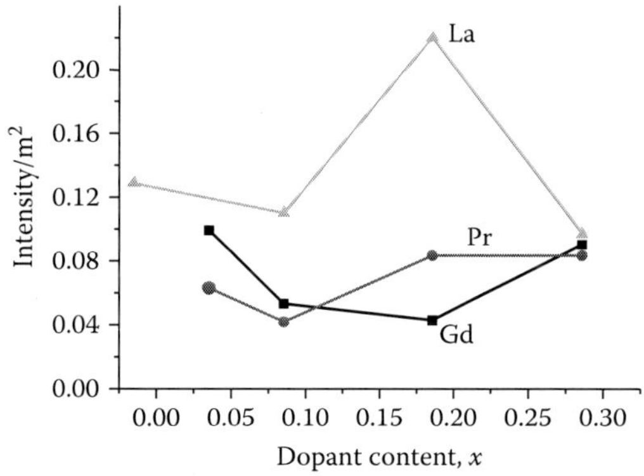

FIGURE 8.8 Intensity of Pt^{2+}-CO band versus dopant content for Pt-supported samples; 16 μmol CO adsorbed at 77K.

All these results clearly demonstrate the utmost importance of extended defects—domain boundaries in complex fluorite-like oxides—in generation of CUS surface sites—active centers in catalysis, as well as in providing strong interaction with the supported metals species.

Surface concentration of transition metal cations in cathode nanocomposites tends to increase with the sintering temperature, while that of Zr or Ce cations decreases (Sadykov et al. 2010c). In agreement with EDX data, this suggests migration of transition metal cations onto the surface of electrolyte domains. Certainly, surface migration can be the most pronounced feature for nanocomposites with developed interfaces.

8.4.4 Oxygen Mobility

In oxygen isotope heteroexchange experiments carried out in a static reactor in the temperature-programmed mode, the oxygen mobility was characterized by a dynamic degree of exchange X_S (amount of oxygen exchanged up to a given temperature) given by $X_S = \lambda_S\{(\alpha^o - \alpha)/(\alpha - \alpha_S^o)$ (Sadykov et al. 2010b, 2011e). Here, $\lambda_S = N/N_S$ (monolayers), N is the number of oxygen atoms in the gas phase; N_S is the number of exchangeable oxygen atoms in the oxide monolayer; α^o and α are the initial and current fractions of ^{18}O atoms in the gas phase, respectively; and α_S^o is the initial fraction of ^{18}O atoms in the oxide. For nanocrystalline ceria-zirconia oxides at 650°C, X_S is below a monolayer (Figure 8.9), so this parameter mainly characterizes the surface/near surface oxygen diffusion.

For ceria-zirconia solid solutions, the decrease of X_S with the Gd or La content (Figure 8.9) certainly correlates with increasing the symmetry and average CN for the Zr-O coordination sphere, contraction of the Ce-O sphere (hence, strengthening of Ce-O bond) and removal of free anion vacancies caused by a complex rearrangement of coordination polyhedra (see Section 8.4.2.2). These results agree with known models explaining enhanced oxygen mobility in ceria-zirconia-based solid solutions by the presence of free anion vacancies and/or Frenkel-type defects, distortion of

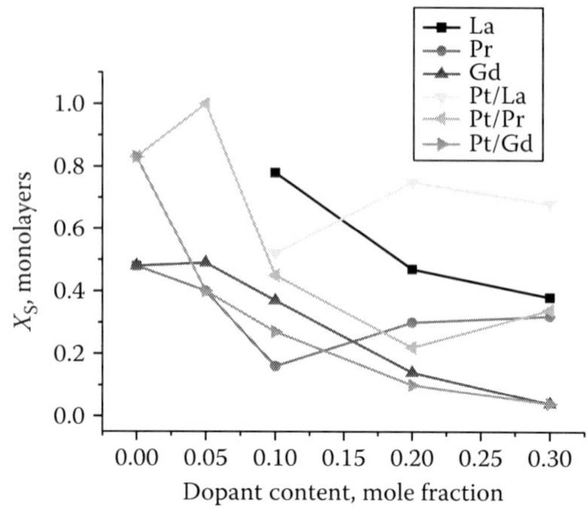

FIGURE 8.9 Dynamic degree of the oxygen heteroexchange at 650°C versus dopant content for doped ceria-zirconia samples with supported 1.4 wt.%Pt (Pt/Ln) or without it (Ln).

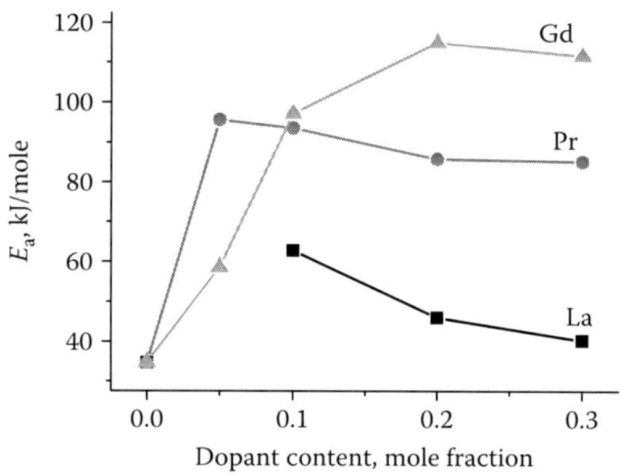

FIGURE 8.10 Activation energy (E_a) of oxygen heteroexchange for $Ln_x(Ce_{0.5}Zr_{0.5})_{1-x}O_2$ samples.

the Zr-O coordination sphere, and weakening of the Ce-O bond (Islam and Balducci 2002; Mamontov et al. 2003). Thus, a higher surface/near surface oxygen mobility in La-Ce-Zr-O samples can be explained by a weaker surface Ce-O bond, as revealed by SIMS studies (Sadykov et al. 2007b,c). This agrees with a smaller exchange activation energy for La-doped ceria-zirconia samples (Figure 8.10). It has been generally observed that E_a of oxygen exchange for samples without Pt decreases with increasing the size of the doping cation and is controlled by the energy barrier for the oxygen migration, which declines with the increase of the fluorite oxide cell size (Islam and Balducci 2002). The increase of X_S at a high Pr content can be explained by increasing the Pr^{4+} share (see Figure 8.9), which favors mixed ionic-electronic conductivity through chains of Pr^{3+}-Pr^{4+} cations, and, hence, oxygen mobility, perhaps, along domain boundaries enriched with Pr cations (Sadykov et al. 2007a–e).

Supporting Pt on doped ceria-zirconia oxides affects the surface/near surface oxygen mobility both through incorporation of Pt^{2+} cations into the surface layer/domain boundaries and by facilitating O_2 molecules dissociation (Sadykov et al. 2006b, 2007b, 2010b; Sadovskaya et al. 2007). A higher surface mobility for La-doped samples is thus explained by a higher ability of more basic La cations to stabilize Pt cations. For samples with a high content of doping cations, the exchange activation energies on Pt-supported samples are practically identical, suggesting the decisive role of the incorporated Pt cations on the activation of oxygen molecules.

A more detailed estimation of the parameters characterizing different stages of oxygen exchange and diffusion is possible by exchange experiments in the steady state isotopic transient kinetic analysis mode and fitting the experimental curves of ^{18}O fraction and fraction of asymmetric $^{18}O^{16}O$ molecules variation with time (Sadovskaya et al. 2007; Sadykov et al. 2011e). Experimental data were satisfactorily described only by taking into account a fast oxygen diffusion along domain boundaries. Moreover, for Pt-supported ceria-zirconia samples, local defects due to Pt incorporation into domain boundaries/subsurface layers of the support also provide a fast path for oxygen diffusion. Comparable rates of surface steps are observed for Pt-supported samples doped with different cations (Table 8.3). However, a much faster oxygen diffusion in the bulk, both within nanodomains and along domain boundaries, is observed for Pr-doped ceria-zirconia, which is expected to be due to positive effects of Pr^{3+}/Pr^{4+} redox pairs on oxygen migration. A partial substitution of Pr for Sm did not affect oxygen mobility in the bulk. The same lattice mobility was estimated for fluorite-like oxides with supported $LaNiO_3$ for the oxide support. Faster surface steps for Pt-supported samples are clearly explained by a high efficiency of Pt in activation of O_2 molecules.

TABLE 8.3
Parameters of Oxygen Heteroexchange and Diffusion for Fluorites-Based Systems

Sample	R_{oxide} (s^{-1})	$R_{Pt/Ni}$ (s^{-1})	R_{sp} (s^{-1})	D_{eff} (s^{-1})	D_{bulk} 10^{-18} m^2/s	$D_{boundaries}$ 10^{-16} m^2/s
Pt/PrCeZrO, $^{18}O_2$	0.25	25	≥60	0.04	4	33
$C^{18}O_2$	7			0.18		2
$LaNiO_3$/	0.28	3	>8	>0.03	3	>25
PrSmCeZrO,$^{18}O_2$				0.008		>5
$C^{18}O_2$						
Pt/LaCeZrO	0.3	10	≥100	Bulk 0.004 Subsurf 0.04	0.4	Bulk 0.45 Subsurf 0.7

Notes: R_{oxide}, $R_{Pt/Ni}$, specific rate of exchange on the oxide sites and metal sites, respectively; R_{sp}, the rate of oxygen spillover from metal to support; D_{eff}, average oxygen diffusion coefficient.

The oxygen isotope heteroexchange experiments using oxygen-labeled $C^{18}O_2$ molecules allow to estimate more precisely the oxygen diffusion parameters due to a much higher rate of the surface reaction. An additional advantage of this technique is that it allows to characterize the oxygen mobility not only for oxidized but also for reduced catalysts, especially after contacting with the reaction feed at high temperatures. Thus, for Pt or $LaNiO_3$-supported samples, the oxygen heteroexchange with $C^{18}O_2$ molecules revealed that after reaching the steady state in the reaction of CH_4 dry reforming in feed 20% CH_4 + 20% CO_2 in He at 700°C, the oxygen diffusion coefficients along domain boundaries somewhat decline but remain high (Table 8.3). This supports the conclusion about the importance of this fast oxygen migration from support domains (on the surface sites where CO_2 molecules are activated) to the metal particles (where CH_4 molecules are activated), in providing high activity and coking stability of these catalysts (Sadykov et al. 2010b, 2011b,e).

The value of X_S exceeds the monolayer for perovskites and cathode nanocomposites even in the intermediate temperature range, due to a much faster bulk oxygen diffusion (Figure 8.11). X_S increases with the sintering temperature, generally exceeding the levels found for separate P or F phases for nanocomposite samples annealed at high T_{sint} (Sadykov et al. 2010c; Kharlamova et al. 2011). Moreover, the specific rate of oxygen heteroexchange characterizing the ability of surface sites to dissociate O_2 molecules also increases with the composite sintering temperature T_{sint}, achieving a level exceeding that for pure perovskites. Since the specific rate of oxygen exchange on ScCeSZ or GDC is lower by two orders of magnitude (Sadykov et al. 2010c), this implies that transfer of transition metal cations onto the surface of electrolyte domains as well as perovskite–fluorite interfaces create sites possessing very high activity in O_2 dissociation.

A satisfactory description of SSITKA results was obtained for GDC-$LSFN_{0.3}$ nanocomposite only in frames of a complex model, suggesting a very fast oxygen exchange between the surface and perovskite–fluorite interfaces in the bulk of

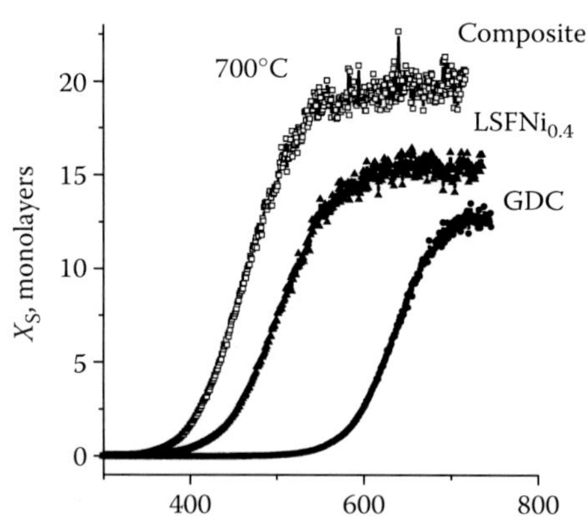

FIGURE 8.11 Dynamic degree of oxygen isotope exchange for samples of $La_{0.8}Sr_{0.2}Fe_{0.6}Ni_{0.4}O_3$ ($LSFN_{0.4}$), $Ce_{0.9}Gd_{0.1}O_{1.95}$ (GDC), and their nanocomposite sintered at 700°C.

TABLE 8.4
The Rate of Surface Oxygen Heteroexchange R^2 and Coefficients of Oxygen Self-Diffusion in the Fluorite Phase D_F, Perovskite Phase D_P, and along Interfaces $D_{interface}$ for LSFN-GDC Nanocomposite

T (°C)	R^2 (Molecules/m²s)	D_F (cm²/s)	D_P (cm²/s)	$D_{interface}$ (cm²/s)
600	0.5×10^{19}	$\geq 6 \times 10^{-14}$	2×10^{-14}	$\geq 1 \times 10^{-8}$
700	1.3×10^{19}	$\geq 30 \times 10^{-14}$	9×10^{-14}	$\geq 5 \times 10^{-8}$

composite particles (Table 8.4). Note that the coefficient of oxygen self-diffusion along perovskite–fluorite interfaces greatly exceeds that for the bulk of perovskite and fluorite domains, and even along domain boundaries of doped ceria-zirconia oxides (compare Table 8.3).

8.4.5 Catalytic Activity

8.4.5.1 Methane Transformation into Syngas

The catalytic activity of Pt-supported doped ceria-zirconia solid solutions in selective methane oxidation/dry reforming/steam reforming in diluted feeds was shown to correlate with the concentration of Pt^{2+} species (and, hence, its dispersion controlled by Pt–support interactions at their interface) within series with the same dopant (Sadykov et al. 2006b, 2007b,c, 2010b). This is explained by a high efficiency for CH_4 molecules activation on dispersed oxidized Pt sites, yielding H_2 as a primary product even in the presence of adsorbed oxygen species (Sadykov et al. 2011e). The difference in activity between series with different dopants changes in the order La > Pr > Gd, correlating with the surface/near-surface oxygen mobility (Sadykov et al. 2006b, 2007c–e, 2010b).

This order of activity is retained in concentrated feeds at low temperatures as well (see, i.e., Figure 8.12 for methane dry reforming). However, at higher temperatures, the highest activity is found for the Pr-doped system, while the activity of La-doped system decreases due to coking. The same trend was observed for partial oxidation of methane (Sadykov et al. 2011e). For active components composed of Pt-(LaNiPt)-promoted fluorite-like oxides supported on structured ceramic/refractory metal substrates, a high and stable performance in partial oxidation of methane into syngas at short contact times was observed in pilot-scale reactors (Sadykov et al. 2009a). Hence, in realistic feeds with a high CH_4 content, a high bulk oxygen mobility is required to prevent deactivation due to coking.

Taking into account the microstructural features of nanocrystalline doped ceria-zirconia samples, a high bulk oxygen mobility is provided by developed domain boundaries with a specific chemical composition structure. Actually, the nature of the supported metal is of importance as well; for instance, for the same fluorite-like support, the highest activity in dry reforming of methane is

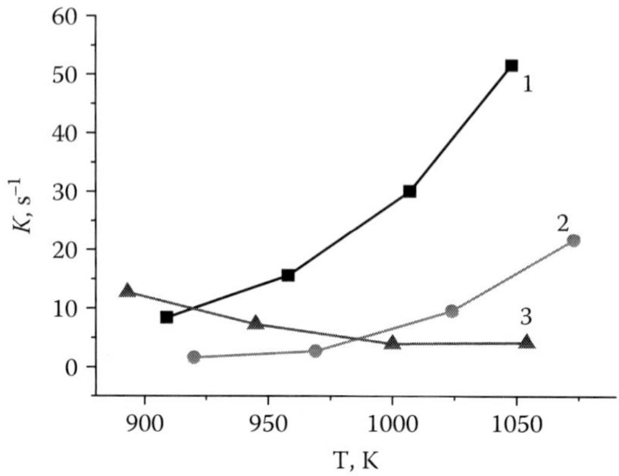

FIGURE 8.12 Rate constants of CH_4 dry reforming over corundum channels with supported 1.4%Pt/LnCeZrO active components doped by Pr (1), Gd (2), and La (3). Feed: 7% CH_4 + 7% CO_2 in He.

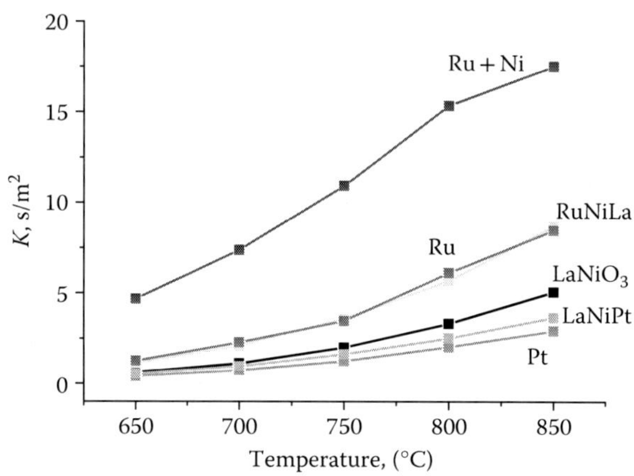

FIGURE 8.13 Temperature dependence of specific first-order rate constant for Ru, Pt, LaNi, and Ni-supported $SmPrCeZrO_2$ catalysts. Feed 10%CH_4 + 10%CO_2 in He, contact time 15 ms.

achieved by supporting Ru + Ni, due to the formation of mixed Ru + Ni clusters strongly interacting with the support, which are highly efficient in CH_4 activation (Figure 8.13).

For anode nanocomposites containing Ni + YSZ and perovskite or fluorite oxide promoters and supported noble metals, a stable performance in CH_4 steam reforming was ensured even in feeds with a small (if any) excess of steam, while unpromoted cermets were either inactive or rapidly deactivated due to coking (Sadykov et al. 2008b, 2009c, 2010c, 2011c; Smirnova et al. 2011). For samples without noble metals, but with the same oxide promoter, the specific rate constant increases with the Ni content in stoichiometric feed (#1 > #4 > #5, Figure 8.14), a fact that can be explained by the increase of the accessible surface area of the active component –Ni

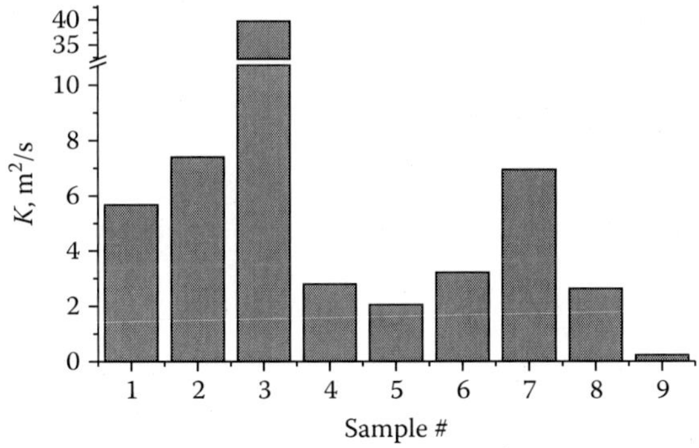

FIGURE 8.14 Specific effective first-order rate constants for samples of anode nanocomposites at 650°C and 10 ms contact time. 8% CH_4 in the feed, $H_2O/CH_4 = 1$. (1) 10%LaMnCrPr + NiO + YSZ; (2) 0.3%Pd/10%LaMnCrPr + NiO + YSZ; (3) 1%Ru/10%LaMnCrPr + NiO + YSZ; (4) 50%LaMnPrCr + 30%NiO + 20% YSZ; (5) 80%LaMnPrCr + 10%NiO + 10% YSZ; (6) 1%Ru + 80%LaMnPrCr + 10%NiO+10% YSZ; (7) 1%Ru + 50%LaMnPrCr + 30%NiO + 20% YSZ; (8) 1.4%Pt 10% SmPrCeZr + 55%NiO + 35%YSZ; (9) 1.4%Pt 80% SmPrCeZr + 10% NiO + 10% YSZ (LaMnCrPr: $La_{0.8}Pr_{0.2}Mn_{0.2}Cr_{0.8}O_3$; SmPrCeZr: $Sm_{0.15}Pr_{0.15}Ce_{0.35}Zr_{0.35}O_2$).

metal-decorated by oxidic species from the support. Activation of H_2O molecules on Ni atoms can be important as well to prevent coking and to ensure a high and stable performance. Promotion of composites with the same content of Ni by noble metals increases the specific activity in stoichiometric feeds: Ru (#3) > Pd (#2) > #1 (Figure 8.14). Combination of Ru with LaPrMnCr clearly provides the highest performance. Specific activity of Ru or Pt-promoted samples increases with the Ni content (Figure 8.14), thus clearly demonstrating synergy of Ni + Ru (Pt) action due to suppression of coking ability of Ni sites, explained by formation of surface alloys. This conclusion agrees with a high efficiency of Ru-LaMnCrPr copromoted NiO/YSZ samples in dissociation of CH_4 molecules even in a completely reduced state, implying generation of loose carbonaceous species unable to block efficiently the Ni surface sites (Sadykov et al. 2011c).

For Ru-LaMnCrPr copromoted Ni/YSZ supported on different types of planar heat-conducting substrates (compressed Ni-Al alloy foam, Fechraloy foil, etc.), a 50% CH_4 conversion was achieved at short contact times (~10 ms) and ~650°C in stoichiometric steam/methane feeds. Middle-term testing of these catalysts revealed an acceptable stability with time-on-stream (Sadykov et al. 2010c, 2011c). For a stack composed of 12 Ni-Al-foam plates and 11 sheets of FeCrAlloy gauzes loaded with $La_{0.8}Pr_{0.2}Mn_{0.2}Cr_{0.8}O_3$ + NiO + YSZ + Ru active component, a high (>45%) concentration of hydrogen in the effluent was ensured at 650–680°C, the CO concentration being approximately 8%.

8.4.5.2 Oxygenates Transformation into Syngas

The catalytic activity of supported Pt or Ru for fuels such as ethanol and acetone was shown to be mainly determined by the oxygen mobility of fluorite-like oxide supports and promoters required to prevent coking, even for diluted feeds (Sadykov et al. 2007c,

2010c, 2011c). For concentrated feeds, the best results were obtained for LaNiPt-supported samples, which provide a high activity in feeds within a broad range of steam/fuel ratios (Figures 8.15 and 8.16). In the autothermal reforming of ethanol, a high and stable performance was demonstrated by $La_{0.8}Pr_{0.2}Mn_{0.2}Cr_{0.8}O_3$ + NiO + YSZ + Ru active component supported on heat-conducting substrates (Figure 8.17). The increase of steam excess in the feed helps to increase the hydrogen yield by increasing ethanol conversion as well as by decreasing the content of by-products such as methane and ethylene.

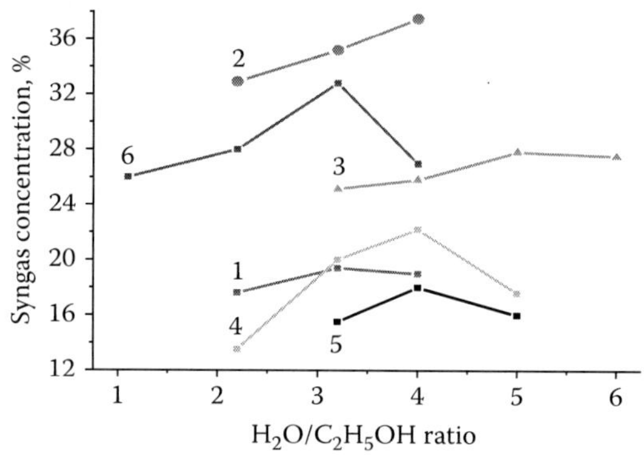

FIGURE 8.15 A sum of CO + H_2 concentration in the effluent from ethanol SR over small pieces of monolithic catalysts on FeCrAlloy substrates. (1) 5.9 wt.% Pr-Ce-Zr-O + 0.92% Pt/FeCrAl gauze; (2) 5.9 wt.% Pr-Ce-Zr-O + 7% LaNi(Pt)O_3/FeCrAl gauze; (3) 6.6 wt.% Pr-Ce-Zr-O + 1.3% Pt + 0.3% Ru/FeCrAl gauze; (4) 5.5 wt.% Pr-Ce-Zr-O + 1.2% Pt/FeCrAl foil; (5) 6.3 wt.% Pr-Ce-Zr-O + 1.0% Pt/FeCrAl foil; (6) 5.4 wt.% La-Ce-Zr-O + 10 LaNi(Pt) O_3/FeCrAl gauze. Feed: 10% EtOH + H_2O in N_2, H_2O/C_2H_5OH = 3.2, 750°C, contact time 0.085 seconds.

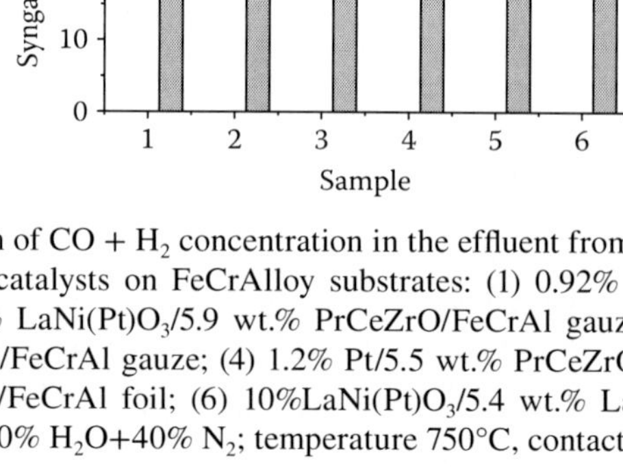

FIGURE 8.16 A sum of CO + H_2 concentration in the effluent from acetone SR over small pieces of monolithic catalysts on FeCrAlloy substrates: (1) 0.92% Pt/5.9 wt.% PrCeZrO/FeCrAl gauze; (2) 7% LaNi(Pt)O_3/5.9 wt.% PrCeZrO/FeCrAl gauze; (3) 1.3% Pt + 0.3% Ru/6.6 wt.% PrCeZrO/FeCrAl gauze; (4) 1.2% Pt/5.5 wt.% PrCeZrO/FeCrAl foil; (5) 1.0% Pt/6.3 wt.% PrCeZrO/FeCrAl foil; (6) 10%LaNi(Pt)O_3/5.4 wt.% LaCeZrO/FeCrAl gauze. Feed 10% acetone + 50% H_2O+40% N_2; temperature 750°C, contact time 0.085 seconds.

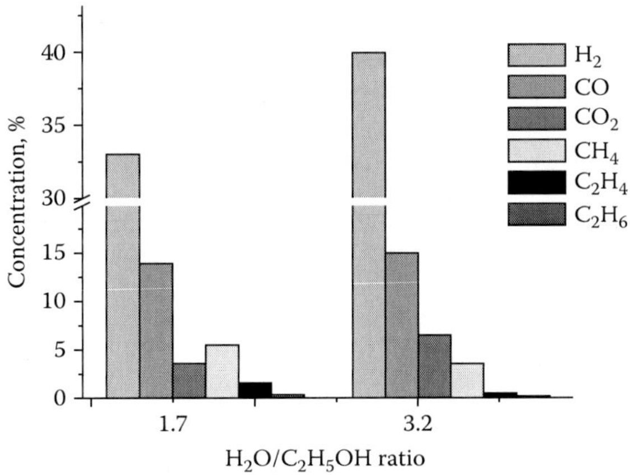

FIGURE 8.17 Concentration of products in the effluent stream of ethanol autothermal reforming on stack comprised of 12 Ni-Al-foam plates and 11 sheets of FeCrAlloy gauzes loaded with $La_{0.8}Pr_{0.2}Mn_{0.2}Cr_{0.8}O_3$ + NiO + YSZ + Ru (volume 34 × 34 × 34 mm^3). Feed composition: 5% O_2+ x% H_2O +25% EtOH + (70 – x)% N_2. Temperature 700°C, contact time 0.3 seconds.

8.5 NANOSTRUCTURE IN MESOPOROUS OXIDES

The key to making the ODH processes industrially viable is to reach good productivities while keeping high selectivity. Conceptually, one should design catalysts with only (or mostly) selective sites, but very active, that is, with the highest possible (specific) concentration of active sites to reach a good productivity. This implies to maximize the active surface area, a goal that can be reached by using nanoparticles or by using high surface area materials, such as mesoporous oxides. Among the vanadium-based catalysts for alkane ODH, the best selectivity is attributed to tetrahedral V^{5+} species, and isolated species have been found to be much more active than oligomeric or polymeric species (Valenzuela and Corberan 2000). This has driven the interest on V-MCM-41 catalysts, which faced the problem of the limited concentration of the heteroatom that can be incorporated into the silica framework when synthesized by the direct hydrothermal methods. The use of the atrane method allows to insert large vanadium contents into the framework of the mesoporous ordered silicas, up to Si/V = 49 in V-MCM-41 (Jia et al. 2004) and even to Si/V = 14 in V-UVM-7 (Huerta et al. 2006), without vanadia phase segregation and keeping the hexagonal order. These catalysts were very active and selective (50%–70% selectivity to total dehydrogenated products) in isobutene ODH. The most remarkable feature of these V-MCM-41 catalysts was the formation of methacrolein, the product of selective oxidation of the olefin, with a relatively high selectivity, with yields per pass up to 5.8%, the highest reported for such a process (Jia et al. 2004). UV-vis diffuse reflectance spectra allowed to identify at least two types of tetragonal vanadium species, isolated and low oligomeric ones. A similar reaction product distribution was obtained when a higher vanadium content could be incorporated into the framework of the mesoporous siliceous material by using the UVM-7 structure. This bimodal hierarchical material can be seen as composed by aggregation of nanoparticles having the ordered mesoporous MCM-41 structure, developing a second type of pores in the interparticle voids. Selective oxidation of isobutene on V-UVM-7 with high V contents (Si/V = 14–40) showed total selectivity to (oxi)dehydrogenation products (58%–80%), formation of methacrolein with a relatively high selectivity; both activity and methacrolein/isobutene ratio increased with the V content (Huerta et al. 2006). Direct formation of the unsaturated aldehyde from the alkane is not observed on nonporous silica-supported vanadium oxide catalysts, for which the only oxygenated products observed (acetone, formaldehyde, etc.) come from oxidative partial degradation. This points to some new functionality provided by the nanosize environment created by the mesopores, whose diameters usually range from 1.5 to 3 nm.

Actually, a similar effect has been recently reported with SBA-15-supported vanadium catalysts (Hess et al. 2006). A 3.3 wt.% V/SBA-15 catalyst, prepared by grafting, was tested for the selective oxidation of propane with water added to the feed (C_3H_8:O_2:N_2:H_2O = 1:2.2:17.9:14.1). Under the reaction conditions used (GHSV: 1200 h^{-1} STP, 400°C), propane conversion was 8%, and the main product was acrylic acid, that is, a product of the selective oxidation of the olefin, with 84% selectivity; propene was also formed (10% selectivity). Formation of the acid instead of the aldehyde might be due to the presence of water in the feed. As in V-MCM-41, two tetrahedral vanadium species were found by UV-vis diffuse reflectance.

Interestingly, methacrolein formation is also observed in the ODH of isobutene on Co-MCM-41 catalysts with Si/Co = 49 and 98 (Cortés Corberán et al. 2004). UV-vis diffuse reflectance reveals that catalysts Co-MCM-41 with Co contents up to Si/Co = 23 contain a unique tetrahedral coordination of the Co(II) centers, a type of species found to be active and selective for the ammoxidation of ethane on Co-containing zeolites with MFI and BEA structures (Li and Armor 1998). It is interesting to notice that mesoporous oxides containing two transition metal cations with so different structural environment and catalytic properties like vanadium and cobalt as active elements produce similar product distributions. This points again to the possibility that mesopores, as micropores in zeolites do, favor the formation of specific nanostructures of the oxide and their cation environment. Moreover, it should be noted that the homologous pure MCM-41 silica sample, where no "active" transition metal is present, showed some activity itself, producing also dehydrogenation products. At 500–525°C, isobutane conversion over Si-MCM-41 was half of that over Co-MCM-41 (Si/Co = 98), but much higher than that observed without any catalyst (homogenous gas phase reaction); in addition, under the conditions tested, both mesoporous systems showed the same selectivity to total dehydrogenated products at isoconversion (Cortés Corberán et al. 2004). This raised the question if mesopores play a role in the overall performance of mesoporous oxides. If they do, some influence of the pore length should be found.

To clarify this point, Huerta et al. (2006) compared the catalytic performance of vanadium-containing mesoporous catalysts with similar V contents and different pore lengths. MCM-41 and UVM-7 share the same hexagonal arrangement of parallel, straight pores, with a total length equal to the particle size in the pore axis direction. Therefore, by comparing the catalytic performance of these two types of materials with similar Si/V ratios, one may get information on the influence of pore

TABLE 8.5
ODH of Isobutane at 475°C on V-Containing Ordered Mesoporous Catalysts and Effect of the Pore Length

Catalyst	V-MCM-41	V-UVM-7
Si/V	49	40
Particle size (magnitude)	microns	Nm
W/F (g.h/mol C4)	12	7.5
Conversion (%)	10.6	9.5
Selectivity (%): Isobutene	35.1	42.6
Methacrolein	28.7	18.0

Source: Adapted from Huerta, L.J. et al., *Catal. Today*, 117, 180–186, 2006.

length, which determines the average residence time inside the pore, on the catalytic behavior. As it can be seen in Table 8.5, the activity per (total) V atom is around 20% higher for V-UVM-7 than for V-MCM-41, and the combined selectivities to dehydrogenation products (isobutene plus methacrolein) at isoconversion are similar, approximately 61%–63%. However, the methacrolein-to-isobutene ratio is almost double in V-MCM-41, where pore length is much higher. This implies that, besides the nature of the vanadium centers, the length of the mesopores also plays a role in determining the product distribution: the longer the channel, the higher the (secondary) conversion of isobutene to methacrolein. This supports the hypothesis of the contribution of heterogeneously initiated, radical gas-phase homogeneous reactions inside the void volume of the mesopores of M-MCM-41 catalysts to its catalytic activity (Cortés Corberán 2005). This effect can be considered to be a new functionality, brought by the nanoscale of the mesopores: the confinement of reactants, intermediates, and products inside the "nanoreactor" pore.

8.6 CONCLUDING REMARKS AND OUTLOOK

Oxide catalysts are intrinsically multifunctional materials whose use involves some specific issues. Application of nanotechnology to master their properties and structures brings both new challenges and opportunities for catalytic applications. Controlling the properties of oxide catalysts at the nanoscale, by a tailored design of the nanostructure opens new ways to improve the catalytic performance of known catalyst compositions as well as to discover new applicable (i.e., useful) materials and even new functionalities. Nanotechnology of materials is opening a broad new range of structures, morphologies, and physicochemical properties of oxides. It requires not only to prepare controlled nanostructures and techniques suitable for studies at the nanoscale but also to discover new materials for current applications as well as identifying new applications for the new materials prepared.

In spite of its potentiality, to date, the research on the application of nanostructured oxides for catalytic redox applications is still scarce.

A question that deserves more attention is the effect of specific (nano) sizes on the catalytic behavior. Usually the effect of the nanosize is studied in a qualitative way, by comparing "bulk" versus "nano" materials, the latter being oxide materials that show sizes within a certain, and relatively broad, dispersion around the reported average value. However, methods to synthesize particles with a narrow size distribution of a number of transition metal oxides of interest for catalytic applications (such as Cu_2O, Mn_3O_4, Co_3O_4, TiO_2, CeO_2, ZrO_2, rare-earth oxides, and some ferrites and perovskites) are already described in the literature (Park et al. 2007). Their application in catalytic studies and real use in catalytic processes are needed to fully understand the variation of the catalytic properties as a function of particle (or domain) size at the nanoscale, as it has been described for metallic (Co, Au) nanoparticle catalysts.

On the other hand, the research on nanostructured oxide materials in many other fields of application, as shown along this book, has generated an increasing interest in the preparation of oxides with a variety of nanostructures and nanomorphologies. They include a range of transition metal oxides well known by their properties as catalysts, or catalytic components, for redox reactions. Some representative examples are ZnO nanorods (Huang et al. 2001), TiO_2 nanotubes (Adachi et al. 2003), or nanotubes and nanowires of a full range of oxides (of Al, Zr, Si, Ti, Nb, V, Co, Ni, W, Ga, etc.) of interest in catalysis (Bae et al. 2008). Oxide nanotubes show promising properties as catalysts, due to their advantages: high SSA, unconventional surface topography (curved surfaces and mouth edges), and confinement within the tubes. They can be also used as precursors of "bulk" solids showing special properties. For instance, nanotubular titania can be used as a support precursor that converts into anatase nanoparticles by calcination at just 400°C (Cortes-Jácome et al. 2007).

A specific feature of oxide catalysts, at a variance of the metallic ones, is that their atoms (anions) may participate in the catalytic oxidation reactions and be inserted in the product molecules. The working state of the oxide catalyst is either in dynamic equilibrium with or irreversibly modified by the reaction mixture. Kinetics of oxidation reactions on oxide catalysts are intrinsically complex and may involve heterogeneous-homogeneous processes. Moreover, process efficiency of oxidation reactions is highly dependent on reactor and process engineering. All these features bring additional complexity to the catalytic processes using oxide catalysts. As a result, convergent approaches from many disciplines and good communication among the different research areas involved are needed to fully understand catalysis by (nanostructured) oxides and to fully exploit the huge potential of their practical application.

REFERENCES

Adachi M, Murata Y, Okada I, Yoshikawa S (2003) Formation of titania nanotubes and applications for dye-sensitized solar cells. *J Electrochem Soc* 150:G488–G493.

Akurati KK, Vital A, Fortunat G, Hany R, Nueesch F et al. (2007) Flame synthesis of TiO_2 nanoparticles with high photocatalytic activity. *Solid State Sci* 9:247–257.

Alconchel S, Sapiña F, Beltran D, Beltran A (1999) A new approach to the synthesis of molybdenum bimetallic nitrides and oxynitrides. *J Mat Chem* 9:749–755.

Anpo M, Takeuchi M (2001) Design and development of second-generation titanium oxide photocatalysts to better our environment—approaches in realizing the use of visible light. *Int J Photoenergy* 3:89–94.

Arai H, Yamada T, Eguchi K, Seiyama T (1986) Catalytic combustion of methane over various perovskite-type oxides. *Applied Catal* 26:265–276.

Armstrong G, Armstrong AR, Canales J, Bruce PG (2005b) Nanotubes with the TiO_2-B structure. *Chem Comm* May 21:2454–2456.

Armstrong AR, Armstrong G, Canales J, Garcia R, Bruce PG (2005a) Lithium-ion intercalation into TiO_2-B. *Adv Mater* 17:862–865.

Asashi R, Morikawa T, Ohwaki T, Aoki K, Taga Y (2001) Visible-light photocatalysis in nitrogen-doped titanium oxides. *Science* 293:269–271.

Atkinson A, Barnett S, Gorte R, Irvine J, Mcevoy A et al. (2004) Advanced anodes for high-temperature fuel cells. *Nat Mater* 3:17–27.

Aw MS, Simovic S, Addai-Menash J, Losic D (2011) Polymeric micelles in porous and nanotubular implants as a new system for extended delivery of poorly soluble drugs. *J Mater Chem* 21:7082–7089.

Bachiller-Baeza B, Rodriguez-Ramos I, Guerrero-Ruiz A (1998) Interaction of carbon dioxide with the surface of zirconia polymorphs. *Langmuir* 14:3360–3556.

Bae C, Yoo H, Kim S, Lee K, Kim J et al. (2008) Template-directed synthesis of oxide nanotubes: Fabrication, characterization, and applications. *Chem Mater* 20:756–767.

Cavani F, Trifirò F (1992) Some innovative aspects in the production of monomers via catalyzed oxidation processes. *Applied Catal* 88:115–135.

Cavani F, Trifirò F (1995) The oxidative dehydrogenation of ethane and propane as an alternative way for the production of light olefins. *Catal Today* 24:307–313.

Chen W, Li F, Yu J (2007) Salt-assisted combustion synthesis of highly dispersed perovskite $NdCoO_3$ nanoparticles. *Mater Lett* 61:397–400.

Cortés Corberán V (2005) Novel approaches for the improvement of selectivity in the oxidative activation of light alkanes. *Catalysis Today* 99:33–41.

Cortés Corberán V, Conesa Cegarra JC (2006). *The CONCORDE NSOCRA White Paper: A Technology Roadmap on Nanostructured Oxide Catalysts for Redox Applications*. ISBN 84-933135-4-8. CONCORDE Consortium, Granada.

Cortés Corberán V, Jia JM, El-Haskouri J, Valenzuela RX, Beltrán-Porter D et al. (2004) Oxidative dehydrogenation of isobutane over Co-MCM-41 catalysts. *Catal Today* 91–92:127.

Cortes-Jácome MA, Morales M, Angeles Chavez C, Ramírez-Verduzco LF, López-Salinas E et al. (2007) WO_x/TiO_2 catalysts via titania nanotubes for the oxidation of benzothiophene. *Chem Mater* 19:6605–6614.

Costa LL, Prado AGS (2009) TiO_2 nanotubes as recyclable catalyst for efficient photocatalytic degradation of indigo carmine dye. *J Photochem Photobiol A: Chemistry* 201:45.

Davies TE, García T, Solsona B, Taylor SH (2006) Nanocrystalline cobalt oxide: a catalyst for selective alkane oxidation under ambient conditions. *Chem Commun* 32:3417–3419.

De Souza Sikora M, Viana Rosario A, Chaves Pereira E, Paiva-Santos CP (2011) Influence of the morphology and microstructure on the photocatalytic properties of titanium oxide films obtained by sparking anodization in H_3PO_4. *Electrochim Acta* 56:3122–3127.

Deng S, Li HQ, Li SG, Zhang Y (2007). Activity and characterization of modified Cr_2O_3/ZrO_2 nano-composite catalysts for oxidative dehydrogenation of ethane to ethylene with CO_2. *J Mol Catal A: Chem* 268:169–175.

Deng S, Li HQ, Zhang Y (2003) Oxidative dehydrogenation of ethane with carbon dioxide to ethylene over nanosized Cr_2O_3 catalysts. *Chinese J Catal* 24:744–755.

Deshpande A, Madras G, Madras NM (2011) Role of lattice defects and crystallite morphology in the UV and visible-light-induced photo-catalytic properties of combustion-prepared TiO_2. *Mater Chem Phys* 126:546–554.

Diamanti MV, Ormeselle M, Marin E, Lanzutti A, Mele A et al. (2011) Anodic titanium oxide as immobilized photocatalyst in UV or visible light devices. *J Hazardous Mater* 186:2103–2109.

Dutoit DCM, Göbel U, Schneider M, Baiker A (1996) Titania-silica mixed oxides: V. Effect of sol-gel and drying conditions on surface properties. *J Catal* 164:433–439.

Efremov D, Pinaeva L, Sadykov VA, Mirodatos C (2008) Original Monte Carlo method for analysis of oxygen mobility in complex oxides by simulation of oxygen isotope exchange data. *Solid State Ionics* 179:847–850.

Fen LB, Han TK, Nee NM, Ang BC, Johan MR (2011) Physico-chemical properties of titania nanotubes synthesized via hydrothermal and annealing treatment. *Appl Surface Sci* 258:431–435.

Ghows N, Enterazi MH (2010) Ultrasound with low intensity assisted the synthesis of nanocrystalline TiO_2 without calcinations. *Ultrasonics Sonochem* 17:878–883.

González A, Martínez Tamayo E, Beltrán Porter A, Cortés Corberán V (1997) Synthesis of high surface area perovskite catalysts by non-conventional routes. *Catal Today* 33:361–369.

González-Reyes L, Hernández-Pérez I, Díax-Barriga Arceo L, Dorantes-Rosales H, Arce-Estrada E et al. (2010) Temperature effects during Ostwald ripening on structural and bandgap properties of TiO_2 nanoparticles prepared by sonochemical synthesis. *Mater Sci Eng B* 175:9–13.

Guo C, Xu J, He Y, Zhang Y, Wang Y (2011) Photodegradation of rhodamine B and methyl orange over one-dimensional TiO_2 catalysts under simulated solar irradiation. *Appl Surface Sci* 257:3798.

He YM, Wu Y, Chen T, Weng WZ, Wan HL (2006) Low-temperature catalytic performance for oxidative dehydrogenation of propane on nanosized Ti(Zr)-Ni-O prepared by modified sol-gel method. *Catal Commun* 7:268–271.

Herrmann J-M (2012) Detrimental cationic doping of titania in photocatalysis: why chromium Cr^{3+}-doping is a catastrophe for photocatalysis, both under UV- and visible irradiations. *New J Chem* 36:883–890.

Hess C, Looi MH, Hamid SBA, Schlögl R (2006) Importance of nanostructured vanadia for selective oxidation of propane to acrylic acid. *Chem Commun* Jan 28:451–453.

Huang MH, Mao S, Feick H, Yan HQ, Wu YY et al. (2001) Room-temperature ultraviolet nanowire nanolasers. *Science* 292:1897–1899.

Huerta LJ, Amorós P, Beltrán-Porter D, Corberán VC (2006) Selective oxidative activation of isobutane on a novel vanadium-substituted bimodal mesoporous oxide V-UVM-7. *Catal Today* 117:180–186.

IEA (2013) Technology roadmap energy and GHG reductions in the chemical industry via catalytic processes. Available at http://www.iea.org/publications/freepublications/publication/Chemical_Roadmap_2013_Final_WEB.pdf.

In S, Orlov A, García F, Tikhov M, Wright DS, Lambert RM (2006) Efficient visible light-active N-doped TiO_2 photocatalysts by a reproducible and controllable synthetic route. *Chem Comm* Oct 28:4236–4238.

Islam MS, Balducci G (2002) Computer simulation studies of ceria-based oxides. In: Trovarelli A (ed) *Catalysis by Ceria and Related Materials*. Imperial College Press, London, UK.

Jia MJ, Valenzuela RX, Amorós P, Beltrán-Porter D, El-Haskouri J et al. (2004) Direct oxidation of isobutane to methacrolein over V-MCM-41 catalysts. *Catal Today* 91–92:43–47.

Johnson DW, Gallagher PK, Schrey F, Rhodes WW (1976) Preparation of high surface area substituted $LaMnO_3$ catalysts. *Am Ceram Soc Bull* 55:520–523, 527.

Kamalov S, Abidov A, Allaberganov BB, Jo SJ, Lee EY et al. (2012) Fabrication and characterization of ordered microsized tubular TiO_2 films by using various anodizing conditions. *Res Chem Intermed* 38:1007–1013.

Kharlamova T, Pavlova S, Sadykov VA, Bespalko Y, Krieger T et al. (2011) Nanocomposite cathode materials for intermediate temperature solid oxide fuel cells. *ECS Transactions* 35:2331–2340.

Kharlamova T, Pavlova S, Sadykov VA, Kriger T, Alikina G et al. (2008a) Perovskite and composite materials for intermediate temperature solid oxide fuel cells. *Mater Res Soc Symp Proc* 1056:HH03-64.1-6.

Kharlamova T, Smirnova A, Sadykov VA, Zarubina V, Krieger T et al. (2008b) Intermediate temperature solid oxide fuel cells based on nano-composite cathode structures. *ECS Transactions* 13:275–284.

Krylov OV, Mamedov AKh, Mirzabekova SR (1995). The regularities in the interaction of alkanes with CO_2 on oxide catalysts. *Catal Today* 24:371–375.

Lee GW, Byeon JH (2009) Effects of ultrasonic processing on phase transition of flame-synthesized anatase TiO_2 nanoparticles. *Mater Characterization* 60:1476–1481.

Lee YN, El-Fadli Z, Sapiña F, Martinez-Tamayo E, Corberán VC (1999) Synthesis and surface characterization of nanometric $La_{1-x}K_xMnO_{3+\delta}$ particles. *Catal Today* 52:45–52.

Lee YN, Sapiña F, Martinez E, Folgado JV, Corberán VC (1997) Catalytic combustion of ethane over high surface area $Ln_{1-x}K_xMnO_{3+\delta}$ (Ln = La, Nd) perovskites: the effect of potassium substitution. *Stud Surf Sci Catal* 110:747–756.

Li C, Hu Y, Yuan W (2010) Nanomaterials synthesized by gas combustion flames: morphology and structure. *Particuology* 8:556–562.

Li YJ, Armor JN (1998) Ammoxidation of ethane to acetonitrile over metal–zeolite catalysts. *J Catal* 173:511–518.

Liu G, Yang HG, Wang X, Cheng L, Pan J et al. (2009a) Visible light responsive nitrogen doped anatase TiO_2 sheets with dominant {001} facets derived from TiN. *J Am Chem Soc* 131:12868–12869.

Liu R, Hsieh CS, Yang WD, Qiang LSS, Wu JF (2011) Applying the statistical experimental method to evaluate the process conditions of TiO_2 nanotube arrays by anodization method. *Current Appl Phys* 11:1294–1298.

Liu Q, Wang LC, Chen M, Cao Y, He HY et al. (2009b) Dry citrate-precursor synthesized nanocrystalline cobalt oxide as highly active catalyst for total oxidation of propane. *J Catal* 263:104–113.

Liu Z, Zhang X, Mishimoto S, Jin M, Tryk DA et al. (2008) Highly ordered TiO_2 nanotube arrays with controllable length for photoelectrocatalytic degradation of phenol. *J Phys Chem C* 112:253–259.

Livage J (1994) *Encyclopaedia of Inorganic Chemistry*. John Wiley & Sons, New York.

Madeira LM, Portela MF, Mazzocchia C (2004) Nickel molybdate catalysts and their use in the selective oxidation of hydrocarbons. *Catal Rev Sci Eng* 46:53–110.

Mamedov EA, Cortés Corberán V (1995) Oxidative dehydrogenation of lower alkanes on vanadium oxide-based catalysts: the present state-of-the-art and outlooks. *Applied Catal* 127:1–40.

Mamontov E, Brezny R, Koranne M, Egami T (2003) Nanoscale heterogeneities and oxygen storage capacity of $Ce_{0.5}Zr_{0.5}O_2$. *J Phys Chem B* 107:13007–130014.

Mamontov E, Egami T, Brezny R, Koranne M, Tyagi S (2000) Lattice defects and oxygen storage capacity of nanocrystalline ceria and ceria-zirconia. *J Phys Chem B* 104:11110–11117.

Mars P, van Krevelen DW (1954) Oxidations carried out by means of vanadium oxide catalysts. *Spec Suppl te Chem Eng Sci* 3:41.

Memarzadeh S, Tolmachoff ED, Phares DJ, Wang H (2011) Properties of nanocrystalline TiO_2 synthesized in premixed flames stabilized on a rotating surface. *Proc Combustion Inst* 33:1917–1924.

Mu Q, Li Y, Zhang Q, Wang H (2011) Template-free formation of vertically oriented TiO_2 nanorods with uniform distribution for organics-sensing application. *J Hazardous Mater* 188:363–368.

Mu R, Xu Z, Li L, Shao Y, Wan H et al. (2010) On the photocatalytic properties of elongated TiO_2 nanoparticles for phenol degradation and Cr(VI) reduction. *J Hazardous Mater* 176:495.

Navas J, Sapiña F, Martínez E, Corberan VC (2007) Nanometric CeO_2–ZrO_2 catalysts for oxidehydrogenation of ethane with CO_2. *Proceedings of the 8th Natural Gas Conversion Symposium (8 NGCS)*, IBP, Brazil 2007, O-34.

Ng J, Wang X, Sun DD (2011) One-pot hydrothermal synthesis of a hierarchical nanofungus-like anatase TiO_2 thin film for photocatalytic oxidation of bisphenol A. *Appl Catal B: Environmental* 110:260–272.

Orlov A, Chan MS, Jefferson DA, Zhou D, Lynch RJ et al. (2006) Photocatalytic degradation of water-soluble organic pollulants on TiO_2 modified with gold nanoparticles. *Environ Technol* 27:747–752.

Oyama ST, Yu CC, Ramanathan S (1999) Transition metal bimetallic oxycarbides: synthesis, characterization, and activity studies. *J Catal* 184:535–549.

Park J, Joo J, Kwon SG, Jang Y, Hyeon T (2007) Synthesis of monodisperse spherical nanocrystals. *Angew Chem Int Ed* 47:4630–4660.

Park JH, Lee TW, Kang MG (2008) Growth, detachment and transfer of highly-ordered TiO_2 nanotube arrays: use in dye-sensitized solar cells. *Chem Comm* Jul 7:2867–2869.

Patzke GR, Zhou Y, Kontic R, Conrad F (2011) Oxide nanomaterials: synthetic developments, mechanistic studies and technological innovations. *Angew Chem Int Ed* 50:826–859.

Pavlova S, Sazonova NN, Sadykov VA, Pokrovskaya S, Kuzmin V et al. (2005) Partial oxidation of methane to synthesis gas over corundum supported mixed oxides: one channel studies. *Catal Today* 105:367–371.

Pavlova SN, Sazonova NN, Sadykov VA, Alikina GM, Lukashevich AI et al. (2007) Study of synthesis gas production over structured catalysts based on $LaNi(Pt)O_x$- and Pt(LaPt)-CeO_2-ZrO_2 supported on corundum. *Stud Surf Sci Catal* 167: 343–348.

Rahiminezhad-Soltani M, Saberyan K, Shahri F, Simchi A (2011) Formation mechanism of TiO_2 nanoparticles in H_2O-assisted atmospheric pressure CVD process. *Powder Technol* 209:15–24.

Rao CNR, Gopalakrishnan J (1987) Synthesis of complex metal oxides by novel routes. *Acc Chem Res* 20:228–235.

Rao CNR, Gopalakrishnan J (1997) *New Directions in Solid State Chemistry*, 2nd edn. Cambridge University Press, Cambridge, UK.

Rao CNR, Gopalakrishnan J, Vidyasagar K, Ganguli AK, Ramanan A et al. (1986) Novel metal oxide prepared by ingenious synthetic routes. *J Mat Res* 1:280–294.

Reddy BM, Han, DS, Jiang NZ, Park SE (2008) Dehydrogenation of ethylbenzene to styrene with carbon dioxide over ZrO2-based composite oxide catalysts. *Catal Surv Asia* 12:56–69.

Reyes de la Torre A, Banda JAM, Alamilla RG, Saldoval-Robles G, Torres Rojas E et al. (2004) Synthesis of supported and unsupported NiMo carbides and their properties for the catalytic hydrocracking of n-octane. *J Phys Condensed Matter* 16:S2329–S2334.

Sadovskaya EM, Ivanova YA, Pinaeva LG, Grasso G, Kuznestova TG et al. (2007) Kinetics of oxygen exchange over CeO_2-ZrO_2 fluorite-based catalysts. *J Phys Chem A* 111:4498–4505.

Sadykov VA, Alikina GM, Lukashevich A, Muzykantov VS, Usoltsev V et al. (2011a) Design and characterization of LSM/ScCeSZ nanocomposite as mixed ionic-electronic conducting material for functionally graded cathodes of solid oxide fuel cells. *Solid State Ionics* 192:540–546.

Sadykov VA, Bobrova L, Pavlova S, Simagina V, Makarshin L et al. (2009a) Syngas generation from hydrocarbons and oxygenates with structured catalysts. In: Kurucz A, Bencik I (eds) *Syngas: Production Methods, Post Treatment and Economics*. Nova Science Publishers, Inc., Hauppauge, NY.

Sadykov VA, Frolova YV, Mezentseva NV, Alikina GM, Lukashevich AI et al. (2006a) Nanocrystalline catalysts based on CeO_2-ZrO_2 doped by praseodymium or gadolinium: synthesis and properties. *Mater Res Soc Symp Proc* 900E:O10.04 1-6.

Sadykov VA, Gubanova EL, Sazonova NN, Pokrovskaya SA, Chumakova NA et al. (2011b) Dry reforming of methane over Pt/PrCeZrO catalyst: kinetic and mechanistic features by transient studies and their modelling. *Catal Today* 171:140–149.

Sadykov VA, Kharlamova T, Batuev L, Mezentseva N, Alikina G et al. (2008a) Design and characterization of nanocomposites based on complex perovskites and doped ceria as advanced materials for solid oxide fuel cell cathodes and membranes. *Mater Res Soc Symp Proc* 1098:HH07-06.1-6.

Sadykov VA, Kriventsov VV, Moroz EM, Borchert YV, Zyuzin DA et al. (2007a) Ceria-zirconia nanoparticles doped with La or Gd: effect of the doping cation on the real structure. *Solid State Phenomena* 128:81–88.

Sadykov VA, Kuznetsova TG, Alikina GM, Frolova YV, Lukashevich AI et al. (2007b) Ceria-based fluorite-like oxide solid solutions promoted by precious metals as catalysts of methane transformation into syngas. In: McReynolds DK (ed) *New Topics in Catalysis Research*. Nova Science Publishers, Inc., Hauppauge, NY.

Sadykov VA, Kuznetsova TG, Frolova YV, Alikina GM, Lukashevich AI et al. (2006b) Fuel-rich methane combustion: role of the Pt dispersion and oxygen mobility in a fluorite-like complex oxide support. *Catal Today* 117:475–483.

Sadykov VA, Mezentseva NV, Alikina GM, Bunina RV, Pelipenko VV et al. (2009d) Nanocomposite catalysts for internal steam reforming of methane and biofuels in solid oxide fuel cells: design and performance. *Catal Today* 146:132–140.

Sadykov VA, Mezentseva NV, Alikina GM, Bunina RV, Pelipenko VV et al. (2011c) Nanocomposite catalysts for steam reforming of methane and biofuels: design and performance. In: Reddy B (ed) *Advances in Nanocomposites: Synthesis, Characterization and Industrial Applications*. INTECH, Vienna, Austria.

Sadykov VA, Mezentseva NV, Alikina GM, Bunina RV, Rogov V et al. (2009c) Composite catalytic materials for steam reforming of methane and oxygenates: combinatorial synthesis, characterization and performance. *Catal Today* 145:127–137.

Sadykov VA, Mezentseva NV, Alikina GM, Lukashevich AI, Borchert YV et al. (2007c) Pt-supported nanocrystalline ceria-zirconia doped with La, Pr or Gd: factors controlling syngas generation in partial oxidation/autothermal reforming of methane or oxygenates. *Solid State Phenomena* 128: 239–248.

Sadykov VA, Mezentseva NV, Alikina GM, Lukashevich AI, Muzykantov VS et al. (2007d) Nanocrystalline doped ceria-zirconia fluorite-like solid solutions promoted by Pt: structure, surface properties and catalytic performance in syngas generation. *Mater Res Soc Symp Proc* 988: QQ06-04.1-6.

Sadykov VA, Mezentseva NV, Alikina GM, Lukashevich AI, Muzykantov VS et al. (2007e) Doped nanocrystalline Pt-promoted ceria-zirconia as anode catalysts for IT SOFC: synthesis and properties. *Mater Res Soc Symp Proc* 1023:JJ02–JJ07.

Sadykov VA, Mezentseva NV, Bunina RV, Alikina GM, Lukashevich AI et al. (2010a) Design of anode materials for IT SOFC: effect of complex oxide promoters and Pt group metals on activity and stability in methane steam reforming of Ni/YSZ (ScSZ) cermets. *J Fuel Cell Sci Technol* 7:011005. 1–6.

Sadykov VA, Mezentseva NV, Bunina RV, Alikina GM, Lukashevich AI et al. (2008b) Effect of complex oxide promoters and Pd on activity and stability of Ni/YSZ (ScSZ) cermets as anode materials for IT SOFC. *Catal Today* 131:226–237.

Sadykov VA, Mezentseva NV, Bunina RV, Pelipenko V, Bobrova L et al. (2008c) Anode materials for IT SOFC based on NiO/YSZ doped with complex oxides and promoted by Pt, Ru or Pd: properties and catalytic activity in the steam reforming of CH_4. *Proc 8th European SOFC Forum*, A0526.1-6 (CD), Lucerne, Switzerland.

Sadykov VA, Mezentseva NV, Muzykantov VS, Efremov D, Gubanova E et al. (2009b) Real structure–oxygen mobility relationship in nanocrystalline doped ceria-zirconia fluorite-like solid solutions promoted by Pt. *Mater Res Soc Symp Proc* 1122:O05-03.1-6.

Sadykov VA, Mezentseva NV, Usoltsev VV, Kharlamova TS, Pavlova SN et al. (2012) Planar thin film solid oxide fuel cells for intermediate temperature operation (IT SOFC): design and performance. In: Liu Z (ed) *Fuel Cell Performance*. Nova Science Publishers, Inc., Hauppauge, NY.

Sadykov VA, Mezentseva NV, Usoltsev V, Sadovskaya E, Ishchenko A et al. (2011d) Solid oxide fuel cell composite cathodes based on perovskite and fluorite structures. *J Power Sources* 196:7104–7109.

Sadykov VA, Muzykantov VS, Bobin A, Batuev L, Alikina GM et al. (2009e) Design and characterization of LSM-ScCeSZ nanocomposite as MIEC material for SOFC cathodes and oxygen-separation membranes. *Mater Res Soc Symp Proc* 1126:S13-03.1-6.

Sadykov VA, Muzykantov VS, Bobin AS, Mezentseva NV, Alikina GM et al. (2010b) Oxygen mobility of Pt-promoted doped CeO_2–ZrO_2 solid solutions: characterization and effect on catalytic performance in syngas generation by fuels oxidation/reforming. *Catal Today* 157:55–60.

Sadykov VA, Pavlova S, Kharlamova T, Muzykantov VS, Uvarov N et al. (2010c) Perovskites and their nanocomposites with fluorite-like oxides as materials for solid oxide fuel cells cathodes and oxygen-conducting membranes: mobility and reactivity of the surface/bulk oxygen as a key factor of their performance. In: Borowski M (ed) *Perovskites: Structure, Properties and Uses*. Nova Science Publishers, Inc., Hauppauge, NY.

Sadykov VA, Sazonova NN, Bobin AS, Muzykantov VS, Gubanova EL et al. (2011e) Partial oxidation of methane on Pt-supported lanthanide doped ceria-zirconia oxides: effect of the surface/lattice oxygen mobility on catalytic performance. *Catal Today* 169:125–137.

Sadykov VA, Sobyanin V, Mezentseva NV, Alikina GM, Vostrikov Z et al. (2010d) Transformation of CH_4 and liquid fuels into syngas on monolithic catalysts. *Fuel* 89:1230–1240.

Sadykov VA, Voronin VI, Petrov AN, Frolova YV, Kriventsov VV et al. (2005) Structure specificity of nanocrystalline praseodymia doped ceria. *Mater Res Soc Symp Proc* 848:231–236.

Schäfer H (1971) Preparative solid state chemistry: the present position. *Angew Chem Int Ed* 10:43–50.

Schwarz JA, Contescu C, Contescu A (1995) Methods for preparation of catalytic materials. *Chem Rev* 95:477–510.

Shelef M (2002) Ceria and other oxygen storage components in automotive catalysts. In: Trovarelli A (ed) *Catalysis by Ceria and Related Materials*. Imperial College Press, London, UK.

Shin JH, Moon JH (2011) Bilayer inverse opal TiO_2 electrodes for dye-sensitized solar cells via post-treatment. *Langmuir* 27:6311–6315.

Sinev MYu, Udalova OV, Tulenina YP, Margolis LYa, Vislovskii VP et al. (2000) Propane partial oxidation to acrolein over combined catalysts. *Catalysis Letters* 69:203–206.

Smirnova AL, Sadykov VA, Mezentseva NV, Bunina RV, Pelipenko VV et al. (2011) Design and testing of structured catalysts for internal reforming of CH_4 in intermediate temperature solid oxide fuel cells (IT SOFC). *ECS Transactions* 35:2771–2780.

Solsona B, Davies TE, Garcia T, Vázquez I, Dejoz A et al. (2008) Total oxidation of propane using nanocrystalline cobalt oxide and supported cobalt oxide catalysts. *Applied Catal B Environ* 84:176–184.

Solsona B, Vázquez I, Garcia T, Davies TE, Taylor SH (2007) Complete oxidation of short chain alkanes using a nanocrystalline cobalt oxide catalyst. *Catal Lett* 116:116–121.

Su Z, Zhou W (2008) Formation mechanism of porous anodic aluminium and titanium oxides. *Adv Mater* 20:3663–3667.

Tejuca LG, Fierro, JLG, Tascon JMD (1989) Structure and reactivity of perovskite-type oxides. *Adv Catal* 36:237–328.

Thambimuthu K, Davison J, Gupta M (2002). CO_2 capture and storage. *Proc. IPCC Workshop on Carbon Dioxide Capture and Storage*, Regina (Canada), November 2002, published by ECN, pp. 31–52.

Thomas JM, Thomas WJ (1996) *Principles and Practice of Heterogeneous Catalysis*. VCH Publishers, Weinham, Germany.

Uyguner CS, Bekbolet M (2007) Contribution of metal species to the heterogeneous photocatalytic degradation of natural organic matter. *Int J Photoenergy* article ID 23156:8.

Valenzuela RX, Bueno G, Corberán VC, Xu Y, Chen Ch (1998). Oxidative dehydrogenation of ethane with CO_2: an analysis of the heterogeneous reaction contribution on CeO_2 catalysts, Abstracts 215th ACS Nat. Meeting, Dallas 1998, ACS, COLL-085.

Valenzuela RX, Bueno G, Corberán VC, Xu Y, Chen C (2000) Selective oxidehydrogenation of ethane with CO_2 over CeO_2-based catalysts. *Catal Today* 61:43–48.

Valenzuela RX, Bueno G, Solbes A, Sapiña F, Martínez E et al. (2001b) Nanostructured ceria-based catalysts for oxydehydrogenation of ethane with CO_2. *Topics Catal* 15:181–188.

Valenzuela RX, Bueno Sobrino G, Vie Giner D, Martínez E, Sapiña F et al. (2001a) Nanostructured cobalt and nickel molybdate catalysts for oxidative dehydrogenation of propane. Book of Extended Abstracts, IV World Congress on Catalytic Oxidation, Berlin-Potsdam, Dechema, II:213–214.

Valenzuela RX, Corberan VC (2000) On the intrinsic activity of vanadium centres in the oxidative dehydrogenation of propane over V-Ca-O and V-Mg-O catalysts. *Topics Catal* 11:153–160.

Vautier M, Guillard C, Herrmann J-M (2001) Photocatalytic degradation of dyes in water: case study of indigo and of indigo carmine. *J Catal* 201:46–59.

Vie D, Martinez E, Sapina F (2004) Freeze-dried precursor-based synthesis of nanostructured cobalt-nickel molybdates $Co_{1-x}Ni_xMoO_4$. *Chem Mater* 16:1697–1703.

Wang Y, Zhou A, Yang Z (2008) Preparation of hollow TiO_2 microspheres by the reverse microemulsions. *Mater Lett* 62:1930–1932.

Wang S, Zhu ZH (2004) Catalytic conversion of alkanes to olefins by carbon dioxide oxidative dehydrogenations: a review. *Energy & Fuels* 18:1126–1139.

West AR (1984) *Solid State Chemistry and Its Applications*. John Wiley & Sons, New York.

Wincewicz KC, Cooper JS (2005) Taxonomies of SOFC material and manufacturing alternatives. *J Power Sources* 140:280–296.

Wu Y, He Y, Chen T, Weng WZ, Wan HL (2006) Low temperature catalytic performance of nanosized Ti–Ni–O for oxidative dehydrogenation of propane to propene. *Applied Surf Sci* 252:5220–5226.

Xie X, Li Y, Liu ZQ, Haruta M, Shen WJ (2009) Low-temperature oxidation of CO catalysed by Co_3O_4 nanorods. *Nature* 458:746–749.

Xin Y, Liu H, Han L, Zhou Y (2011) Comparative studies of photocatalytic and photoelectrocatalytic properties of alachlor using different morphology TiO_2/Ti photoelectrodes. *J Hazardous Mater* 192:1812–1818.

Xu WW, Kershaw R, Dwight K, Wold A (1990) Preparation and characterization of TiO_2 films by a novel spray pyrolysis method. *Mat Res Bull* 25:1385–1392.

Yang XF, Jin CJ, Liang CL, Chen DH, Wu MM et al. (2011) Nanoflower arrays of rutile TiO_2. *Chem Comm* 47:1184–1186.

Yoneyama M, Will RK, Müller S, Yang W (2010) Catalysts: petroleum and chemical process. SRI International. Abstracts available at: http://www.ihs.com/products/chemical/planning/scup/catalysts-petroleum-and-chemical.aspx?pu=1&rd=chemihs.

Yoriya S, Bao N, Grimes CA (2011) Titania nanoporous/tubular structures via electrochemical anodization of titanium: effect of electrolyte conductivity and anodization voltage on structural order and porosity. *J Mater Chem* 21:13909–13912.

Yu Y, Xu D (2007) Single-crystalline TiO_2 nanorods: highly active and easily recycled photocatalysts. *Appl Catal B: Environmental* 73:166–171.

Zhai Y, Gao Y, Liu F, Zhang Q, Gao G (2007) Synthesis of nanostructured TiO_2 particles in room temperature ionic liquid and its photocatalytic performance. *Mater Lett* 61: 5056–5058.

Zheng Q, Kang H, Yun J, Lee Y, Park JH et al. (2011) Hierarchical construction of self-standing anodized titania nanotube arrays and nanoparticles for efficient and cost-effective front-illuminated dye-sensitized solar cells. *ACS Nano* 5:5088–5093.

Zhou L, Xu J, Miao H, Wang F, Li XQ (2005) Catalytic oxidation of cyclohexane to cyclohexanol and cyclohexanone over Co_3O_4 nanocrystals with molecular oxygen. *Applied Catal A: General* 292:223–228.

Zhou Q, Yang X, Zhang S, Han Y, Ouyang G et al. (2011) Rutile nanowire arrays: tunable surface densities, wettability and photochemistry. *J Mater Chem* 21:15806–15812.

9 Solar Photocatalytic Drinking Water Treatment for Developing Countries

John Anthony Byrne and Pilar Fernandez-Ibañez

CONTENTS

9.1 THE NEED FOR SAFE DRINKING WATER

On July 28, 2010, the UN General Assembly adopted Resolution 64/292 recognizing that safe and clean drinking water and sanitation is a human right essential to the full enjoyment of life and all other human rights. It called on United Nations Member States and international organizations to offer funding, technology, and other resources to help poorer countries scale up their efforts to provide clean, accessible, and affordable drinking water and sanitation for everyone (WHO/UNICEF 2011).

Water is the most important natural resource in the world, and the availability of safe drinking water is a high priority issue for human existence and quality of life. Unfortunately, water resources are coming under increasing pressure due to population growth, over-use of resources, and climate change. Because of the adoption of the Millennium Development Goals, the WHO/UNICEF Joint Monitoring Programme for Water Supply and Sanitation has reported on progress toward achieving Target 7c: reducing by half the proportion of people without sustainable access to safe drinking water and basic sanitation. As of 2010, the target for drinking water has been met; however, it still remains that 780 million people are without access to an improved drinking water source and many more are forced to rely on sources that are microbiologically unsafe, leading to a higher risk of contracting and therefore risk of contracting waterborne diseases, including typhoid, hepatitis A and E, polio, and cholera (Burch and Thomas 1998; Clasen and Edmondson 2006; WHO 2007; WHO/UNICEF 2012).

Water safety is affected by geogenic contamination of groundwater; pollution from industry and wastewater; poor sanitation; weak infrastructure; unreliable services; and the need for collection, transportation, and storage in the home. Although the Millennium Development Goal (MDG) for drinking water target refers to sustainable access to safe drinking water, the MDG indicator—"use of an improved drinking water source"—does not include a measurement of either drinking water safety or sustainable access. This means that accurate estimates of the proportion of the global population with sustainable access to *safe* drinking water are likely to be significantly lower than estimates of those reportedly using improved drinking water sources. It is estimated that, at the current rate of progress, 672 million people will not use improved drinking water sources in 2015. It is likely that many hundreds of millions more will still lack sustainable access to safe drinking water.

Of the estimated 884 million people who did not have access to improved sources of drinking water in 2008, 37% lived in Sub-Saharan Africa, 25% in Southern Asia, 17% in Eastern Asia, and 9% in South-Eastern Asia (Figure 9.1). The use of piped water on premises was lowest in Sub-Saharan Africa, Southern Asia, and South-Eastern Asia. Of particular concern among those people without access to improved drinking water sources are those who rely on surface water sources. Such sources

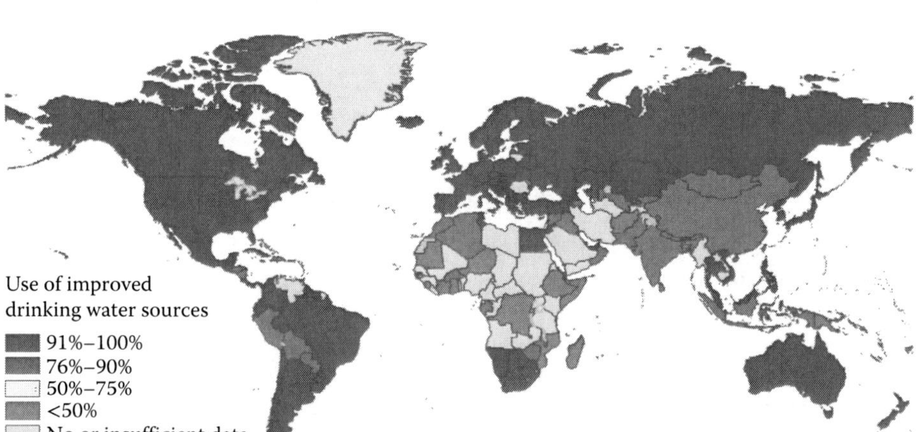

FIGURE 9.1 Worldwide use of improved drinking water sources in 2008. (From UNICEF and World Health Organization, Drinking Water Equity, safety and sustainability, JMP Thematic Report on Drinking Water 2011. With permission.)

include rivers, dams, lakes, ponds, and canals, and are often the most susceptible to pollution and more likely to have poor water quality. As ever, the poor are the worst affected and, in developing countries, 50% of the population are exposed to polluted water sources that, along with inadequate supplies of water for personal hygiene and poor sanitation, are the main contributors to an estimated 4 billion cases of diarrhea each year. This causes an estimated 2.2 million deaths annually, and most of these are children under the age of five (Clasen and Edmondson 2006).

The provision of piped-in water supplies is an important long-term goal; however, the WHO and the United Nations Children's Fund (UNICEF) acknowledge that we are unlikely to meet the MDG target of halving the proportion of the people without sustainable access to safe drinking water and basic sanitation by 2015. The proxy indicator used in the global survey methodology—"use of improved drinking water sources"—does not guarantee that the quality of drinking water consumed by people meets the standards for safe drinking water as proposed in the WHO Guidelines for Drinking water Quality (WHO 2011).

Although conventional interventions to improve water supplies at the source (point of distribution) have long been recognized as effective in preventing diarrhea, more recent reviews have shown household-based (point-of-use) interventions to be significantly more effective than those at the source. As a result, there is increasing interest in such household-based interventions that can deliver the health gains of safe drinking water at lower cost (Clasen and Haller 2008).

There has been rapid growth in the use of boreholes and tubewells, especially in Southern Asia where 310 million more people used boreholes in 2008 than in 1990, and the limited water quality compliance for this technology is of significant concern (Figure 9.2).

Household water treatment and safe storage (HWTS) is one option for improving the quality of water for consumption within the home, especially where water handling and storage is necessary and recontamination is a real risk between the point of collection

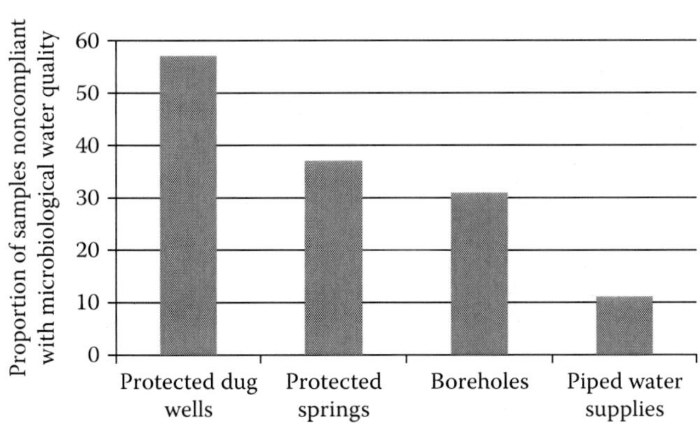

FIGURE 9.2 Noncompliance with microbiological water quality guideline values by improved drinking water source type. The survey was conducted by rapid assessment of drinking water quality in five countries, that is Ethiopia, Jordan, Nicaragua, Nigeria, and Tajikistan. Proportion refers to percentage. (From UNICEF and World Health Organization, Drinking Water Equity, safety and sustainability, JMP Thematic Report on Drinking Water 2011. With permission.)

and point of use. Access to a distant source only, unreliable piped supplies, and reliance on rainwater are all factors that make household storage a necessity. Living conditions in many humanitarian crises also call for effective HWTS. The practice of HWTS can help improve water quality at the point of consumption, especially when drinking water sources are distant, unreliable, or unsafe. However, HWTS is a stop-gap measure only and does not replace the obligation of a service provider to supply access to safe drinking water. It is intended for people who have no access to improved drinking water sources at all, for people with access to improved sources outside of their home or premises (i.e., when contamination can occur during transport and storage), for people with unreliable piped supplies who have to store water to bridge the gaps between deliveries, and for people in emergency situations. People relying on unimproved drinking water sources who apply an appropriate household water treatment (HWT) method are still not considered to have sustainable access to safe drinking water. Doing so would absolve the providers of their responsibility to provide safe drinking water and in effect transfer this responsibility to consumers (UNICEF and WHO 2011).

HWTS can serve as an effective means to remove pathogens and reduce diarrheal diseases associated with ingested water, even when drinking water is collected from an unimproved or unsafe source. The prevalence of appropriate HWT is relatively high where drinking water is piped into the dwelling, suggesting that consumers may not trust the quality of their tap water. In contrast, less than one quarter of those using unprotected dug wells and unprotected springs use appropriate HWT. Appropriate HWT is practiced by over 50% of people using protected wells but only by 23% of those using unprotected wells. This indicates that many households with the poorest drinking water quality and most in need of HWT do not use such technologies.

Appropriate HWT methods include boiling, filtration, adding chlorine or bleach, and solar disinfection. Straining water through a cloth or letting it stand and settle are not considered appropriate methods. Households are at least four times more likely to use boiling than any other HWT method (Figure 9.3).

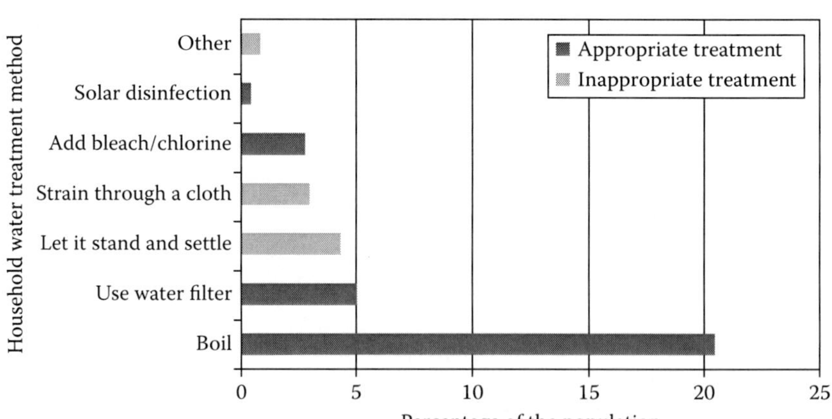

FIGURE 9.3 Prevalence of household water treatment methods reported across selected countries. (From UNICEF and World Health Organization, *Drinking Water Equity, safety and sustainability*, WHO/UNICEF, 2011. With permission.)

In 2008, Clasen and Haller reported on the cost and cost effectiveness of household-based interventions to prevent diarrhea (Clasen and Haller 2008). They compared chlorination using sodium hypochlorite following the *Safe Water System* developed and promoted by the U.S. Centers for Disease Control and Prevention (CDC); gravity filtration using either commercial *candle* style gravity filters or locally fabricated pot-style filters developed by Potters for Peace; solar disinfection following the SODIS method in which clear 2-L PET bottles are filled with raw water and then exposed to sunlight for 6–48 hours; and flocculation disinfection using Procter & Gamble's PUR® sachets, which combine an iron-based flocculent with a chlorine-based disinfectant to treat water in 10-L batches. They concluded that household-based chlorination was the most cost-effective method. Solar disinfection was only slightly less cost-effective, owing to its almost identical cost but lower overall effectiveness. Given that household-based chlorination requires the distribution of sodium hypochlorite, solar disinfection has a major advantage in terms of nonreliance on chemical distribution.

Sunlight is widely and freely available on Earth and the combined effects of infrared (IR), visible, and ultraviolet (UV) energy from the sun can inactivate pathogenic organisms present in water. There are, however, a number of parameters that affect the efficacy of the solar disinfection (SODIS) process, including the solar irradiance, ambient temperature, the quality of the water to be treated (turbidity, suspended solids, etc.), and the nature of contamination (as some pathogens are more resistant to SODIS than others). Furthermore, SODIS enhancement technologies may improve the effectiveness without increasing the cost substantially. One approach to SODIS enhancement is the use of heterogeneous photocatalysis. Photocatalysis uses light along with a semiconductor material to produce highly oxidative species that destroy organic pollutants in water and inactivate pathogenic microorganisms (Bahnemann 2004; Augugliaro et al. 2006; Fujishima et al. 2008; Gaya and Abdullah 2008; Malato et al. 2009). The process operates at atmospheric pressure and ambient temperature, and does not require consumable chemicals except oxygen from the air. Photocatalysis may provide a relatively low cost to the purification of water in developing regions where solar irradiation can be used. Solar photocatalysis is truly a *Clean Technology* and could provide a sustainable, low-cost, and low-carbon approach to the remediation of contaminated water.

9.2 SOLAR DISINFECTION OF WATER

Solar disinfection of water (SODIS) is a simple and low-cost technique that can be used to disinfect contaminated drinking water. UV-vis transparent bottles (preferably polyethylene terephthalate (PET)) are filled with contaminated water and placed in direct sunlight for a minimum of 6 hours. Following exposure, the water is safer to drink as the viable pathogen load can be significantly decreased. This protocol is referred to as SODIS. Simple guidance for the use of SODIS is given in Figure 9.4. SODIS is recognized and promoted by the WHO and there are an estimated 4.5 million regular users worldwide, predominantly in Africa, Latin America, and Asia (SODIS 2010a).

SODIS harnesses light and heat from the sun to inactivate pathogens via a synergistic mechanism (McGuigan et al. 1998). It is well known that UV light and thermal

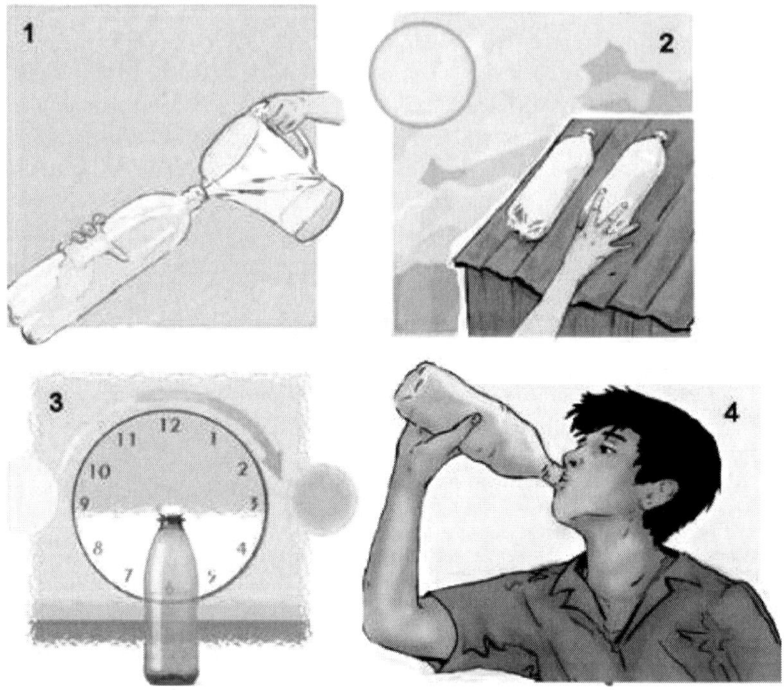

FIGURE 9.4 SODIS process. (Courtesy of Kevin McGuigan, Royal College of Surgeons in Ireland, drawn by Maria Boyle.)

treatment can inactivate pathogenic microorganisms. However, only around 4%–6% of the solar spectrum reaching the surface of Earth is in the UV domain, with maximum reported value of around 50 W/m^2 (Goswami et al. 2000). UV radiation (200–400 nm) can be classified as UVA (320–400 nm), UVB (280–320 nm), and UVC (200–280 nm). The highly energetic UVC radiation is absorbed by the ozone layer along with a proportion of the UVB. Therefore, UVA represents the main fragment of solar ultraviolet radiation reaching the earth's surface, with a small proportion of UVB.

Disinfection of water using solar energy has been carried out since Egyptian times. The process was first studied and reported in scientific literature by London-based scientists Downes and Blunt in the late 1870s (Downes and Blunt 1877). The process was effectively rediscovered as a low-cost water disinfection method by Acra et al. (1980, 1984) in the late 1970s. The main findings of their work were that *Escherichia coli* was more resistant to SODIS than other organisms tested, and that *E. coli* should be used as an indicator of SODIS efficiency. Indeed the presence of viable fecal coliforms is used as an indicator of the effectiveness of conventional disinfection processes. Furthermore, it is the UV component of sunlight, and to a lesser extent, the blue end of the visible spectrum, that is mainly responsible for the biocidal action. Wegelin et al. (1994) from the Swiss Federal Institute of Aquatic Science and Technology (Eawag) reported on SODIS in terms of the scope of the process and the analysis of radiation experiments. They concurred with the conclusions of Acra et al., that is, the bactericidal effect of UVA and violet light, and their combined effects. A 3-log reduction in the number of *E. coli* required a fluence of 555 W h m^{-2} (350–450 nm). The same dose was found to inactivate bacteriophage f2, whereas picornavirus was twice as resistant. Water temperatures above 50°C increased the rate of bacteria inactivation significantly, but below this temperature there was no synergistic effect. However, the virus inactivation rate was found to increase with increasing temperatures above 20°C. They concluded, among other things, that *E. coli* and fecal coliforms are good indicators of SODIS efficiency. Eawag promotes the use of SODIS via an online information network and the publication and distribution of SODIS literature (SODIS 2010b).

Laboratory studies have demonstrated the effects of key operational parameters such as light intensity and wavelength, solar exposure time, availability of oxygen, turbidity, and temperature (Reed 2004; Berney et al. 2006a). Solar disinfection has been mostly explained by the synergetic effect of the light and temperature (Berney et al. 2006a). The SODIS mechanism is understood to involve a number of biocidal pathways based on absorption of UVA radiation and thermal inactivation. Direct UVA exposure can induce cellular membrane damage and delay microbial growth (Hamamoto et al. 2007). The biocidal action of UVA has also been attributed to the production of reactive oxygen species (ROS) that are generated from dissolved oxygen in water (Khaengraeng and Reed 2005) and the photosensitization of any molecule in the cell and/or any naturally occurring dissolved organic matter that can absorb photons of wavelengths between 320 and 400 nm to induce photochemical reactions (Oates et al. 2003). The thermal effect has been attributed to the high absorption of red and IR photons by water. At temperatures below 40°C, the thermal effect is negligible with UVA inactivation mechanisms dominating the inactivation process. Significant bactericidal action is evident at temperatures above 40°C–45°C with a synergistic SODIS process observed at temperatures above 45°C (McGuigan et al. 1998; Navntoft et al. 2008; Blanco et al. 2009). Furthermore, detailed genetic assessment of the microorganisms has been used to probe the biocidal mechanism of SODIS (Berney et al. 2006b,c). Researchers have shown SODIS to be effective against a wide range of microorganisms responsible for diarrheal illness (Kehoe et al. 2004; Lonnen et al. 2005; Boyle et al. 2008). The inactivation of resistant protozoa has also been reported (Heaselgrave et al. 2006; McGuigan et al. 2006; Gómez-Couso et al. 2009; Mtapuri-Zinyowera et al. 2009).

Field trials have demonstrated significant health benefits from consumption of SODIS-treated water as compared to untreated water (Conroy et al. 1996, 1999). The effectiveness of SODIS against cholera was also demonstrated in a Kenyan health impact assessment, where an 86% reduction of cholera cases was observed in households regularly using SODIS (Conroy et al. 2001). Recent contributions show that dysentery and nondysentery diarrhea were significantly reduced by solar disinfection in field trials, carried out in Kenya and Cambodia, with children aged under 5 years (Du Preez et al. 2011; McGuigan et al. 2011).

Studies to improve the efficiency of the SODIS process by using low-cost, commonly available materials have been conducted (Sommer et al. 1997; Kehoe et al. 2001; Mania et al. 2006; Fisher et al. 2008); however, the simple approach of exposing a 2-L PET bottle to full sun for a minimum of 6 hours is most commonly practiced.

9.3 ENHANCEMENT TECHNOLOGIES FOR SODIS

There are several drawbacks of conventional SODIS technology. The use of bottles allows for only small volumes to be treated (2–3 L). Of course, one can deploy as many bottles as one sees fit. The process efficiency, however, is dependent on a range of environmental parameters including the solar irradiance (which depends on the latitude, time of day, and atmospheric conditions), the initial water quality (e.g., organic loading, turbidity), and the level and nature of the bacterial contamination. Different microorganisms display varying levels of resistance to solar disinfection and this leads to variation in the required treatment times. Malato et al. (2009) reviewed the reported inactivation time required for a range of microorganisms using SODIS under ca. 1 kW/m^2 global irradiance. These vary enormously from 20 minutes, for *Campylobacter jejuni*, to 8 hours, for *Cryptosporidium parvum* oocysts. For *Bacillus subtilis* endospores, no inactivation was observed after 8 hours of SODIS treatment. Furthermore, SODIS is user dependent in that it requires the user to time the exposure and as such there is no quality assurance for the process, and lack of compliance with the recommended protocol is a major issue (Du Preez et al. 2010).

There are a number of ways to improve or enhance the conventional SODIS process and these include the design of SODIS bags where the solar dose per volume is increased (Dunlop et al. 2011); the use of UV dosimetric sensors that indicate to the user when the desired dose has been received by the water (Bandala et al. 2011); thermal enhancement (mild heat, i.e., up to 50°C) so that disinfection is more efficient; the use of solar flow reactors that have been designed to maximize the solar dose, for example, using compound parabolic collectors (CPCs) and may include UV feedback sensors for automated control; and the use of semiconductor photocatalysis to enhance the treatment efficacy (Byrne et al. 2011). All of these enhancements have been focused on one or more of the following objectives: (1) maximizing the collection of solar energy dose, (2) enhancing the disinfecting efficacy especially against resistant waterborne pathogens; (3) increasing the output of treated water in given solar exposure time; (4) reducing the user dependence of the process; and (5) finding as cheap and robust as possible disinfection systems, which may also be constructed with local materials without sophisticated technological needs.

Since this chapter deals with the solar photocatalytic disinfection of water, we shall concentrate on two areas that are important and complimentary, that is, the use of nanophotocatalytic materials and enhancement of solar reactors using CPCs.

9.4 SOLAR REACTORS AND CPC COLLECTORS FOR DRINKING WATER DISINFECTION

Some authors have worked on the improvement produced by increasing temperature using solar thermal systems (Fjendbo et al. 1998; Saitoh and El-Ghetany 2002). Other researchers have focused on the development of recirculating batch flow reactors, which increase the optical component of sunlight inactivation by using different solar collectors (Vidal 1999; McLoughlin 2004; Fernández et al. 2005; Navntoft et al. 2008), whereas others have used TiO_2 photocatalyst either immobilized or suspended (Dunlop et al. 2002; Fernandez et al. 2005; Fernández-Ibáñez et al. 2009).

SODIS relies heavily on the solar UVA which, as received at sea level, is composed of roughly similar portions of both direct and diffuse electromagnetic radiation. The diffuse component is greater on cloudy days or when the atmosphere is polluted with particles. The SODIS bottles are only illuminated on the upper side when exposed to sunlight so that a large fraction of the available radiation cannot reach the water.

Given the diffuse nature of the UVA and the cylindrical shape of bottles, the use of concentrating systems based on nonimaging optics with a low concentrating factor has obvious potential compared to imaging optics-based systems. CPCs are nonimaging concentrating systems with a diffuse focus. The concentrated rays are homogeneously distributed in the absorber. Their main advantage is that they concentrate diffuse radiation. Hence, they do not rely solely on direct solar radiation and are effective even on cloudy days. In addition, they concentrate radiation independently of the direction of sunlight and do not require sun tracking in contrast to direction-dependent image forming systems (Figure 9.5).

McLoughlin et al. (2004) analyzed three types of static solar collectors for the disinfection of water containing *E. coli*. They tested three collector geometries—compound parabolic, parabolic, and V-groove—in lab-scale solar photoreactors with aluminum reflectors under natural sunlight. Water disinfection was improved in all cases, although the CPC was shown to be the most efficient system tested. They also proved that bacterial deactivation rates using sunlight alone can be enhanced by low concentrations of titanium dioxide suspended in the water. They also suggested a detrimental effect of intermittent light during the solar treatment on the disinfection efficacy, which was also confirmed by other researchers.

Navntoft et al. (2008) investigated the CPC-concentrating optics to enhance the radiation reaching the bottles in SODIS. The authors reported improved solar disinfection results for suspensions of *E. coli* in well water using CPC for batch reactors under sunlight in cloudy and clear sky conditions. They demonstrated that the use of a CPC improved disinfection efficacy of SODIS, and the total treatment time to disinfect the water was reduced in all cases studied (Figure 9.6).

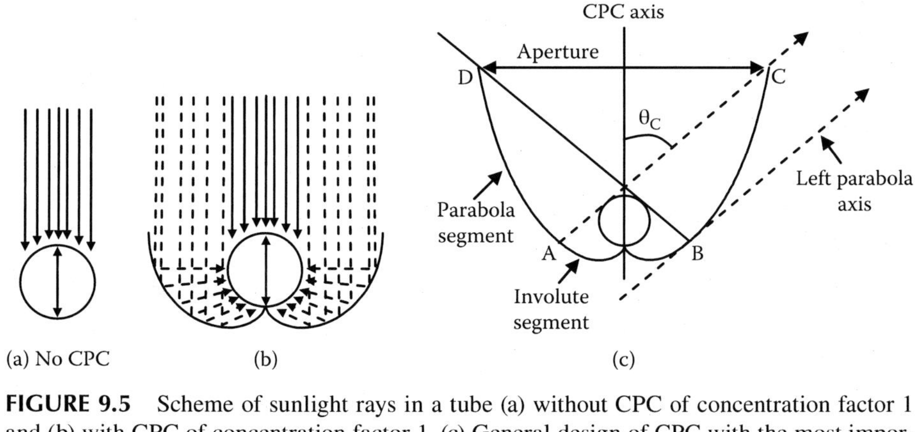

FIGURE 9.5 Scheme of sunlight rays in a tube (a) without CPC of concentration factor 1 and (b) with CPC of concentration factor 1. (c) General design of CPC with the most important geometric parameters.

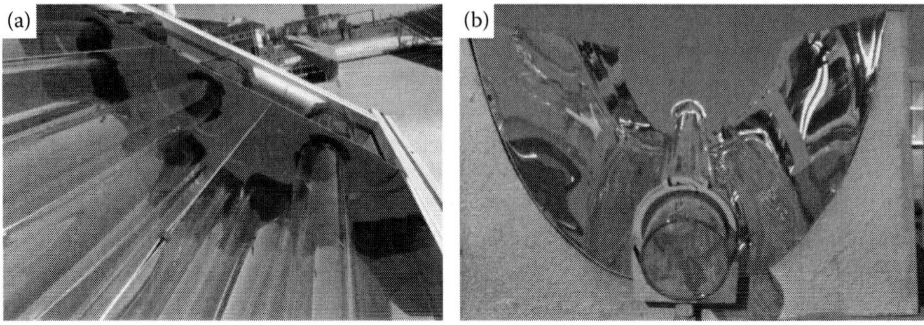

FIGURE 9.6 Photographs of CPC aluminum mirrors with concentration factor equal to 1 (a) and 1.9 (b) placed in solar photoreactors located in the facilities of Plataforma Solar de Almería (Spain).

SODIS has also clear limitations related with total volume of treated water. When the process is brought to larger scale, the use of modular flow systems is preferred; however, unless the system is properly designed, poor disinfection efficiencies may be observed. These problems have been addressed in recent contributions (Ubomba et al. 2009; Polo-Lopez et al. 2011).

Ubomba et al. (2009) reported a study with important implications for scaling-up SODIS through the use of pumped, recirculatory, continuous flow reactors. They studied the effect of the total volume of treated water and the flow rate on inactivation using solar CPC flow- and static-batch reactors. They observed that increasing flow rate has a negative effect on inactivation of bacteria, irrespective of the long exposure time. They attributed this behavior to the need of a certain exposure time to UV (or a given solar UV dose) for bacteria to ensure inactivation as compared to having bacteria exposed to sublethal doses over a long period. They found that complete inactivation of *E. coli* in well water occurred at all solar intensities studied, as long as an uninterrupted solar-UV dose > 108 kJ m^{-2} was reached. Complete inactivation of bacteria was only observed for static systems; that is, the water was constantly illuminated and hence the required uninterrupted UV dose was delivered and complete inactivation (to the detection limit) took place. Continuous flow systems only deliver the lethal dose to the bacteria in an intermittent manner, resulting in a residual bacteria population remaining after the treatment. According to this study, if the operational parameters are set such that the microbial pathogens are repeatedly exposed to sublethal doses of solar radiation followed by a period within which the cells have an opportunity to recover or repair, complete inactivation may not be achieved.

This problem may be solved in continuous flow systems only if the lethal UV dose is delivered to the bacteria in one solar treatment before collection and storage (Polo-Lopez et al. 2011). To address these practical problems associated with SODIS, a novel sequential batch photoreactor was designed and constructed with the aim of avoiding residual viable bacteria in the treated water, decreasing the treatment time required, and reducing user dependency (Figure 9.7).

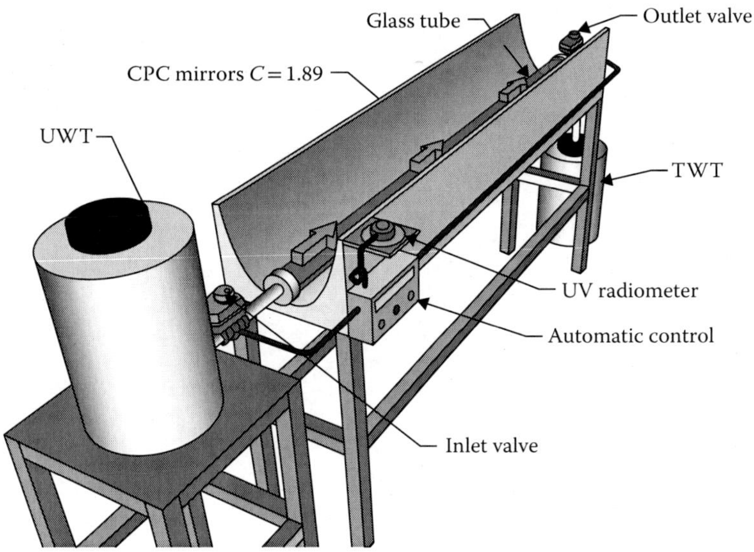

FIGURE 9.7 Schematic of the sequential batch system, where UWT is the Untreated Water Tank; TWT is the Treated water Tank, and the solar CPC collector has a concentration factor of 1.89.

This photoreactor incorporated two major improvements over traditional solar reactors: (1) it reduced the solar exposure time required to receive the lethal UVA dose increasing the concentration factor of the CPC from 1.00 to 1.89 and (2) the treatment time was automatically controlled by an electronic UVA sensor. The feedback sensor system controlled the gravity filling of the reactor from an untreated water reservoir and controlled the discharge of the treated water into a clean reservoir tank following receipt of the predefined UVA dose. The reactor was successfully proven for the inactivation of *E. coli* under real sun conditions (Polo-Lopez et al. 2011).

Another way to address the limitation relating to the volume that can be treated is the design of larger volume static batch systems. A recent contribution presented a new SODIS-enhanced batch reactor design fitted with a CPC for the disinfection of 25 L of water in ≥6 hours of strong sunlight (Ubomba-Jaswa et al. 2010). This prototype avoids the need for a constant supply of PET bottles. Such a large volume batch system should (1) be constructed from materials of minimum cost, (2) be robust in nature so that it can withstand adverse environmental conditions, and (3) require very little maintenance. The reactor consisted of a cylindrical container placed along the linear focus of the CPC. The use of flow was avoided in the reactor. Furthermore, flow-through systems might require higher maintenance and power costs to operate the system. The use of the CPC provides an enhancement to the disinfection process. The authors studied the microbial inactivation efficacy of the reactor using *E. coli* K-12 as the model organism in 25 L volumes of natural well water, with different turbidities. They demonstrated that the CPC enhanced the solar disinfection in this system (Figure 9.8).

(a)

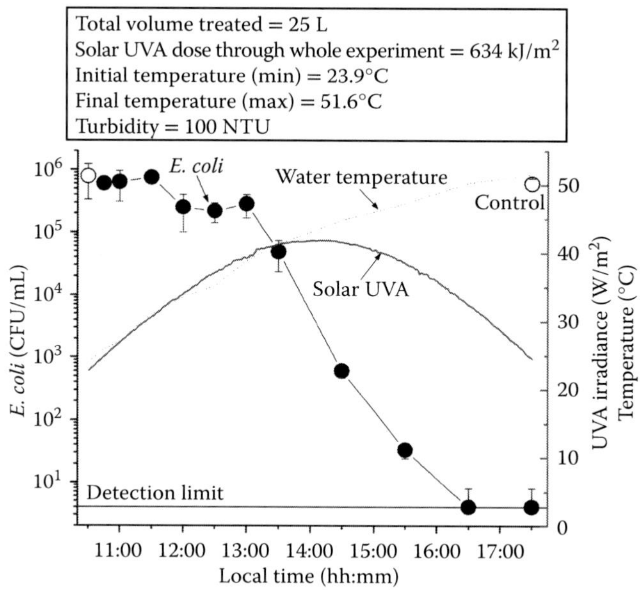

(b)

FIGURE 9.8 (a) Picture of the SODIS-enhanced batch reactor of 25 L, (b) Disinfection curve of well water contaminated with *E. coli* K-12 and turbidity of 100 NTU.

9.5 SEMICONDUCTOR PHOTOCATALYSIS

9.5.1 Mechanism of Photocatalysis

Photocatalysis is the acceleration of a photoreaction in the presence of a catalyst. When a semiconductor is illuminated with light of wavelength equal or greater than its band gap, the radiation energy is absorbed and electrons are promoted from the valence band to the conduction band giving rise to the formation of electron-hole pairs (e^- and h^+). These charge carriers can recombine with the energy being re-emitted as light or heat, or they may migrate to the catalyst surface. If they survive to reach the semiconductor/water interface, they may participate in redox reactions (Cassano and Alfano 2000). If an electron acceptor (A) and electron donor (D) are adsorbed or close to the surface of the semiconductor particle, the following reactions may occur (Litter 1999):

$$TiO_2 \xrightarrow{h\nu \geq E_{bg}} TiO_2(e^-_{cb}, h^+_{vb}) \Leftrightarrow \text{recombination} \tag{9.1}$$

$$TiO_2(h^+_{vb}) + H_2O_{ads} \longrightarrow TiO_2 + HO^{\bullet}_{ads} + H^+ \tag{9.2}$$

$$TiO_2(h^+_{vb}) + HO^-_{ads} \longrightarrow TiO_2 + HO^{\bullet}_{ads} \tag{9.3}$$

$$TiO_2(h^+_{vb}) + D_{ads} \longrightarrow TiO_2 + D^+_{ads} \tag{9.4}$$

$$HO^{\bullet} + D_{ads} \longrightarrow D_{oxid} \tag{9.5}$$

$$TiO_2(e^-_{cb}) + A_{ads} \longrightarrow TiO_2 + A^-_{ads} \tag{9.6}$$

$$TiO_2(e^-_{cb}) + O_{2ads} + H^+ \longrightarrow TiO_2 + HO_2^{\bullet} \Leftrightarrow O_2^{-\bullet} + H^+ \tag{9.7}$$

$$HO_2^{\bullet} + TiO_2(e^-_{cb}) + H^+ \longrightarrow H_2O \tag{9.8}$$

$$2HO_2^{\bullet} \longrightarrow H_2O_2 + O_2 \tag{9.9}$$

$$H_2O_2 + h\nu \longrightarrow 2HO^{\bullet} \tag{9.10}$$

$$H_2O_2 + TiO_2(e^-_{cb}) \longrightarrow HO^{\bullet} + HO \tag{9.11}$$

The ROS that are produced are very active, indiscriminate oxidants, especially the hydroxyl radical (Lee et al. 2006). The ROS not only can destroy a large variety of chemical contaminants in water but also can cause fatal damage to microorganisms (Blake et al. 1999). Figure 9.9 shows a schematic for the mechanism of semiconductor photocatalysis. In most cases, the final products of the photocatalytic degradation of organic compounds (given long enough) are CO_2 and H_2O, and respective inorganic acids or salts. For more detailed information on the mechanisms of semiconductor photocatalysis, the reader is referred to one of the many reviews (e.g., Bahnemann 2004; Augugliaro et al. 2006; Gaya and Abdullah 2008; Fujishima et al. 2008; and Malato et al. 2009).

9.5.2 Photocatalytic Materials

Several compounds have been investigated as potential semiconductor photocatalysts for water treatment such as metal oxides (TiO_2, ZnO, ZrO_2, V_2O_5, Fe_2O_3, and SnO_2) and metal sulfides (CdS, ZnS) (Herrmann 1999; Karunakaran et al. 2004). Among

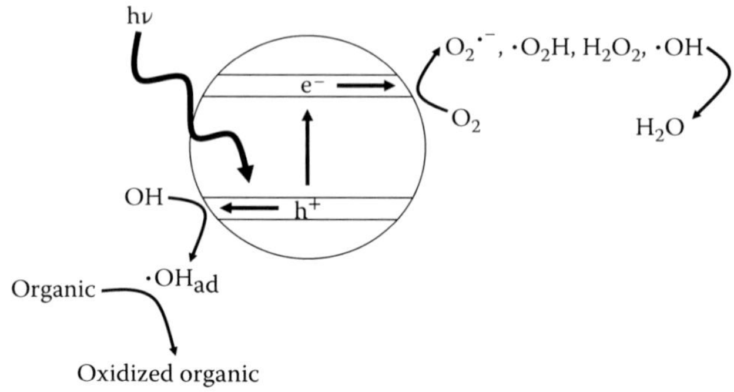

FIGURE 9.9 Schematic representation of photocatalysis mechanism on a titanium dioxide particle. Oxidation of hydroxyl ion by the valence band hole yields hydroxyl radical. Reduction of molecular oxygen yields superoxide radical anion that can be reduced again to yield hydrogen peroxide.

these, the most popular photocatalyst used for water treatment applications is titanium dioxide (TiO_2). $TiO_2$is a wideband semiconductor and the band gap energy is 3.2 eV for anatase meaning it requires UV excitation. For solar applications, visible-light-active materials are desirable. However, the smaller band gap, while absorbing a greater number of solar photons, gives a narrower voltage window to drive the redox reactions at the interface. Furthermore, metal sulfide semiconductors, which absorb in the visible region of the spectrum, tend to undergo photo-anodic corrosion (Kanade 2006). Considering cost, chemical and photochemical stability, availability, and lack of toxicity, the most suitable catalyst reported to date for the treatment of water is TiO_2 (Cassano and Alfano 2000).

9.5.3 Immobilized Versus Suspended Photocatalyst

The photocatalyst can be used in aqueous suspension or it may be immobilized on a supporting solid substrate. Most previous studies have reported that suspension reactors are more efficient due to large surface area available for the reaction (Byrne et al. 1998). The main drawback of using nano- or microparticles in suspension is the requirement for posttreatment separation and recycling of the catalyst, potentially making the treatment more complex and expensive. Therefore, treatment reactors using immobilized TiO_2 have gained more attention. Unfortunately, immobilizing TiO_2 on a solid substrate will reduce the surface area available for reaction and limit the mass transfer of reactants to the photocatalyst surface (McMurray et al. 2004). There are a wide range of methodologies available for the preparation of immobilized photocatalyst films on a range of supporting substrates (Byrne et al. 1998).

9.6 PHOTOCATALYTIC DISINFECTION OF WATER

Matsunaga et al. (1985) reported the first application of TiO_2 photocatalysis for the inactivation of bacteria in 1985. Since then, there have been a large number of research publications dealing with the inactivation of microorganisms including bacteria, viruses, protozoa, fungi, and algae. Blake et al. (1999) carried out an extensive review of the microorganisms reported to be inactivated by photocatalysis. In 2007, McCullagh et al. (2007) reviewed the application of photocatalysis for the disinfection of water contaminated with pathogenic microorganisms. In 2009, Malato et al. published an extensive review on the decontamination and disinfection of water by solar photocatalysis (Malato et al. 2009) and, in 2010, Dalrymple et al. reviewed the modeling and mechanisms of photocatalytic disinfection (Dalrymple et al. 2010).

In most studies concerning photocatalytic disinfection, the hydroxyl radical is suggested to be the primary species responsible for microorganism inactivation. Some articles reported other ROS such as H_2O_2, $O_2^{\bullet -}$ to be responsible for inactivation process (Huang et al. 2000; Wainwright 2000; Sunada et al. 2003; Cho et al. 2004). These reactive species can cause fatal damage to microorganisms by disruption of the cell membrane or by attacking DNA and RNA (Blake et al. 1999). Other modes of action of TiO_2 photocatalysis have been proposed including damage to the respiratory system within the cells (Rincon et al. 2001) or loss of fluidity and increased ion permeability in the cell membrane (Wainwright 2000). Many researchers have attributed cell death to lipid peroxidation of bacterial cell membrane (Huang et al. 2000; Wainwright 2000; Sunada et al. 2003). The peroxidation of the unsaturated phospholipids that are contained in the bacterial cell membrane causes loss of respiratory activity (Rincon et al. 2001) and/or leads to a loss of fluidity and increased ion permeability (Wainwright 2000). This is suggested to be the main reason for cell death. Other researchers suggested that the cell membrane damage can open the way for further oxidative attack of internal cellular components, ultimately resulting in cell death (Rincon and Pulgarin 2004).

Research at the University of Ulster has been mainly concerned with the use of immobilized TiO_2 films prepared by the deposition of Degussa P25 onto a range of supporting substrates, including borosilicate glass, ITO glass, and titanium metal. For example, Alrousan et al. (2009) reported on the photocatalytic inactivation of *E. coli* in surface water using immobilized nanoparticle TiO_2 films. In this work, the photocatalyst (Degussa P25) was immobilized onto borosilicate glass using a dip coating method. The photoreactor was a custom-built stirred tank reactor that has excellent mixing and good mass transfer properties. The catalyst was irradiated in a back-face configuration, that is, the light passes through the glass to excite the photocatalyst on the surface. It was found that the rate of photocatalytic inactivation of *E. coli* was more efficient with UVA-TiO_2 than direct photolytic inactivation with UVA alone, both for distilled water and a real surface water (see Figure 9.10). The optimum catalyst loading for the inactivation of *E. coli* was determined to be 0.5 mg cm^{-2}, approximately half that reported for photocatalytic degradation of formic acid and atrazine in the same reactor, under the same incident light intensity.

The organic and inorganic content of the surface water led to a reduction in the rate of photocatalytic disinfection in comparison to that observed in distilled water. The effect of selected individual constituents present in the surface water was examined to identify the main constituent responsible for the reduction in the rate of photocatalytic disinfection. The presence of inorganic ions, that is sulfate and nitrate, reduced the rate of photocatalytic inactivation, with sulfate having a more pronounced effect than nitrate. The presence of organic matter was found to be the dominating parameter responsible for the decrease in the rate of photocatalytic

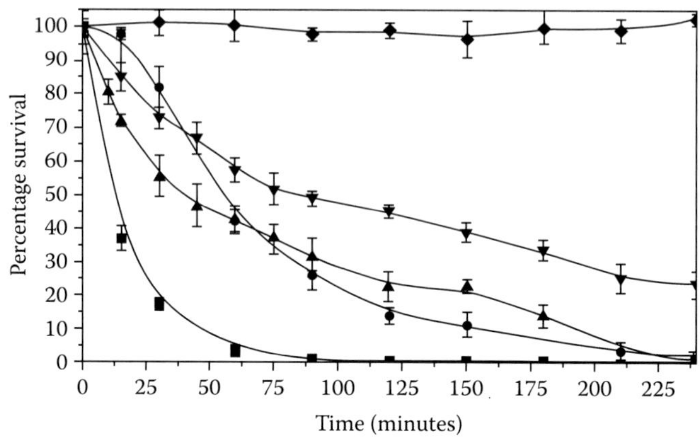

FIGURE 9.10 Comparison between photocatalytic and photolytic inactivation of *E. coli* in surface and distilled water. (TiO_2, UV, Distilled water) ■, (TiO_2, UV, Surface water) ●, (UV, Distilled water) ▲, (UV, Surface water) ▼, Dark control ◆. (From Alrousan, D.M.A., Dunlop, P.S.M., McMurray, T.A., Byrne, J.A., *Water Res*, 43, 47–54, 2009. With permission.)

disinfection. The important points to note are that the presence of inorganic ions will lead to a reduction in efficiency either by absorption of light, competing for ROS, or by adsorption to the catalyst surface. Organic matter will compete for ROS and may compete for photon absorption. The efficacy of photocatalytic disinfection will be strongly dependent on the initial water quality. Therefore, where the water to be treated is of poor quality, a pretreatment stage, for example simple filtration or settling, may be desirable.

Bacterial cells have been described as a rather easy target for disinfectants, with bacterial spores and protozoa suggested as more robust target organisms. *Clostridium perfringens* spores have been reported to be chlorine resistant at levels used in potable water supplies. Dunlop et al. (2008) reported on the photocatalytic inactivation of *Clostridium perfringens* spores on TiO_2 electrodes. The TiO_2 electrodes were made using electrophoretic immobilization of commercially available TiO_2 powders onto conducting supports, that is indium-doped tin oxide-coated glass, titanium metal, and titanium alloy. The photocatalytic inactivation of *E. coli* and *Clostridium perfringens* spores in water was observed on all immobilized TiO_2 films under UVA irradiation. The rate of photocatalytic inactivation of *E. coli* was found to be one order of magnitude greater than that of *Clostridium perfringens* spores, demonstrating the greater resistance of the spores to environmental stress. In this work, it was shown that the application of an external electrical bias (electrochemically assisted photocatalysis) significantly increased the rate of photocatalytic disinfection of *Clostridium perfringens* spores. The effect of incident light intensity and initial spore loading were investigated and disinfection kinetics was determined to be pseudo-first order. This work demonstrated that UVA photocatalysis is effective against bacterial spores that are more resistant to environmental stress, including UVA irradiation.

Cryptosporidium species are waterborne, protozoan parasites that infect a wide range of vertebrates. The life cycle involves the production of an encysted stage (oocyst), which is discharged in the feces of their host. The disease, cryptosporidiosis, in humans usually results in self-limited watery diarrhea but has far more devastating effects on immunocompromised patients (e.g., AIDS patients) and can be life threatening as a result of dehydration caused by chronic diarrhea. Owing to their tough outer walls, the oocysts are highly resistant to disinfection and can survive for several months in standing water. *Cryptosporidium* oocysts, therefore, present as an excellent challenge for disinfection technologies. Sunnotel et al. (2010) reported on the photocatalytic inactivation of *Cryptosporidium parvum* oocysts on nanostructured titanium dioxide films (see Figure 9.11).

The photocatalytic inactivation of the oocysts was shown to occur in Ringers buffer solution (78.4% after 180 minutes) and surface water (73.7% after 180 minutes). Scanning electron microscopy confirmed cleavage at the suture line of oocyst cell walls, revealing large numbers of empty (ghost) cells after exposure to photocatalytic treatment. Importantly, no significant inactivation was observed in the oocysts exposed to UVA radiation alone in the timescale of the experiments.

It is clear from the research literature, that in lab-scale reactors, photocatalysis under UVA irradiation is more efficient than UVA irradiation alone for the disinfection of water contaminated with pathogenic microorganisms. Therefore, it is essential that photocatalytic reactors are tested for disinfection under real sun conditions either on small-scale batch (personal use) or at pilot scale (aimed at household or small community use), and using real water sources.

Gelover et al. (2006) studied the small-scale batch disinfection of spring water naturally polluted with coliform bacteria in plastic bottles with and without TiO_2. The bottles were mounted in simple home-made solar collectors (see Figure 9.12). Two-liter PET bottles were filled with spring water and exposed to direct sunlight. The wild strain content was typically 2.5×10^3 MPN/100 mL (most probable number/100 mL) of total coliforms and 9.0×10^2 MPN/100 mL of fecal coliforms.

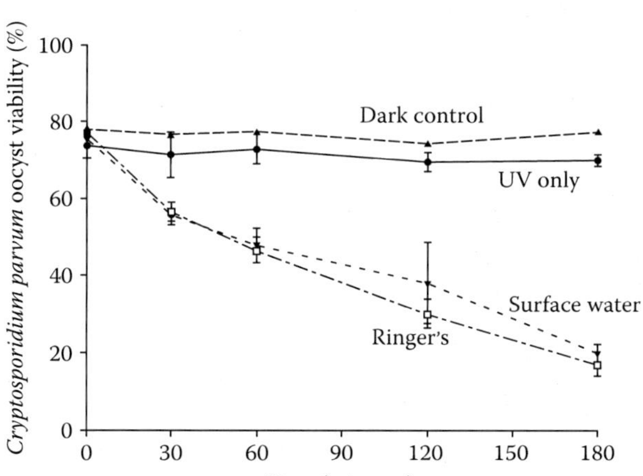

FIGURE 9.11 Photocatalytic inactivation of *Cryptosporidium parvum* oocysts suspended in Ringer's solution and surface water. (From Sunnotel, O., Verdoold, R., Dunlop, P.S.M., Snelling, W.J., Lowery, C.J., Dooley, J.S.G., Moore, J.E., Byrne, J.A., *J Water Health*, 8, 83–91, 2010. With permission.)

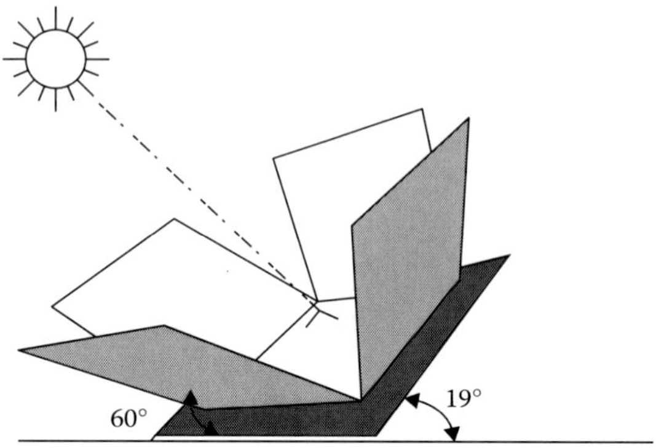

FIGURE 9.12 Simple solar collector made from five wooden sheets covered with aluminum foil. (From Gelover, S., Gomez, L.A., Reyes, K., Leal, M.T., *Water Res*, 40, 3274–3280, 2006. With permission.)

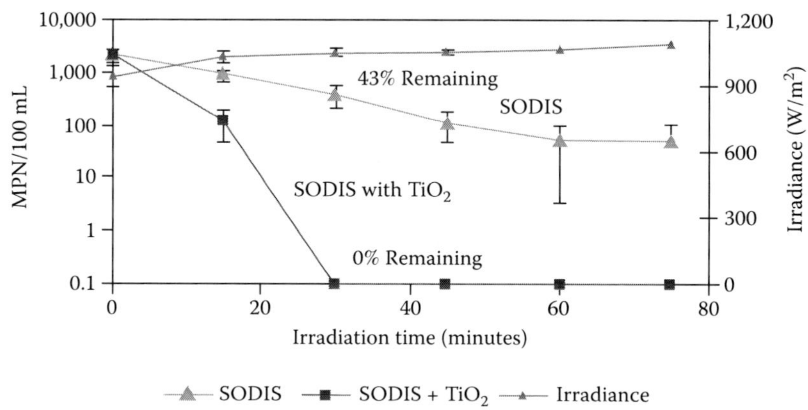

FIGURE 9.13 Showing decrease in fecal coliforms in natural spring water exposed to direct sunlight, comparing SODIS to TiO_2-enhanced SODIS. (From Gelover, S., Gomez, L.A., Reyes, K., Leal, M.T., *Water Res*, 40,3274–3280, 2006. With permission.)

TiO_2 was coated onto small Pyrex glass cylinders, using a sol–gel method, and these were placed inside each bottle. It was found that photocatalytically enhanced SODIS was by far more effective than SODIS alone for the inactivation of both the total coliforms and the fecal coliforms (see Figure 9.13). They tested bacterial regrowth following treatment and found that regrowth was observed with SODIS alone, but not with photocatalytically enhanced SODIS. This is an important finding, as bacteria have repair mechanisms, which allow recovery following stress, and demonstrates the differences in the kill mechanisms involved.

With respect to large-scale systems, Fernandez-Ibanez et al. (2009) reported on the pilot scale photocatalytic disinfection of water under real sun conditions using a photoreactor with CPC. The experiments were carried out under sunlight at the Plataforma Solar de Almeria (PSA) in southern Spain using compound parabolic concentrators (CPCs). The pilot plant consisted of three CPCs, and each collector consisted of two Pyrex tubes, which had a 0.25 m^2 collector surface. The photoreactor volume was 5.4 L and the total plant volume 11 L. Experiments were performed with suspended TiO_2 and, following modification to the reactor, with supported TiO_2. In both cases, Degussa P25 was used as the photocatalyst. For the immobilized study, the catalyst was immobilized on glass fiber using SiO_2 as an inorganic binder. *E. coli* was used as the model microorganism suspended in distilled water. It was found that the photocatalytic suspension reactor was most efficient, followed by the immobilized photocatalytic reactor, with the solar irradiation alone being the least efficient (see Figure 9.14).

Further work undertaken by the group at PSA investigated the effect of UV intensity and dose on SODIS and photocatalytic SODIS of bacteria and fungi (Sichel et al. 2007). The aim of the work was to study the dependence on solar irradiation conditions under natural sunlight. This dependency was evaluated for solar photocatalysis with TiO_2 and solar-only disinfection of three microorganisms, a pure *E. coli* K-12 culture and two wild strains of the *Fusarium* genus, *F. solani* and *F. anthophilum*. Photocatalytic disinfection experiments were carried out with TiO_2 supported on a paper matrix around concentric tubes, in compound parabolic collectors (CPCs) or with TiO_2 as slurry in bottle reactors, under natural solar irradiation at PSA. The experiments were performed with different illuminated reactor surfaces, in different seasons of the year, and under changing weather conditions (i.e., cloudy and sunny days). All results showed that once the minimum solar dose had been received, the photocatalytic disinfection efficacy was not particularly enhanced by any further increase. The solar-only disinfection turned out to be more susceptible to changes in solar irradiation, and therefore, only took place at higher irradiation intensities.

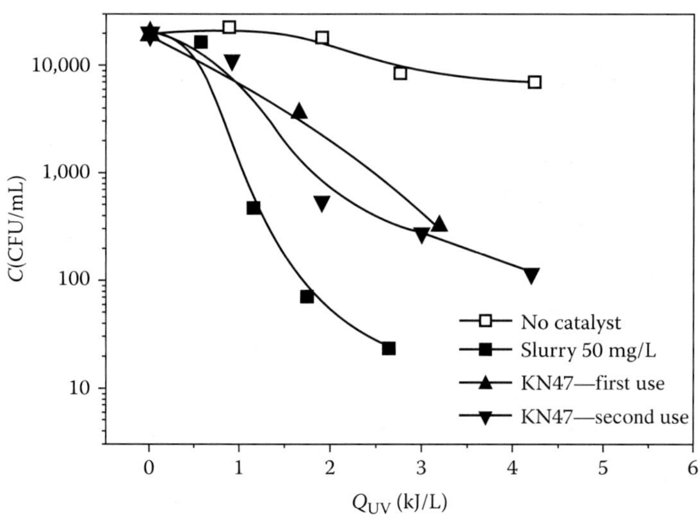

FIGURE 9.14 *E. coli* deactivation (C_0 = 104 CFU/mL) vs. Q_{UV} in the CPC solar photoreactor. Comparison of TiO_2 slurry, 50 mg/L (solid squares), with supported photocatalyst KN47 (19.3 g TiO_2/m^2) in two sequential uses (solid triangles) and only solar light (open squares). Q_{UV} is the accumulated energy (per unit of volume, kJ/L) incident on the reactor for each sample taken during the experiment. (From Fernandez, P., Blanco, J., Sichel, C., Malato, S., *Catal Today*, 101, 345–352, 2005. With permission.)

9.7 ISSUES TO BE ADDRESSED

9.7.1 Photoreactor Design

The SODIS process depends mainly on the UV-A wavelengths present in sunlight. Solar UV at sea level is composed of roughly similar portions of both direct and diffuse electromagnetic radiation. Without cloud cover, the solar UV-A spectrum is approximately 60% direct and 40% diffuse. Therefore, the use of concentrating systems based on non-imaging optics with low concentrating factor has a clear potential compared to imaging optics-based systems. The major advantage with compound parabolic collectors (CPCs) is that concentration factor remains constant for all values of sun zenith angle within the acceptance angle limit. Therefore, CPC enhancement can be used in the design of large-scale solar disinfection systems that can be used for household or small community use. A photoreactor for use in developing countries should have the following attributes:

- High illuminated volume to total volume ratio
- Operate under a low flow rate when using immobilized photocatalyst to maximize the residence time in flow systems
- Include a UVA dose indicator as the efficiency of both SODIS and photocatalytic disinfection are UV dose dependent
- Include a CPC of good quality, that is, high UVA reflectivity (for aluminum UVA reflectivity is ca. 87%–90%) and that decreases minimally following environmental exposure
- A reaction vessel with high (90%) UVA transmission, for example, borosilicate glass
- Robust under environmental conditions
- Low as possible life-cycle cost; low environmental impact
- Low maintenance requirements and easy access to replacement parts if necessary
- Low power requirements

Furthermore, in photocatalytic disinfection, the electron acceptor is normally dissolved oxygen, which is easily available from the air. With static batch systems, the concentration of dissolved oxygen will be rapidly depleted and must be replenished to maintain photocatalytic activity. Also, the solubility of oxygen in water is reduced by temperature. This must be taken into consideration if solar photocatalytic disinfection in developing countries is to be considered as the temperature within solar irradiated reactors can reach 55°C. Novel reactor design must address the need for replenishment of dissolved oxygen in photocatalytic disinfection systems. An alternative may be to introduce other oxidants, for example H_2O_2; however, this would give rise to a dependence on consumable chemicals, which is undesirable.

9.7.2 Photocatalyst Longevity

For application in remote locations and developing regions, the treatment system must be robust, noncomplex, and require only low-level maintenance. Therefore, photocatalyst regeneration stages are undesirable. There needs to be more research into the longevity of the photocatalyst under working conditions. To reduce the complexity of the treatment system, immobilized photocatalyst systems are preferred; however, catalyst stripping may be a problem if the immobilization protocol does not produce a robust hard wearing coating. Also, catalyst fouling by inorganic species present in the water can lead to a reduction in the photocatalytic efficiency over time. Miranda-García et al. (2010), based at PSA, investigated the degradation of 15 emerging contaminants in a photocatalytic pilot plant using TiO_2 immobilized onto glass beads. The CPC plant consisted of two modules of 12 Pyrex glass tubes mounted on a fixed platform tilted 37° (local latitude). Two of the glass tubes were packed with the TiO_2-coated glass spheres. The total illuminated area was 0.30 m^2 and total volume was 10 L of which 0.96 L was the irradiated volume. The system operated in recirculating batch mode with a flow rate of 3.65 L min^{-1}. The photoactive layer of TiO_2 was deposited on glass spheres using a sol–gel method. They found that the degradation of the organic pollutants was achieved under solar irradiation, and, importantly, after 5 cycles of photocatalysis, the photocatalyst activity was not decreased significantly. However; it would appear that distilled water was used as the water source in the experiments.

9.7.3 Visible-Light-Active Photocatalyst Materials

The overall efficiency of TiO_2 under natural sunlight is limited to the UV-driven activity (for anatase $\lambda \leq 400$ nm), accounting only to ca. 4% of the incoming solar energy on Earth's surface. Therefore, there has been a substantial amount of research effort toward shifting of the absorption spectrum of TiO_2 toward the visible region of the electromagnetic spectrum. Different approaches have been attempted including doping the TiO_2 with metal ions (Hamilton et al. 2008). According to the literature, one of the more promising approaches to achieve visible-light activity is doping with nonmetal elements including N and S. Since Asahi et al. (2001) reported the visible-light photoactivity of TiO_2 with nitrogen doping, many groups have demonstrated that anion doping of TiO_2 extends the optical absorbance of TiO_2 into the visible-light region. However, the number of publications concerning the photocatalytic activity of these materials for the inactivation of microorganisms is limited.

Li et al. (2008) based at the University of Illinois, reported on the inactivation of MS2 phage under visible-light irradiation using a palladium-modified nitrogen-doped titanium oxide (TiON/PdO) photocatalytic fiber, synthesized on a mesoporous activated carbon fiber template by a sol–gel process. Dark adsorption led to virus removal and subsequent visible-light illumination (wavelengths greater than 400 nm and average intensity of 40 mW/cm^2) resulted in additional virus removal of 94.5%–98.2% within 1 hour of additional contact time. By combining adsorption and visible-light photocatalysis, TiON/PdO fibers reached final virus removal rates of 99.75%–99.94%. Electron paramagnetic resonance (EPR) measurements confirmed the production of ·OH radicals by TiON/PdO under visible-light illumination. Wu et al. (2009) from the University of Illinois also reported on the visible-light-induced photocatalytic inactivation of bacteria by composite photocatalysts of palladium oxide and nitrogen-doped TiO_2. The PdO/TiON catalysts were tested for visible-light-activated photocatalysis using gram-negative organisms, that is *E. coli* and *Pseudomonas aeruginosa*, and a gram-positive organism *Staphylococcus aureus*. Their disinfection data showed that the PdO/TiON photocatalysts had a much better visible photocatalytic activity than either palladium-doped (PdO/TiO_2) or nitrogen-doped titanium oxide (TiON). The photocatalytic reactor was rather basic, using a petri dish that was stirred periodically. The light source was a metal halogen desk lamp with a low UV output (<0.01 mW/cm^2 for $\lambda \leq 400$ nm). While these photocatalysts show promise for truly *visible*-light activity, there was no comparison with undoped TiO_2 for photocatalytic activity under solar simulated light (which has around 5% UV).

In many cases, the UV activity of undoped TiO_2 is much greater than the visible-light activity of the doped material. Therefore, for solar applications, the photocatalysts should be tested under simulated solar irradiation or under real sun conditions. Indeed, Rengifo-Herrera and Pulgarin (2010) recently reported on the photocatalytic activity of N,S co-doped and N-doped commercial anatase (Tayca TKP 102) TiO_2 powders toward phenol oxidation and *E. coli* inactivation under simulated solar light irradiation. Their proposed mechanism is shown in Figure 9.15.

However, these novel materials did not present an enhancement for the photocatalytic degradation of phenol or the photocatalytic inactivation of *E. coli* under simulated solar light, as compared to Degussa P-25. They suggest that while the N or N,S co-doped TiO_2 may show a visible-light response, the localized states responsible for the visible-light absorption do not play an important role in the photocatalytic activity.

More research is required to determine if visible-light active materials can deliver an increase in the efficiency of photocatalysis under solar irradiation.

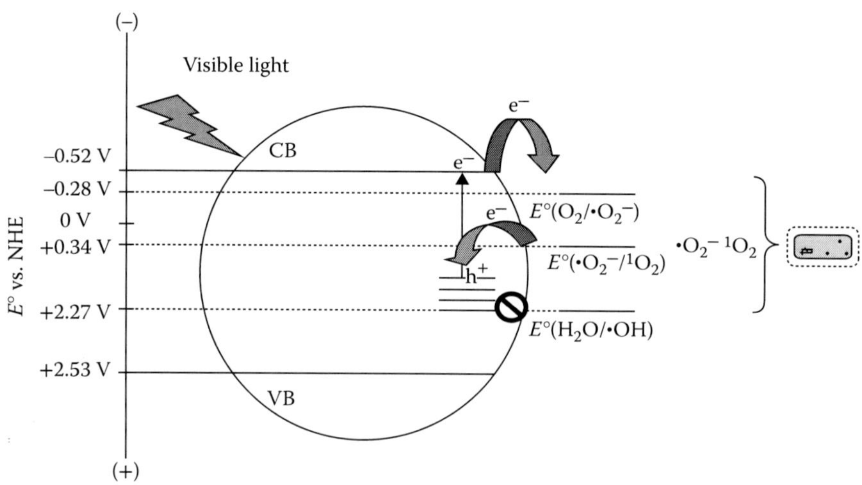

FIGURE 9.15 Possible mechanism for the photocatalytic action of N,S co-doped TiO_2 exposed to visible light, including singlet oxygen photogeneration. (From Rengifo-Herrera, J.A., Pierzchała, K., Sienkiewicz, A., Forro, L., Kiwi, J., Pulgarin, C., *Appl Catal B: Environ*, 88, 398–406, 2009. With permission.)

9.8 CONCLUSIONS

SODIS is a simple and low-cost technique used to disinfect contaminated drinking water. Transparent bottles (preferably PET) are filled with contaminated water and placed in direct sunlight for a minimum of 6 hours. Following exposure, the water is safer to drink as the viable pathogen load can be significantly decreased. The process has approximately 4.5 million regular users, predominately in Africa, Latin America, and Asia and is recognized and promoted by the WHO. However; there are several drawbacks of *conventional* SODIS technology. The use of PET bottles allows for only small volumes to be treated (2–3 L), and the process efficiency is dependent on a range of environmental parameters including solar irradiance (which depends on the latitude, time of day, and atmospheric conditions) and initial water quality, for example organic loading, turbidity, level, and nature of the bacterial contamination. There are a number of ways to improve or enhance the conventional SODIS process and these include the design of SODIS bags where the solar dose per volume is increased; the use of UV dosimetric indicators, which measure the UV dose and indicate to the user when the desired dose has been received by the water; design of customized SODIS treatment systems, which maximize the solar dose and may include UV feedback sensors for automated control; and the use of semiconductor photocatalysis to enhance the treatment efficacy. Semiconductor photocatalysis has been shown to be effective for the inactivation of a wide range of microorganisms at lab scale and under real sun conditions for both small-scale and large-scale applications. The use of CPC reactors enhances the efficiency of solar disinfection and photocatalytically enhanced solar disinfection. Nevertheless, there are a number of issues to be addressed before photocatalytically enhanced solar disinfection can be effectively deployed in developing regions. These include improvements in photoreactor design and assessment of photocatalyst longevity under real operating conditions. Future developments in relation to visible-light active photocatalytic materials may lead to more efficient solar photocatalysis for the disinfection of water.

ACKNOWLEDGMENTS

We acknowledge Dr. Kevin McGuigan, Royal College of Surgeons in Ireland, and Dr. Cathy McCullough, Robert Gordon University (UK), for reviewing this chapter and their helpful suggestions and comments.

REFERENCES

Acra A, Karahagopian Y, Raffoul Z, Dajani R (1980) "Disinfection of oral rehydration solutions by sunlight," *Lancet*, 316:1257–1258.

Acra A, Raffoul Z, Karahagopian Y (1984) *Solar Disinfection of Drinking Water and Oral Rehydration Solutions*, UNICEF, Paris.

Alrousan DMA, Dunlop PSM, McMurray TA, Byrne JA (2009) "Photocatalytic inactivation of *E. coli* in surface water using immobilised nanoparticle TiO_2 films," *Water Res*, 43:47–54.

Asahi R, Morikawa T, Ohwaki T, Aoki T, Taga K (2001) "Visible-light photocatalysis in nitrogen-doped titanium oxides," *Science*, 293:269–271.

Augugliaro V, Litter M, Palmisano L, Soria J (2006) "The combination of heterogeneous photocatalysis with chemical and physical operations: a tool for improving the photoprocess performance," *J Photochem Photobiol C*, 7:127–144.

Bahnemann D (2004) "Photocatalytic water treatment: solar energy applications," *Solar Energy*, 77:445–459.

Bandala ER, González L, de la Hoz F, Pelaez MA, Dionysiou DD, Dunlop PSM, Byrne JA, Sanchez JL (2011) "Application of azo dyes as dosimetric indicators for enhanced photocatalytic solar disinfection (ENPHOSODIS)," *J Photochem Photobiol, A*, 218:185–191.

Berney M, Weilenmann H, Simonetti A, Egli T (2006a) "Efficacy of solar disinfection of *Escherichia coli, Shigella flexneri, Salmonella typhimurium* and *Vibrio cholerae*," *J Appl Microbiol*, 101:828–836.

Berney M, Weilenmann HU, Egli T (2006b) "Flow-cytometric study of vital cellular functions in *Escherichia coli* during solar disinfection (SODIS)," *Microbiology*, 152:1719–1729.

Berney M, Weilenmann HU, Egli T (2006c) "Gene expression of *Escherichia coli* in continuous culture during adaptation to artificial sunlight," *Environ Microbiol*, 9:1635–1647.

Blake DM, Maness PC, Huang Z, Wolfrum EJ, Huang J, Jacoby WA (1999) "Application of the photocatalytic chemistry of titanium dioxide to disinfection and the killing of cancer cells," *Sep Purif Methods*, 28:1–50

Blanco J, Malato S, Fernández-Ibañez P, Alarcón D, Gernjak W, Maldonado MI (2009) "Review of feasible solar energy applications to water processes," *Renew Sust Energ Rev*, 13:1437–1445.

Boyle M, Sichel C, Fernandez-Ibanez P, Arias-Quiroz GB, Iriarte-Puna M, Mercado A, Ubomba-Jaswa E, McGuigan KG (2008) "Bactericidal effect of solar water disinfection under real sunlight conditions," *Appl Environ Microbiol* 74:2997–3001.

Burch JD, Thomas K.E (1998) "Water disinfection for developing countries and potential for solar thermal pasteurization," *Solar Energy* 64:87–97.

Byrne JA, Eggins BR, Brown NMD, McKinney B, Rouse M (1998) "Immobilisation of TiO_2 powder for the treatment of polluted water," *Appl Catal B: Environ*, 17:25–36.

Byrne JA, Fernandez-Ibanez PA, Dunlop PSM, Alrousan DMA, Hamilton JWJ (2011) "Photocatalytic enhancement for solar disinfection of water: a review," *Int J Photoenergy*, 12. DOI:10.1155/2011/798051.

Cassano AE, Alfano OM (2000) "Reaction engineering of suspended solid heterogeneous photocatalytic reactors,"*Catal Today*, 58:167–197.

Cho M, Chung H, Choi W, Yoon J (2004) "Linear correlation between inactivation of *E. coli* and OH radical concentration in TiO_2 photocatalytic disinfection," *Wat Res*, 38:1069–1077.

Clasen T, Edmondson P, (2006) "Sodium dichloroisocyanurate (NaDCC) tablets as an alternative to sodium hypochlorite for the routine treatment of drinking water at a household level," *Int J Hyg Environ Health*, 209:173–181.

Clasen TF, Haller L (2008) *Water Quality Interventions to Prevent Diarrhoea: Cost and Cost-Effectiveness*, WHO, Geneva, Switzerland.

Conroy RM, Elmore-Meegan M, Joyce T, McGuigan KG, Barnes J (1996) "Solar disinfection of drinking water and diarrhoea in Maasai children: a controlled field trial," *Lancet*, 348:1695–1697.

Conroy RM, Elmore-Meegan M, Joyce T, McGuigan KG, Barnes J (1999) "Solar disinfection of water reduces diarrhoeal disease: an update," *Arch Dis Child*, 81:337–338.

Conroy RM, Meegan M, Joyce T, McGuigan KG, Barnes J (2001) "Solar disinfection of drinking water protects against cholera in children under 6 years of age," *Arch Dis Child*, 85:293–295.

Dalrymple OK, Stefanakos E, Trotz MA, Goswami DY (2010) "A review of the mechanisms and modeling of photocatalytic disinfection," *Appl Catal B: Environ*, 98:27–38.

Downes A, Blunt TP (1887) "The influence of light upon the development of bacteria," *Proc Roy Soc Med*, 26:488–500.

Du Preez M, Conroy RM, Ligondo S, Hennessy J, Elmore-Meegan M, Soita A, McGuigan KG (2011) *Environ Sci Technol*, 45:9315–9323.

Du Preez M, McGuigan KG, Conroy RM (2010) "Solar disinfection of drinking water in the prevention of dysentery in South African children aged under 5 years: the role of participant motivation," *Environ Sci Technol*, 15:8744–8749.

Dunlop PSM, Byrne JA, Manga N, Eggins BR (2002) "The photocatalytic removal of bacterial pollutants from drinking water," *J Photochem Photobiol A*, 148:355–363.

Dunlop PSM, Ciavola Rizzo ML, Byrne JA (2011) "Inactivation and injury assessment of *Escherichia coli* during solar and photocatalytic disinfection in LDPE bags," *Chemosphere*, 85:1160–1166.

Dunlop PSM, McMurray TA, Hamilton JWJ, Byrne JA (2008) "Photocatalytic inactivation of *Clostridium perfringens* spores on TiO_2 electrodes," *J Photochem Photobiol A*, 196:113–119.

Fernandez P, Blanco J, Sichel C, Malato S (2005) *Catal Today*, "Water disinfection by solar photocatalysis using compound parabolic collectors," 101:345–352.

Fernández-Ibáñez P, Sichel C, Polo-López I, de Cara-García M, Tello JC (2009) "Photocatalytic disinfection of natural well water contaminated by *Fusarium solani* using TiO_2 slurry in solar CPC photo-reactors," *Catal Today*, 144:62–68.

Fisher MB, Keenan CR, Nelson KL, Voelker BM (2008) "Speeding up solar disinfection (SODIS): effects of hydrogen peroxide, temperature, pH, and copper plus ascorbate on the photoinactivation of *E. coli*," *J Water Health*, 6:35–51.

Fjendbo Jorgensen AJ, Nohr K, Sorensen H, Boisen F (1998) "Decontamination of water by direct heating in solar panels," *J Appl Microbiol*, 85:441–447.

Fujishima A, Zhang XT, Tryk DA (2008) "TiO_2 photocatalysis and related surface phenomena," *Surf Sci Rep*, 63:515–582.

Gaya UI, Abdullah AH (2008) "Heterogeneous photocatalytic degradation of organic contaminants over titanium dioxide: a review of fundamentals, progress and problems," *J Photochem Photobiol C*, 9:1–12.

Gelover S, Gomez LA, Reyes K, Leal MT (2006) "A practical demonstration of water disinfection using TiO_2 films and sunlight," *Water Res*, 40:3274–3280.

Gómez-Couso H, Fontán-Saínz M, Sichel C, Fernández-Ibáñez P, Ares Mazás E (2009) "Efficacy of the solar water disinfection method in turbid waters experimentally contaminated with *Cryptosporidium parvum* oocysts under real field conditions," *Trop Med Int Health*, 14:620–627.

Goswami YD, Kreith F, Kreider JF (2000) *Principles of Solar Engineering*, 2nd ed., Taylor & Francis, Philadelphia, PA.

Hamamoto A, Mori M, Takahashi A, Nakano M, Wakikawa N, Akutagawa M, Ikehara T, Nakaya Y, Kinouchi Y (2007) "New water disinfection system using UVA light-emitting diodes," *J Appl Microbiol*, 103:2291–2298.

Hamilton JWJ, Byrne JA, McCullagh C, Dunlop PSM (2008) "Electrochemical investigation of doped titanium dioxide," *Int J Photoen*, 8. DOI:10.1155/2008/631597.

Heaselgrave W, Patel N, Kilvington S, Kehoe SC, McGuigan KG (2006) "Solar disinfection of poliovirus and *Acanthamoeba polyphaga* cysts in water—a laboratory study using simulated sunlight," *Lett Appl Microbiol*, 43:125–130.

Herrmann J (1999) "Heterogeneous photocatalysis: fundamentals and applications to the removal of various types of aqueous pollutants," *Catal Today*, 53:115–129.

Huang Z, Maness P, Blake DM, Wolfrum EJ, Smolinski SL, Jacoby WA (2000) "Bactericidal mode of titanium dioxide photocatalysis," *J Photochem Photobiol A*, 130:163–170.

Kanade KG, Baeg J, Mulik UP, Amalnerkar DP, Kale BB (2006) "Nano-CdS by polymer-inorganic solid-state reaction: visible light pristine photocatalyst for hydrogen generation," *Mater Res Bul*, 41:2219–2225.

Karunakaran C, Senthilvelan S, Karuthapandian S, Balaraman K (2004) "Photooxidation of iodide ion on some semiconductor and non-semiconductor surfaces," *Catal Commun*, 5:283–290.

Kehoe SC, Barer MR, Devlin LO, McGuigan KG (2004) "Batch process solar disinfection is an efficient means of disinfecting drinking water contaminated with *Shigella dysenteriae* type I," *Lett Appl Microbiol*, 38:410–414.

Kehoe SC, Joyce TM, Ibrahim P, Gillespie JB, Shahar RA, McGuigan KG (2001) "Effect of agitation, turbidity, aluminium foil reflectors and container volume on the inactivation efficiency of batch-process solar disinfectors," *Water Res*, 35:1061–1065.

Khaengraeng R, Reed RH (2005) "Oxygen and photoinactivation of *Escherichia coli* in UVA and sunlight," *J Appl Microbiol*, 99:39–50.

Lee C, Lee Y, Yoon J (2006) "Oxidative degradation of dimethylsulfoxide by locally concentrated hydroxyl radicals in streamer corona discharge process," *Chemosphere*, 65:1163–1170.

Li Q, Page MA, Marinas BJ, Shang JK (2008) "Treatment of coliphage MS2 with palladium-modified nitrogen-doped titanium oxide photocatalyst illuminated by visible light," *Environ Sci Technol*, 42:6148–6153.

Litter MI (1999) "Heterogeneous photocatalysis: transition metal ions in photocatalytic systems," *Appl Catal B*, 23:89–114.

Lonnen J, Kilvington S, Kehoe SC, Al-Touati F, McGuigan KG (2005) "Solar and photocatalytic disinfection of protozoan, fungal and bacterial microbes in drinking water," *Water Res*, 39:877–883.

Malato S, Fernandez-Ibanez P, Maldonado MI, Blanco J, Gernjak W (2009) "Decontamination and disinfection of water by solar photocatalysis: recent overview and trends," *Catal Today*, 147:1–59.

Mani SK, Kanjura R, Singha ISB, Reed RH (2006) "Comparative effectiveness of solar disinfection using small-scale batch reactors with reflective, absorptive and transmissive rear surfaces," *Water Res*, 40:721–727.

Matsunaga T, Tomoda R, Nakajima T, Wake H (1985) "Photoelectrochemical sterilization of microbial cells by semiconductor powders," *FEMS Microbiol Letters*, 29:211–214.

McCullagh C, Robertson JM, Bahnemann DW, Robertson PJK (2007) "The application of TiO_2 photocatalysis for disinfection of water contaminated with pathogenic microorganisms: a review," *Res Chem Intermed*, 33:359–375.

McGuigan KG, Joyce TM, Conroy RM, Gillespie JB, Elmore-Meegan M (1998) "Solar disinfection of drinking water contained in transparent plastic bottles: characterizing the bacterial inactivation process," *J Appl Microbiol*, 84:1138–1148.

McGuigan KG, Mendez-Hermida F, Castro-Hermida JA, Ares-Mazas E, Kehoe SC, Boyle M, Sichel C et al (2006) "Batch solar disinfection inactivates oocysts of *Cryptosporidium parvum* and cysts of *Giardia muris* in drinking water," *J Appl Microbiol*, 101:453–463.

McGuigan KG, Samaiyar P, du Preez M, Conroy RM (2011) "Randomized intervention study of solar disinfection of drinking water in the prevention of dysentery in Kenyan children aged under 5 years," *Environ Sci Technol*, 45:7862–7867.

McMurray TA, Byrne JA, Dunlop PSM, Winkelman JGM, Eggins BR, McAdams ET (2004) "Intrinsic kinetics of photocatalytic oxidation of formic and oxalic acid on immobilised TiO2 films." *Appl Catal A*, 262:105–110.

McLoughlin OA, Fernández P, Gernjak W, Malato S, Gill LW (2004) "Photocatalytic disinfection of water using low cost compound parabolic collectors," *Solar Energy*, 77:625–633.

Miranda-García N, Maldonado MI, Coronado JM, Malato S (2010) "Degradation study of 15 emerging contaminants at low concentration by immobilized TiO_2 in a pilot plant," *Catal Today*, 151:107–113.

Mtapuri-Zinyowera S, Midiz N, Muchaneta-Kubera CE, Simbini T, Mduluza T (2009) "Impact of solar radiation in disinfecting drinking water contaminated with *Giardia duodenalis* and *Entamoeba histolytica/dispar* at a point-of-use water treatment," *J Appl Microbiol*, 106:847–852.

Navntoft C, Ubomba-Jaswa E, McGuigan KG, Fernández-Ibáñez P (2008) "Effectiveness of solar disinfection using batch reactors with non-imaging aluminium reflectors under real conditions: natural well-water and solar light," *J Photochem Photobiol B*, 93:155–161.

Oates PM, Shanahan P, Polz MF (2003) "Solar disinfection (SODIS): simulation of solar radiation for global assessment and application for point-of-use water treatment in Haiti," *Water Res*, 37:47–54.

Polo-López MI, Fernández-Ibáñez P, Ubomba-Jaswa E, Navntoft C, Garcia-Fernandez I, Dunlop PSM, Schmid, M et al. (2011) "Elimination of water pathogens with solar radiation using an automated sequential batch CPC reactor," *J Hazard Mater*, 196:16–21.

Reed RH (2004) "The inactivation of microbes by sunlight: solar disinfection as a water treatment process," *Adv Appl Microbiol*, 54:333–365.

Rengifo-Herrera JA, Pierzchała K, Sienkiewicz A, Forro L, Kiwi J, Pulgarin C (2009) "Abatement of organics and *Escherichia coli* by N, S co-doped TiO_2 under UV and visible light. Implications of the formation of singlet oxygen (O-1(2)) under visible light,"*Appl Catal B: Environ*, 88:398–406.

Rengifo-Herrera JA, Pulgarin C (2010) "Photocatalytic activity of N, S co-doped and N-doped commercial anatase TiO_2 powders towards phenol oxidation and *E. coli* inactivation under simulated solar light irradiation," *Sol Energy*, 84:37–43.

Rincon A, Pulgarin C (2004) "Field solar *E-coli* inactivation in the absence and presence of TiO_2: is UV solar dose an appropriate parameter for standardization of water solar disinfection?" *Sol Energy*, 77:635–648.

Rincon AG, Pulgarin C, Adler N, Peringer P (2001) "Interaction between *E-coli* inactivation and DBP-precursors - dihydroxybenzene isomers - in the photocatalytic process of drinking-water disinfection with TiO_2," *J Photochem Photobiol A*, 139:233–241.

Saitoh TS, El-Ghetany HH (2002) "A pilot solar water disinfecting system: performance analysis and testing" *Sol Energy*, 72:261–269.

Sichel C, Tello J, de Cara M, Fernández-Ibáñez P (2007) "Effect of UV solar intensity and dose on the photocatalytic disinfection of bacteria and fungi," *Catal Today*, 129:152–160.

SODIS (2010a) Newsletter No. 1, Swiss Federal Institute of Aquatic Science and Technology.

SODIS (2010b) Available at http://www.sodis.ch, accessed August 24, 2010.

Sommer B, Mariño A, Solarte Y, Salas ML, Dierolf C, Valiente C (1997) "SODIS—an emerging water treatment process," *J Water Supply Res Technol-Aqua*, 46:127–137.

Sunada K, Watanabe T, Hashimoto K (2003) "Studies on photokilling of bacteria on TiO_2 thin film," *J Photochem Photobiol A*, 156:227–233.

Sunnotel O, Verdoold R, Dunlop PSM, Snelling WJ, Lowery CJ, Dooley JSG, Moore JE, Byrne JA (2010) "Photocatalytic inactivation of *Cryptosporidium parvum* on nanostructured titanium dioxide films," *J Water Health*, 8:83–91.

Ubomba-Jaswa E, Navntoft C, Polo-López MI, Fernández-Ibáñez P, McGuigan KG (2009) "Solar disinfection of drinking water (SODIS): an investigation of the effect of UV-A dose on inactivation efficiency," *Photochem Photobiol Sci*, 8(5):587–595.

Ubomba-Jaswa E, Fernández-Ibáñez P, Navntoft C, Polo-López MI, McGuigan KG (2010) "Investigating the microbial inactivation efficiency of a 25 L batch solar disinfection (SODIS) reactor enhanced with a compound parabolic collector (CPC) for household use," *J Chem Tech Biotech***,** 85:1028–1037.

Vidal A, Díaz AI, El Hraiki A, Romero M, Muguruza I, Senhaji F, González J (1999) "Solar photocatalysis for detoxification and disinfection of contaminated water: pilot plant studies," *Cat Today*, 54:183–290.

Wainwright M (2000) "Methylene blue derivatives—suitable photoantimicrobials for blood product disinfection?" *Int J Antimicrob Agents*, 16:381–394.

Wegelin M, Canonica MS, Mechsner K, Fleischmann T, Pesaro F, Metzler A (1994) "Solar water disinfection: scope of the process and analysis of radiation experiments," *J Water Supply: Res Technol - Aqua*, 43:154–169.

WHO (2007) *Economic and health effects of increasing coverage of low cost household drinking-water supply and sanitation interventions to countries off-track to meet MDG target 10*. World Health Organisation, Geneva, Switzerland.

WHO (2011) *Guidelines for Drinking-water Quality* (4th Edition). World Health Organization, Geneva.

WHO/UNICEF (2011) *Drinking Water Equity, safety and sustainability*, JMP Thematic Report on Drinking Water 2011. World Health Organization, Geneva, Switzerland, and UNICEF, New York.

WHO/UNICEF (2012) *Progress on Sanitation and Drinking-water: 2012 Update*. World Health Organization, Geneva, Switzerland, and UNICEF, New York.

Wu P, Xie R, Imlay JA, Shang JK (2009) "Visible-light induced photocatalytic inactivation of bacteria by composite photocatalysts of palladium oxide and nitrogen-doped titanium oxide," *Appl Catal B: Environ*, 88:576–581.

10 Applications of Nanotechnology in the Building Industry

Michael B. Cortie, Nicholas Stokes, Gregory Heness, and Geoffrey B. Smith

CONTENTS

10.1 INTRODUCTION

A growing number of materials technologies that deliberately exploit some aspect of nanotechnology are available to the building industry. Surface coatings of nanoscale thickness or comprising some nanomaterials are a common example. The interest in using these nanotechnologies is driven mainly by two factors: a desire to reduce maintenance costs and/or a need to reduce the energy consumption of a building to the minimum consistent with reasonable comfort for its occupants. Here, we provide an overview of these technologies and offer some comments regarding their performance and relevance within the context of the building industry. Only a somewhat abbreviated account is provided; greater technical detail may be found in the references provided or in the comprehensive work by Smith and Granqvist (2011).

The coatings technologies may be roughly grouped into those that control the optical properties of a surface and those with some chemical or other physical functionality.

A spectrally selective coating on a window is a classic example of the first kind, whereas a coating that can control either hydrophobicity or corrosion resistance of a surface is an example of the second kind. Some crossovers between these functionalities in the same product are also possible. These technologies are already well established. As we will see, however, there are also some other interesting new products or ideas that are still somewhat underexploited in the building industry. For example, there are several substances that can be used to prepare photocatalytic surfaces for use in bathrooms or on windows. These surfaces consequently acquire an active functionality and, for example, can clean themselves, destroy microorganisms, or purify indoor air of contaminants such as formaldehyde.

Surface coatings of various kinds were a prominent feature of the prototypical NanoHouse demonstrator designed by the Australian architect James Muir and built in 2004 in Sydney, Australia (Muir et al. 2004; Cortie et al. 2005). The NanoHouse (Figure 10.1) made extensive use of coated glass of various types, ultraviolet (UV)-resistant coatings for furniture, and coatings of silver nanoparticles for antibacterial surfaces. It was also designed to draw some of its electrical energy from photovoltaic panels and exploited several energy-saving ideas of a non-nanotech nature, such as passive cooling. Over 70,000 visitors passed through the NanoHouse while it was on display, and it was noted that the interior of the structure was at a pleasant temperature even on a hot Sydney afternoon. This achievement is in part attributable to the extensive use of spectrally selective coatings.

In this chapter, we discuss some of the nanotechnologies that were used in the NanoHouse and several that were not. However, this chapter excludes topics such as photovoltaics and other forms of energy generation.

10.2 COATINGS FOR SPECTRAL SELECTIVITY

Coatings that control optical properties are very accessible to designers because a wide range of coating products and coated products are already on the market.

FIGURE 10.1 The UTS-CSIRO NanoHouse, on display on the forecourt of the Sydney Opera House. This experimental construction made extensive use of spectrally selective coatings and other energy-saving technologies (photograph by the authors).

10.2.1 Passive Coatings on Windows and Glass

10.2.1.1 General

The consumption of electrical energy for lighting, heating, cooling, and ventilating the environment within buildings represents a significant fraction of the electrical energy drawn in industrialized countries. When the operation of indoor electrical appliances is also included, it has been estimated that about 40% of all electrical energy is used for these purposes (Granqvist 2003). Much of this energy is released as heat within the building and, although this may be welcome in winter, will obviously create discomfort in summer or in tropical climates.

In hot climates, the problem is exacerbated by the ingress through windows of additional heat from the outdoors. The sun can irradiate surfaces with approximately 500 $W{\cdot}m^{-2}$ over a wide range of latitudes, and about 90% of this will pass through a normal pane of glass. In addition, solid-state conduction of heat takes place from the hot air outside, through the glass, and into a cooler interior environment. Therefore, windows can generate a significant heat load on the adjacent interior space in warm climates. Of course, in cold climates or in winter the problem is reversed. In this case, the priority may be to slow the transfer of heat through windows and out into the cold external environment.

It has been estimated that in the United States alone 4×10^{18} J of energy use may be directly attributed to the presence of windows, and the extra cost to the consumer is on the order of $40 billion (Anon 2006). It is worth noting that heat is also readily transferred between indoors and outdoors by the movement of air, so the aforementioned comments about windows are only relevant if the building is not excessively ventilated. Control of the exchange of air between interior and exterior is therefore a key preparatory step before any of the energy-saving nanotechnologies can be applied.

Heat transfer through a window takes place by a combination of radiation, convection, and conduction. Considerable restriction of the latter two mechanisms can be achieved using double glazed assemblies, but this is not primarily a nanotechnological solution and will therefore not be discussed further here. However, heat transfer by radiation through the nominally transparent window remains an important issue, particularly in the tropics (Dai 2001), and this can be readily controlled by using nanoscale coatings of various kinds.

The basic requirement for such coatings is that they should block as much of the near-infrared radiation received as possible while transmitting most of the visible wavelengths. Since perceived color is an important consideration, the spectrum of the transmitted light is very important, and a neutral color balance is required. The challenge in the field has been to find material and geometry combinations that maximize this combination of effects, at a realistic cost. The available coating technologies are of two rather different types: reflective and absorptive.

10.2.1.2 Coatings That Reflect Infrared Radiation

In general, a continuous film of a metal has the reflectivity desired to block infrared wavelengths, but unfortunately in most cases the high reflectivity also extends into the visible range, thereby also blocking the desired visible light. Only a few materials have sufficient free electron densities combined with specific dielectric functions that simultaneously allow both reflection of the infrared and transmission of much of the visible. Gold; silver; and the compounds titanium nitride, tin dioxide, and indium oxide are examples of such materials. Currently, nanoscale coatings of silver, in conjunction with dielectric compounds that protect and are antireflective in the visible, dominate this aspect of the architectural glass market. These are manufactured by magnetron sputtering, and the individual layers are from 3 to 40 nm thick.

Although reflective coatings are very efficient from an optical and energy perspective, they are relatively expensive to manufacture. This is due to the high cost of the vacuum equipment and the batch nature of the manufacturing process. Another problem with reflective coatings is that the light and heat that they reflect may in turn be shed onto neighboring buildings. The town planning legislation of some jurisdictions around the world now specifically controls or prohibits this possibility.

10.2.1.3 Coatings That Absorb Infrared Radiation

Given the issues with reflective coating systems, there is a growing interest in an alternative type of coating in which near-infrared radiation is absorbed rather than reflected. In these technologies, either a nanoparticle with suitable spectral selectivity or a dye molecule in solid solution within a polymer is applied as a coating to the window. Suitable nanoparticulate materials in commercial use for this purpose include iron oxide, indium tin oxide, and lanthanum hexaboride (Takeda et al. 2001; Smith et al. 2002; Schelm and Smith 2003). Gold nanospheres and gold nanorods have also been trialed but are not yet competitive with commercially available products on the basis of performance (Xu et al. 2004; Stokes et al. 2010). Surprisingly, it remains feasible that these precious metal coatings could compete on the basis of cost if only their performance could be improved. This is because of the very small mass of gold used per square meter of glass (Xu et al. 2006) and is an illustration of the capability of nanotechnology to radically reduce the amount of material required to achieve a particular effect.

It is particularly convenient to blend the absorptive nanoparticles into the polymer interlayer of safety glass, since that is part of the construction of the windowpane anyway. In addition, this protects the nanoparticles and polymer matrix from adverse chemical interactions with the atmosphere. Unfortunately, an absorptive coating will, in principle, always be less efficient than a reflective one because about half of the energy that is absorbed by the coating will be transmitted to the interior. (The other half, roughly speaking, will be shed to the outside environment by convection and radiation.) Solar control by a combination of reflection and double glazing can reduce the heat input through a sunlit window to as little as 20%–30% of its normal value, whereas the combination of absorption and double glazing can typically reduce the energy input only by about 50%. However, absorptive coatings are potentially cheaper to manufacture and do not reflect infrared radiation onto adjacent buildings, so they are being increasingly specified. These coatings do have problems of their own, however, with cracking of the glass due to the generation of thermal stresses being a possibility (Chowdhury and Cortie 2007).

The energy balance over three types of glazing systems is shown schematically in Figure 10.2. It is clear that the reflective type of coating is superior in terms of minimizing solar heat gain.

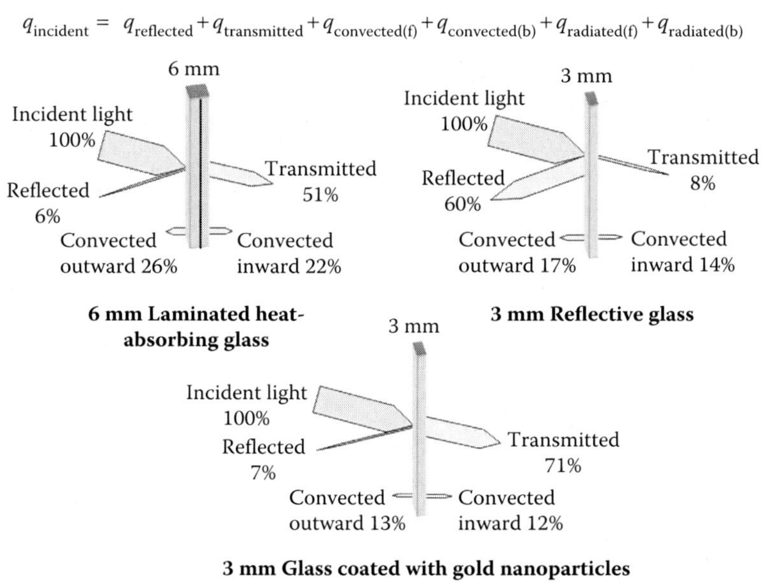

FIGURE 10.2 Energy balance over three kinds of glazing systems, subscript b refers to the back surface of the glass, subscript f to the front surface. (Reproduced from Chowdhury 2007. With permission.)

10.2.1.4 Angularly Selective Coatings

These provide another technology that has good potential for reducing heat transfer into the interior environment. They would be useful in situations where it is desired that sunlight at low angles (early morning and late afternoon and more in winter than summer) passes through a window but light at higher incident angles is blocked or attenuated. This type of design allows people on the inside to see out horizontally and below the horizontal, as shown in Figure 10.3, and also allows in adequate daylight without glare. Here, we only consider angularly selective coatings of nanoscale dimensions. (Obviously, many macroscale solutions are also there, including the classic venetian blind.)

Angular selectivity can be achieved at the nanoscale by controlling conditions during the deposition of coatings by magnetron sputtering, for example. An opaque material, for example, chromium metal, may be deposited in the form of columns inclined at some angle to the plane of the glass, produced by the process of glancing angle deposition. Maximum transmittance occurs when incident light is parallel to the columns, allowing clear viewing below the horizontal. Typical overall transmittance could be approximately 68% for a 90 nm thick coating (Bell and Matthews 1998). Another type of angularly selective coating can be produced by depositing a substance onto a templating monolayer of polymer microspheres (Liu et al. 2006), a process sometimes known as nanosphere lithography. One complication with all these types of coatings is that the polarization of the incident light is also a factor (Bell and Matthews 1998); another is that they are exceedingly expensive to produce.

10.2.2 Active or *Smart* Coatings on Windows or Glass

10.2.2.1 General

There is widespread interest in systems that would provide a switchable optical property, especially for windows. In particular, a means to trigger a change from transparent to opaque, or transparent to reflective, is of interest for reasons of privacy or energy conservation. These possibilities have given rise to the concept of the smart window. A variety of systems of varying degrees of complexity are commercially available to meet this need, but several problems still need to be overcome

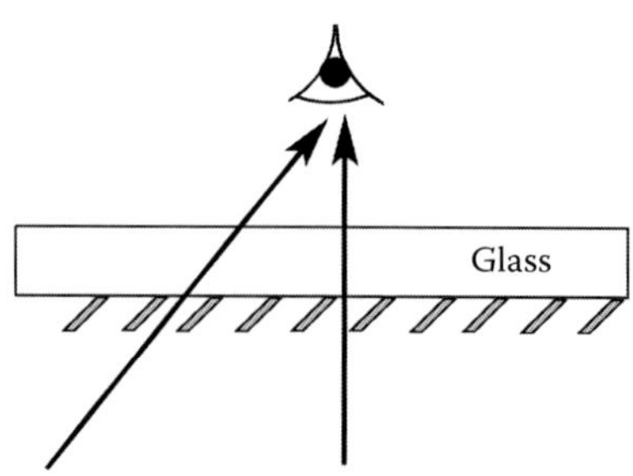

FIGURE 10.3 Schematic of angular selective films showing venetian blind–like microcolumns, typically 45°–60° from perpendicular and significantly reducing the light from higher angles.

before a wider uptake is likely. One issue is the high cost of such systems; another is that uniformity of color is only available from relatively small panes of glass; and, finally, there is the question of durability. Architectural applications probably require 20 years or more of service, and many of the current switchable windows suffer from a gradual decline in performance over the years.

The change in optical property can be obtained by various means, and these are briefly discussed in Sections 10.2.2.2 through 10.2.2.6.

10.2.2.2 Active Liquid Crystal Systems

Liquid crystal–based systems currently only allow visual transmission when a voltage is applied, so there is a continual consumption of power, albeit at a low level. Their key nanotechnology is the transparent conducting oxide that is required for their function. However, they may only be suitable for internal and small-sized glass walls and are extremely expensive. They are not discussed further in this chapter.

10.2.2.3 Electrochromic Systems

The essential property of an electrically activated chromogenic material is that it exhibits a large change in its optical properties upon either a change in the applied electric field or the injection or ejection of a charge. This change in optical properties is exhibited in the form of a change in absorbance, reflectance, or scattering (Lampert 1998). The change depends on the type of material and can form a highly transmissive state or a partly reflecting or absorbing state; the change can also occur over part or all of the solar spectrum (Lampert 1998). Most designs draw electric power only during the switching event, and a fine control over the resulting tint can be obtained (Wittkopf 1997). Electrochromic systems may currently be the most popular of the switchable coating technologies. Electrochromic materials may be divided into two groups: those that develop an absorption band and those in which the effective free electron density in the material is modulated. These phenomena are also called *absorptance modulation* and *reflectance modulation*, respectively (Bell and Matthews 1998). The most common electrochromic material currently in use is tungsten oxide (WO_3), and it is particularly noteworthy that WO_3 can be both absorption modulating with amorphous WO_3 films and reflectance modulating with crystalline WO_3 (Bell and Matthews 1998).

Realizing the electrochromic effect in glazing is relatively complex and requires the deposition of several layers (Figure 10.4). The transparent conducting oxide, typically either indium- or fluorine-doped tin oxide, produces the required electric field when a potential difference is applied between the two transparent conductors. An electrochromic device also requires an ion-containing material (the electrolyte) in close proximity to the electrochromic layer, allowing the device to shuttle ions into and out of the electrochromic layer with an applied potential (Bell and Matthews 1998; Lampert 1998). There are several advantages of electrochromic materials over other switchable devices, including the following: power is required only during switching, only moderate voltages are required to switch (1–5 V), they have continuous dimming, and several designs have long-term memory (12–48 hours) (Lampert 1998). Unfortunately, they respond slowly to changes in the applied electric field (requiring a few minutes to switch), they have a limited cycle lifetime, and

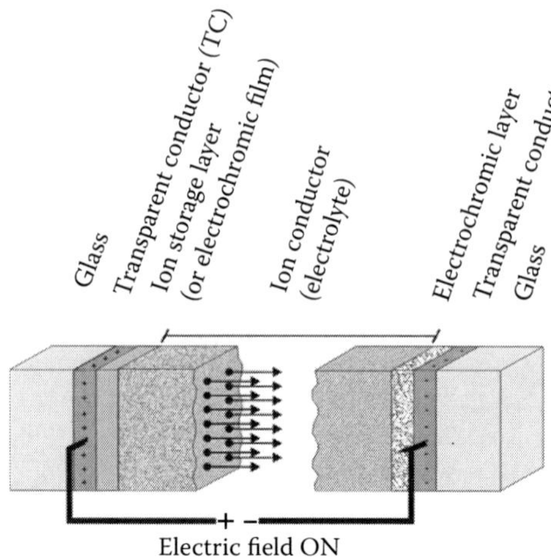

FIGURE 10.4 Schematic of an electrochromic glazing: TCO, transparent conducting oxide. (Reproduced from Seeboth et al., *Sol. Energy Mater. Sol. Cells*, 60, 263–277, 2000. With permission.)

larger panels can exhibit an *iris* effect where the tint changes from the edges inward (Benson and Branz 1995; Lampert 1998; Anon 2006). It is interesting from a broader perspective that many of the better chromogenic nanomaterials have much in common with new materials under development for advanced batteries, supercapacitors, and hydrogen storage materials.

10.2.2.4 Gaschromic Systems

A gaschromic system can be achieved by coating one of the internal surfaces in a double glazing unit with a nanoscale film of WO_3 plus a thin coating of platinum catalyst (Lampert 2004). The change in color of the window would then be induced by opening a valve to pass a gas through the interior of the double glazed unit. The gas could be simply air when oxidation of WO_x is required to make a clear state or a gas mixture comprised of nitrogen-1% hydrogen when reduction is needed to impart a dark blue tint. The gas infrastructure required would probably scale most economically in a large building project.

10.2.2.5 Thermochromic Systems

Thermotropic glazing and thermochromic glazing, as their names suggest, undergo a phase transformation at some temperature, with the low- and high-temperature forms having rather different optical properties. Some conditions need to be met for thermotropic glazing to be viable. First, the transmittance of one of the phases should be greater than 85% and the other less than 15%, with the high-transmittance phase associated in most applications with the low-temperature state. The temperature at which phase change occurs should be controllable so that a temperature suitable for local climate and window orientation can be selected. Probably, a switching temperature of 30°C or a little higher is ideal for warm climate conditions (Seeboth et al. 2000). The change in transmission in thermotropic materials is caused by a change in scattering, primarily backscattering in this case, causing the materials to have a high solar reflectance (Bell and Matthews 1998).

Figure 10.5 shows data for a commercial thermochromic material, Cloud-Gel, which consists of a thermochromic layer sandwiched between two plastic films. The thermochromic layer consists of hydrocarbon copolymers dissolved in water, where at temperatures below the switching point the polymers are elongated (and transparent) but at temperatures above it the polymer coagulates, forming structures larger than the wavelength of light, causing a massive increase in the amount of scattering. Altering the component concentrations allows the switching point to be controlled between 10°C and 70°C (Beck et al. 1995).

The thermochromic effect can also be achieved by laminating a polymer film containing a temperature-sensitive dye between two sheets of glass. When the structure exceeds the switching temperature, the dyes change color and darken, thereby blocking radiation.

Vanadium dioxide has been frequently mooted as a possible component of a thermochromic coating system for the building industry, either as a nanoscale coating (Granqvist 1990; Manning and Parkin 2004) or as a pigment added to a paint system (Guinneton et al. 2005; Bai et al. 2009). This material undergoes a metal/insulator transition at about 69°C, being metallic and reflective above 69°C and semiconducting and partially transparent below that temperature. Under ideal conditions, a significant degree of switching can be obtained from thin films coatings on glass (Figure 10.6). Switching has also been demonstrated in coatings containing VO_2 nanoparticles, but the extent of the property change has been poor in the past. Recently, however, Ji et al. (2011) reported a coating of VO_2 nanoparticles in which the transmittance in the near infrared (NIR) could be switched from 20% to 70%. Nevertheless, despite these intriguing characteristics, commercial uptake of this idea has been very slow. One reason is that 69°C is too high a switching temperature for convenient application. The switching temperature can be brought down to about 30°C by alloying the VO_2 with W or Mo; but when this is done, the extent of change

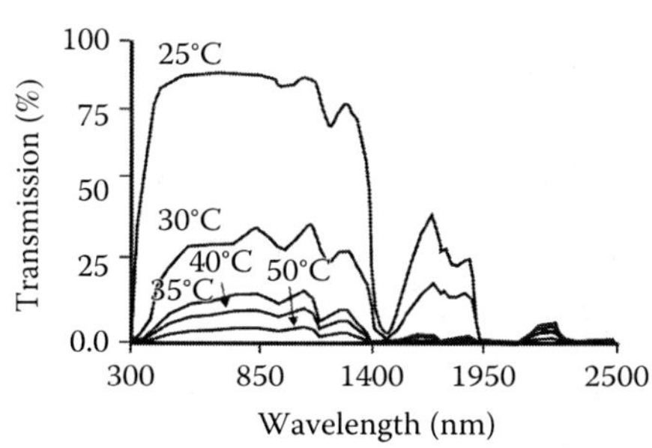

FIGURE 10.5 Spectral transmittance of a commercial thermochromic window glazing at different temperatures. (Reproduced from Seeboth et al., *Sol. Energy Mater. Sol. Cells*, 60, 263–277, 2000. With permission.)

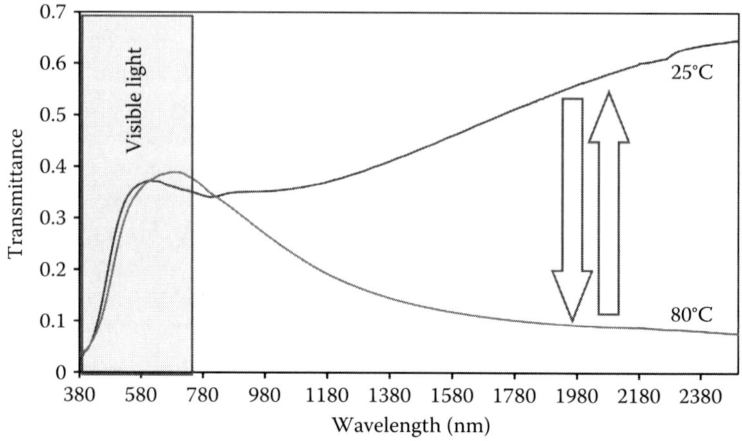

FIGURE 10.6 Switching of the transmittance of glass coated with a 50-nm-thick film of VO_2. (Courtesy of Dr A. Maaroof; see Gentle, Maaroof et al. [2007] for a related discussion.)

in optical properties becomes muted and, hence, less useful. Another problem is that VO_2 coatings have a yellow color in transmission when in the semitransparent low-temperature form. This attribute would work against its more widespread use even if good switching could be obtained at a climatically relevant temperature. However, it is feasible that doping and/or the application of antireflection coatings can improve the color.

A significant advantage of thermotropic glazing and thermochromic glazing is that transmission is regulated passively by temperature, so no external field or wiring is required. This also leads to a disadvantage as occasionally an occupant may desire independent control over visibility, which is not possible with these systems.

10.2.2.6 Systems Based on Dispersed Particles

Dispersed particles have been used in windows where long, thin particles, possibly magnetic, are suspended in an organic fluid or gel. The position and orientation of these particles are initially random and can absorb significant quantities of light. Under the influence of an electric field, the particles align along the field lines increasing the light transmission. This change in transmission between electrically aligned (ON) and randomly orientated (OFF) is illustrated in Figure 10.7.

Dispersed particle glazing generally consists of five layers: the first and last layers are generally a transparent conductor like indium tin oxide–coated or fluorine-doped tin oxide–coated glass, they are followed by a dielectric as the second and fourth layers to prevent current flow, and the middle layer is dispersed particles in a gel or fluid. Some problems associated with suspended particle glazing include long-term stability, cyclic durability, particle settling and agglomeration, and gap spacing control. However, this has not prevented several companies from producing commercial products in architectural glass, automotive mirrors, and flat panel displays (Lampert 1998).

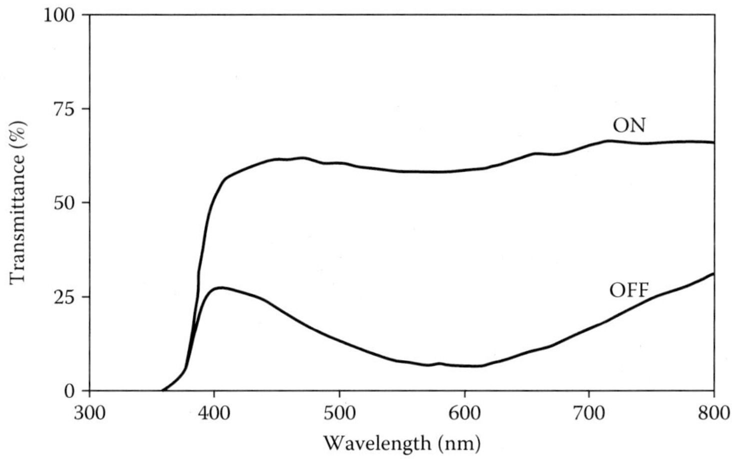

FIGURE 10.7 Spectral transmittance of a smart coating for glass based on dispersed particles. (From Lampert, C., *Sol. Energy Mater. Sol. Cells*, 52, 207–221, 1998.)

10.2.3 Other Spectrally Selective Coatings on Buildings

10.2.3.1 Nanotechnology in Paint Systems

It is normally accepted that dark-colored paints absorb the NIR and heat up in sunlight to a greater degree than light-colored paints. This is not, however, inevitably true. In principle, it is possible to design a paint that is dark colored (in the visible) and yet reflects the NIR, or vice versa. This can be achieved by using pigment particles that are spectrally selective with respect to absorption, scattering, or reflection. It is important to note the implications of Kirchoff's law, which is that a highly reflective surface is necessarily simultaneously one of very low emissivity and vice versa. So, for example, a paint that is designed purely to minimize solar-induced heating should possess maximum reflectivity across the strongest wavelengths of the incident solar spectrum ($\lambda < 2500$ nm) and minimum reflectivity (hence maximum emissivity) across the thermal wavelengths of the mid infrared ($\lambda > 2500$ nm) (Smith and Granqvist 2011).

When visible color is an important constraint, a transition in spectral selectivity at about 750 nm, the limit of human color perception, is helpful. It is possible to design paints that have similar colors to the eye but very different thermal performance. For example, a paint designed for the exterior of homes in cold climates should have minimum reflectivity in the NIR but maximum reflectivity (lowest emissivity) in the thermal infrared wavelengths. In contrast, a paint designed for the exterior of a house in hot climates should have maximum reflectivity in the NIR but the lowest reflectivity (hence maximum emissivity) in the thermal region of the infrared. *Whiteness* does not guarantee high solar reflectance as some white paints absorb a significant amount of invisible solar NIR radiation.

Commercial exploitation of these principles is still at a relatively early stage. Pigments with high absorptivity include mixed Fe-Mn-Cu oxides, or mixed Cu-Cr oxides. Low emissivity (and hence high reflectivity) can be provided by placing a layer of Al flake particles below the oxide pigments. Such a composite system has a high absorbance of the NIR from the metal oxides but a low emissivity in the thermal wavelengths due to the Al flake (Smith and Granqvist 2011). Color with high solar reflectance can be achieved using pigments with narrow visible absorption bands mixed with or on top of layers of otherwise high solar-reflecting materials such as TiO_2, which is the commonest base paint pigment before the addition of color.

Finally, it is possible in principle to formulate a smart paint with a thermochromic pigment, such as the VO_2 discussed previously. However, this idea has not yet been proved practically in the laboratory, so its commercial exploitation has not even been tried to our knowledge.

10.2.3.2 *Sky Cooling* of Buildings

An intriguing idea that has been mooted from time to time (Granqvist 2003; Gentle and Smith 2010) is that of a nanoparticle-based coating that has high radiative emissivity in the *sky window* but low absorptivity (and hence, by Kirchoff's law, emissivity) outside of it. The sky window is a region of the electromagnetic spectrum, roughly from 8 to 13 μm, in which the atmosphere is relatively more transparent. The idea here is that the roof of a building can be painted with such a coating and on clear nights this surface will radiate a significant amount of the building's heat into deep space (which is nominally at a radiative temperature of 4 K at these wavelengths, although, due to the unavoidable heating of atmospheric gases, the ostensible temperature of the sky window may be more like 180 K [Gentle and Smith 2009]). Very conveniently, the blackbody radiation of a body at about room temperature peaks at about 10 μm. The success of this strategy depends critically on identifying materials with enhanced emissivity at these wavelengths but low emissivity at other wavelengths. One way to achieve this is to ensure the development of an optical resonance (analogous to the localized plasmon resonance of precious metal nanoparticles), one that is based on a photon–phonon interaction rather than a photon–electron interaction (Gentle and Smith 2010). Promising candidate materials for such nanoparticles include SiC, SiO_2, TiO_2, $BaSO_4$, or their mixtures (Granqvist 2003; Gentle and Smith 2010). It has been calculated that up to 90 $W{\cdot}m^{-2}$ can be shed by these means, which could lead to a depression of the temperature of the radiating surface of up to 30°C provided that reverse heat flow from the environment (by convection or conduction) can be restricted to 1 $W{\cdot}m^{-2}{\cdot}K^{-1}$ or less (Granqvist 2003). In simpler, low-cost structures, a more attainable value for the reverse heat flow might be 2–2.5 $W{\cdot}m^{-2}{\cdot}K^{-1}$, which would give a ΔT of 12–17 K below ambient, corresponding to a net cooling power of 40–50 $W{\cdot}m^{-2}$ (Gentle and Smith 2010). The associated coolth could obviously be transferred to a fluid in a heat exchanger and used later in a building for cooling purposes. However, it is important to note that the operation of these devices requires the driest possible air column between radiating surface and deep space due to the highly absorptive nature of H_2O, which is a potent greenhouse gas. Therefore, this scheme would work best in desert regions with clear skies.

10.2.3.3 Ultraviolet Resistance in Wood, Paints, and Fabrics

Nanoparticles of compounds such as TiO_2 or ZnO are too small to scatter visible light but are still strong absorbers of UV wavelengths. Since UV light is very damaging to polymers and fabrics, there has been some interest in using these compounds in coating systems. A transparent varnish for exterior timber, for example, can be made opaque to UV radiation by the incorporation of such nanoparticles. The coating remains transparent in the visible wavelengths, however, which is what is normally required by the end user for aesthetic reasons.

10.3 PRODUCTS WITH NONOPTICAL FUNCTIONALITY

10.3.1 Catalytically Active Building Materials

The idea of coating the surfaces of a building with a catalyst capable of purifying indoor air, for example, has been mooted many times. Titanium dioxide is an obvious candidate for this because it is photocatalytic and, when irradiated by suitable wavelengths of light, it can oxidize most kinds of organic molecules adhering to its surface (Figure 10.8). In principle, this can remove hydrocarbons and similar organic contaminants, but this can also degrade materials such as organic paint binders and polymer sheeting. (The TiO_2 pigment particles used in paints and plastic sheets are deliberately coated with an isolating layer of SiO_2 to prevent the occurrence of this effect.) Unfortunately, UV radiation is required to make this effect work in ordinary TiO_2, so its application indoors is not likely to be particularly efficacious except in air conditioning ducts containing UV-emitting lamps (note, however, that conventional fluorescent tubes do emit a small amount of UV light, so there can still be a limited amount of photocatalysis indoors). Many attempts have been made to modify TiO_2 so that its photocatalytic effect would work in the visible range (Osterloh 2008). Doping with nitrogen or silver atoms is considered a promising route (El Saliby et al. 2011). However, it appears fair to say that the results have not yet been so attractive as to make these ideas commercially viable.

A more successful technology is that of *self-cleaning* glass based on a coating of TiO_2 nanoparticles. This was introduced into the market a decade or so ago as Aktiv glass, a trademark of Pilkington, St. Helens, United Kingdom, and several

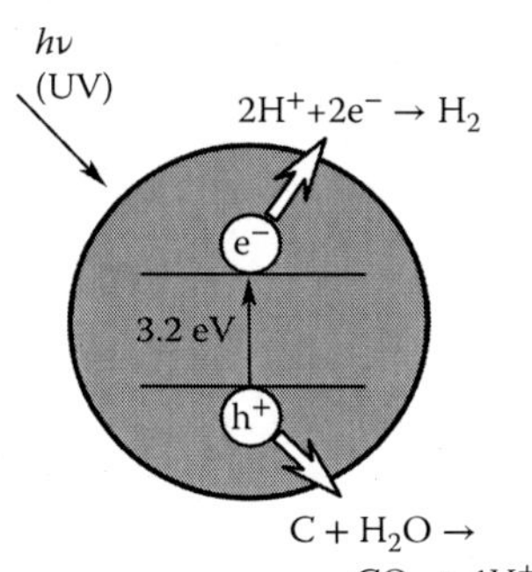

FIGURE 10.8 Schematic illustration of the principle of photocatalysis, showing how an incoming photon of ultraviolet (UV) light is absorbed by TiO_2 with the energy causing the promotion of an electron into the conduction band. The hole (h^+) left behind in the valence band can oxidize an adjacent carbonaceous species, shown here for simplicity as one atom of carbon.

allied products are now available from other suppliers. This glass makes use of two different properties of TiO_2 nanoparticles to achieve its effect. The first is the photocatalysis effect referred to earlier. In this case, the TiO_2 coating will oxidize urban-style deposits of carbonaceous substances when illuminated by UV light. The second effect is that the contact angle of water on TiO_2 drops to almost 0° when it is illuminated with UV light, a phenomenon known as superhydrophilicity. This means that any water on the surface of the window spreads out to make a thin film that then readily drains off the glass, hopefully carrying the residual material generated by photocatalysis with it. Because the water drains off as a sheet, there is also a reduced probability of the water forming unsightly streaks on the glass. Of course, successful operation of this system requires occasional rain (or perhaps a quick hose down), and some sunlight, to stay clean and clear.

10.3.2 Antimicrobial and Antifungal Coatings

The use of silver nanoparticles in antimicrobial coatings is well known and has found application in medical dressings, clothing, and the surfaces of white goods (e.g., refrigerators). In principle, such coatings can also be applied to suitable surfaces in buildings; however, as far as the authors know this has not been attempted yet. Zinc oxide nanoparticles are useful in this role too, primarily as fungistats. These are substances that retard the growth of fungi, for example, in the form of mildew growing on painted surfaces exposed to the external environment or mold growing on the grouting between bathroom tiles. Today, the use of antifungal mercury compounds is discouraged or prohibited in most paint and grouting formulations, so they may use ZnO instead as one component of the overall antifungal formulation.

10.3.3 Corrosion Resistance

Corrosion of metallic structures is accelerated when water and chloride ions are present, and this obviously presents problems in the building industry for reasons of both appearance and, ultimately, structural integrity. One solution is to select alloys that are reasonably corrosion resistant in the first place; another is to apply a coating that somehow imparts an improved resistance to corrosive attack. Most of the coating processes, for example, anodizing or galvanizing, apply a layer of micrometer-scale thickness and so obviously lie outside the scope of this chapter. However, in principle, nanoscale layers of cathodic materials such as nickel can be applied by techniques such as electroplating or electroless deposition, and these have some potential to reduce corrosive attack (Rabizadeh et al. 2010) provided that they can be applied in a continuous and undamaged form.

Monolayers of tightly attached organic molecules can also provide resistance to corrosion, and this idea is exploited in various kinds of corrosion inhibitors. Generally, these are not yet widely exploited in the building industry, except in packaging and closed-loop water reticulation circuits.

Sol gel coatings are another of the nanotechnologies with the potential to improve corrosion resistance of building materials. These consist of very thin films of mixed oxides that are deposited onto, for example, a steel substrate by a wet chemical process (Duran, Castro et al. 2007). Sol gel coatings can be very durable, relatively cheap to apply, and transparent. Of course, like other inert coatings, any scratches, pinholes, or other defects will become the sites of corrosive attack, so care needs to be taken in the specification and application of sol gel coatings. One solution is to include a corrosion inhibitor in the sol gel formulation. For example, laboratory trials have shown that the corrosion of aluminum can be reduced by coating it with a SiO_2 sol gel containing the cerium nitrate corrosion inhibitor (Tavandashti et al. 2011). Another idea is to design *self-healing* coatings that can somehow compensate for being scratched (Hikasa et al. 2004).

10.3.4 Thermal Insulation

Aerogels are very-low-density materials that can be produced by the supercritical evaporation of, for example, colloidal suspensions of SiO_2 nanospheres. The resulting solid will consist of an interconnected skeleton of SiO_2 nanospheres surrounded by air. If carefully prepared the density of this solid may only be small percentage that of bulk SiO_2 (Granqvist 2003). Although these materials are exceedingly fragile, they are interesting because they have a low heat conductivity (~1 $W{\cdot}m^{-2}{\cdot}K^{-1}$). They may also be relatively transparent (Granqvist 2003). However, aerogel sheeting has yet to find widespread use in the building industry due to cost and handling problems. Its use in granule form has shown improved take-up recently as it costs less and enables filling from a hopper into gaps between clear or opaque sheeting.

10.3.5 Products for Control of Surface Physical Properties

The contact angle of water on the surfaces of structural materials in common use varies widely. A material that is wetted by liquid water has a contact angle less than 90° and is termed hydrophilic. Exceedingly hydrophilic surfaces have a contact angle close to 0° and are termed *superhydrophilic*. In this case, water spreads out on them as a very thin film. In contrast, a surface with very little affinity for water will have a contact angle greater than 90° and is termed hydrophobic. A surface with a contact angle greater than 140° is *superhydrophobic*. In this latter case, any water droplets tend to form spheres on the surface. Obviously, there is also a continuum of surface types between the two extremes, with examples of intermediate contact angle and hence intermediate wettability. Although the surface properties can be controlled to some extent by selecting the appropriate bulk material, it may be more convenient in practice to do so by applying a surface coating. In many cases, the thickness of such coatings is in the nanometer range and, hence, they do qualify as nanotechnologies.

Sites where controlling the wettability might be considered useful include tiles and walls of humid or wet rooms such as bathrooms, saunas, or kitchens; surfaces of windows or mirrors; floor tiles; carpets; windows; skylights; conservatories; sunrooms; or outdoors paving.

If a greater degree of wetting, that is, a hydrophilic surface, is desired, then a surfactant may be applied to the surface. The adhering surfactant layer may be only one or two molecules thick, but the net result is to lower the surface tension of the substrate and thereby cause any droplets of water to exhibit such a low contact angle that they spread out to become a continuous film. One application of this in the household is as an antifog agent, which, when applied to mirrors or windows, prevents the formation of the semiopaque film of water droplets that is associated with fogging. In contrast, ordinary hydrophobicity is promoted by having a surface comprising weakly bonded molecules such as fluoro- or hydrocarbons. Attachment to the substrate can be via physisorption or by a covalent bond. Superhydrophobicity is quite different again and is promoted by having surfaces that are so rough on the nanoscale that some air is entrained between them and the drops of water.

Although superhydrophilic and superhydrophobic surfaces have completely different interactions with water, both can be considered to be self-cleaning, although with very different mechanisms. As mentioned earlier in this section, superhydrophilic surfaces tend to resist the formation of stains and run marks. On the other hand, if a superhydrophobic surface is inclined at a sufficient angle then droplets of water on it will run downward and tend to drag any loosely attached dirt particles with them. Therefore, hydrophobic surfaces tend to stay clean too. This is an especially desirable attribute for locations that are difficult to reach for cleaning, such as, for example, the exterior windows on high-rise buildings or the transparent roofs of conservatories or sunrooms. Of course, this technology may also be welcome in much more accessible locations, such as the walls of shower cubicles, because it may provide a degree of labor saving.

An interesting biomimetic variation has been inspired by the process by which Namibian desert beetles of the genus *Stenocara* obtain water. The beetle's carapace is coated with a structure of microscale hydrophobic and hydrophilic features (Parker and Lawrence 2001). Water droplets in the morning coastal fog coalesce onto hydrophilic protrusions and, once they become big enough, drain away to the hydrophobic surface, where they are rapidly aggregated into water droplets of 4 or 5 mm diameter, which the beetle drinks. The artificial mimic uses spaced-out hydrophilic nanofeatures to condense the water vapor on top of smooth hydrophobic surfaces, which drain the water away (Thickett et al. 2011). One proposal is to coat textile meshes with this structure and use this to collect sufficient water for human consumption. However, this technology would only be useful in environments that receive fog or in which the temperature is below the dew point.

10.4 NANOTECHNOLOGY IN THE LIGHTING OF BUILDINGS

Since lighting fixtures are generally installed at the time of construction of buildings, their technology is certainly relevant to a chapter on the building industry. Several nations have enacted legislation to discourage the use of ordinary incandescent lamps (which are based on a heated tungsten filament). The argument is that such lighting systems generate 10–18 lumens (lm) of light output per watt of electrical power consumed (10–18 $lm{\cdot}W^{-1}$), compared to the 60–80 $lm{\cdot}W^{-1}$ or so produced by fluorescent lights, for example. Of course, the excess heat generated by an incandescent during winter may be a useful fraction of the building's heating, and its loss would have to be replaced by some other source of heat. However, in summer such heat is generally inconvenient and may even need to be counteracted by air conditioning. The problem in the marketplace at the moment is that the old-style fluorescent tubes are perceived by many consumers to be suitable only for commercial or retail-type applications. Consumer resistance is based on their flickering start-up, less desirable color balance, content of mercury, slight humming noise, and large size. The last of these problems has been partially addressed by the introduction of compact fluorescent lamps (CFLs). However, there is currently a huge opportunity in the building industry for the introduction of other lighting technologies.

Fluorescent lamps could be considered to exploit nanotechnology insofar as their phosphors are concerned, but this is a specialized topic. It is worth mentioning that a CFL contains up to 5 mg of mercury, a very toxic element, so breakage or disposal of these devices presents some problems. Another issue is that they have a short life if used in applications where they are frequently switched on and off, and they are in general rapidly destroyed if used in a dimmer circuit. This latter point can be an expensive irritation to homeowners given that 50% of home lighting in some countries is on a dimmer circuit (Tonzani 2009). And if these figures are not problematic enough, it has been reported that an analysis of manufacturing practice showed that making a halogen incandescent used only half of the energy required to manufacture a CFL and generated only 5 g of nonhazardous waste, compared to the 108 g of waste generated by the CFL of which 78 g could be considered hazardous (Van Tichelen 2009).

More recently, light-emitting diodes (LEDs) have been promoted as the solution to energy-efficient lighting. These may be considered to exploit nanotechnology too, primarily by virtue of the fact that they comprise very precisely deposited layers of nanoscale semiconductors and the light-emitting process itself takes place within a nanoscale volume of material. Since LED chips generate light internally and have smooth layers, much light cannot escape due to total internal reflection unless special surface structures or chip shapes are used. Nanostructured surfaces can be used to enhance light output, including coatings based on plasmonic nanoparticles. Better light extraction has been one of the main sources of the recent gains in efficiency. At the time of writing this chapter, LED-based lighting products with efficiencies in the range of 10 to 60 $lm \cdot W^{-1}$ are available for purchase at retail outlets. Unfortunately, these are still expensive and of course their output still falls short of fluorescent lamps in overall luminous efficiency (Tonzani 2009). On the other hand, LEDs last an order of magnitude longer and are continually being improved. Their theoretical limit is about 240 $lm \cdot W^{-1}$, so there is plenty of room for improvement. Current LEDs consume relatively scarce elements such as gallium or indium; but there is some optimism in the field that an LED technology based on zinc oxide can be developed, either as an all-ZnO homojunction using *n*- and *p*-type ZnO or as a heterojunction of *n*-type ZnO coupled to some other *p*-type semiconductor (Choi et al. 2010). Zinc oxide is a ubiquitous and relatively cheap industrial chemical that is now mainly used in the rubber and ceramics industries. However, one issue that has slowed progress is the lack of a way to produce stable *p*-type ZnO with desired properties.

Organic light-emitting diodes (OLEDs) are another potential competitor to the compact fluorescent bulb. Their performance is steadily improving and was recently reported (Reineke et al. 2009) to have exceeded 90 $lm \cdot W^{-1}$. However, the lifetime of these devices is still considered by many to be inadequate (Reineke et al. 2009; Tonzani 2009).

Only time will tell whether LED-based systems (or indeed one of the other lighting technologies that we have not had the space to mention here) can eventually displace fluorescent ones in mainstream building applications or not.

10.5 CONCLUSIONS

Several nanotechnologies have commercial applications, or credible prospects of applications, in the building industry. Although the basic structural elements of a building are still completely dominated by macroscale materials engineering, nanoscale coatings of various kinds offer benefits in terms of spectral selectivity or controlled surface properties. The motivation for applying a nanotechnology in the building industry is most commonly based on improved appearance, reduced maintenance, or greater efficiency of energy use. Coatings for the glass used in windows may be the biggest and best developed of the building-industry nanotechnologies. Passive windows can be designed to admit visible wavelengths and reject the NIR, and these are already in widespread use. Active window coatings (also known as smart or switchable windows) are still in their infancy but can be expected to become increasingly important due to the incredible convenience that they offer. Nanotechnology is also exploited for controlling the wettability of building materials, and their durability. Finally, the issue of indoor lighting and nanotechnology is closely intertwined, with all of the alternatives to incandescent lighting being based on some key nanotechnology.

REFERENCES

Anon (2006) Zero energy window prototype. US Department of Energy Fact Sheet. Available at http://windows.lbl.gov/adv_Sys/hi_R_insert/ZeroEnergyWindowDOE-FactSheet.pdf, accessed January 24, 2014.

Bai, H, Cortie, MB et al. (2009). Preparation of plasmonically resonant VO_2 thermochromic pigment. *Nanotechnology* **20**: 085607.

Beck, A, Korner, W et al. (1995). Control of solar insolation via thermochromic light-switching gels. *Sol. Energy Mater. Sol. Cells* **36**: 339–347.

Bell, JM and Matthews, JP (1998). Glazing materials. *Mater. Forum* **22**: 1–24.

Benson, DK and Branz, HM (1995). Design goals and challenges for a photovoltaic-powered electrochromic window covering. *Sol. Energy Mater. Sol. Cells* **39**: 203–211.

Choi, YS, Kang, JW et al. (2010). Recent advances in ZnO-based light-emitting diodes. *IEEE Trans. Electron Dev.* **57**(1): 26–41.

Chowdhury, H and Cortie, MB (2007). Thermal stresses and cracking in absorptive solar glazing. *Constr. Build. Mater.* **21**(2): 464–468.

Chowdhury, HA 2007. Heat transfer and thermally induced stresses in window glass coated with optically active nano-particles, MSc thesis, University of Technology Sydney, Sydney, Australia.

Cortie, M, Muir, J et al. (2005). Materials engineering and design in the UTS-CSIRO Nanohouse. *Mater. Aust.* **38**(3): 10–11.

Dai, Y (2001). Solar control film retrofitted energy efficient windows for tropical climate. *Proc. Glass Processing Days 2001*, Tampere, Finland, Tamglass Ltd Oy., 156–161.

Duran, A, Castro, Y et al. (2007). Protection and surface modification of metals with sol-gel coatings. *Int. Mater. Rev.* **52**(3): 175–192.

El Saliby, I, Shon, HK et al. (2011). Visible-light active doped titania for water purification: nitrogen and silver doping. *J. Indus. Eng. Chem.* **17**(2): 358–363.
Gentle, A, Maaroof, AI et al. (2007). Nanograin VO_2 in the metal phase: a plasmonic system with falling dc resistivity as temperature rises. *Nanotechnology* **18**(2): 025202.
Gentle, AR and Smith, GB (2009). Angular selectivity: impact on optimised coatings for night sky radiative cooling. *Nanostructured Thin Films II*, San Diego, SPIE, article 74040J.
Gentle, AR and Smith, GB (2010). Radiative heat pumping from the Earth using surface phonon resonant nanoparticles. *Nano Lett.* **10**(2): 373–379.
Granqvist, CG (1990). Window coatings for the future. *Thin Solid Films* **193/194**: 730–741.
Granqvist, CG (2003). Solar energy materials. *Adv. Mater.* **15**(21): 1789–1803.
Guinneton, F, Sauques, L et al. (2005). Role of surface defects and microstructure in infrared optical properties of thermochromic VO_2 materials. *J. Phys. Chem. Solids* **66**: 63–73.
Hikasa, A, Sekino, T et al. (2004). Preparation and corrosion studies of self-healing multi-layered nano coatings of silica and swelling clay. *Mat. Res. Innovat.* **8**(2): 84–88.
Ji, S, Zhang, F et al. (2011). Preparation of high performance pure single phase VO_2 nanopowder by hydrothermally reducing the V_2O_5 gel. *Sol. Energy Mater. Sol. Cells* **95**: 3520–3526.
Lampert, C (1998). Smart switchable glazing for solar energy and daylight control. *Sol. Energy Mater. Sol. Cells* **52**: 207–221.
Lampert, CM (March, 2004). Chromagenic smart materials. *Mater. Today* **7**: 28–35.
Liu, J, Cankurtaran, B et al. (2006). Anisotropic optical properties of semitransparent coatings of gold nanocaps. *Adv. Funct. Mater.* **16**(11): 1457–1461.
Manning, TD and Parkin, IP (2004). Atmospheric pressure chemical vapour deposition of tungsten doped vanadium(IV) oxide from $VOCl_3$, water and WCl_6. *J. Mater. Chem.* **14**: 2554–2559.
Muir, J, Smith, G et al. (2004). The Nanohouse™—an Australian initiative for the future of energy efficient housing. In: *Nanotechnology in Construction*. P. J. M. Bartos, J. J. Hughes, P. Trtik and W. Zhu (eds.). Cambridge, The Royal Society of Chemistry: 291–304.
Osterloh, FE (2008). Inorganic materials as catalysts for photochemical splitting of water. *Chem. Mater.* **20**: 35–54.
Parker, AR and Lawrence, CR (2001). Water capture by a desert beetle. *Nature* **414**(6859): 33–34.
Rabizadeh, T, Allahkaram, SR et al. (2010). An investigation on effects of heat treatment on corrosion properties of Ni-P electroless nano-coatings. *Mater Design* **31**(7): 3174–3179.
Reineke, S, Lindner, F et al. (2009). White organic light-emitting diodes with fluorescent tube efficiency. *Nature* **459**: 234–238.
Schelm, S and Smith, GB (2003). Dilute LaB_6 nanoparticles in polymer as optimized clear solar control glazing. *Appl. Phys. Lett.* **82**(24): 4346–4348.
Seeboth, A, Schneider, J et al. (2000). Materials for an intelligent sun protecting glazing. *Sol. Energy Mater. Sol. Cells* **60**: 263–277.
Smith, GB, Deller, CA et al. (2002). Nanoparticle-doped polymer foils for use in solar control glazing. *J. Nanopart. Res.* **4**: 157–165.
Smith, GB and Granqvist, CG (2011). *Green Nanotechnology.* CRC Press, Boca Raton, Florida.
Stokes, NL, Edgar, JA et al. (2010). Spectrally selective coatings of gold nanorods on architectural glass. *J. Nanopart. Res.* **12**(8): 2821–2830.
Takeda, H, Yabuki, K et al. (2001). Coating solution for forming a film for cutting off solar radiation and the film formed therefrom. United States Patent. 6319613.
Tavandashti, NP, Sanjabi, S et al. (2011). Evolution of corrosion protection performance of hybrid silica based sol–gel nanocoatings by doping inorganic inhibitor. *Mater. Corros.* **62**(5): 411–415.
Thickett, SC, Neto, C et al. (2011). Biomimetic surface coatings for atmospheric water capture prepared by dewetting of polymer films. *Adv. Mater.* **23**: 3718–3722.

Tonzani, S (2009). Lighting technology: time to change the bulb. *Nature* **459**: 312–314.

Van Tichelen, P (2009). Lot 19: domestic lighting, Flemish Institute for Technological Research NV. Study for European Commission DGTREN unit D3.

Wittkopf, H (1997). Electrochromics for architectural glazing applications. *Proc. Glass Processing Days 1997*. Tampere, Finland: 299–303.

Xu, X, Gibbons, T et al. (2006). Spectrally-selective gold nanorod coatings for window glass. *Gold Bull.* **39**(4): 156–165.

Xu, X, Stevens, M et al. (2004). In situ precipitation of gold nanoparticles onto glass for potential architectural applications. *Chem. Mater.* **16**(1): 2259–2266.

11 Anticipatory Life-Cycle Assessment of SWCNT-Enabled Lithium Ion Batteries

Ben A. Wender and Thomas P. Seager

CONTENTS

11.1 INTRODUCTION

Until recently, the environmental impacts of developing technologies were neither explored nor regulated until after commercialization. Thus, technological innovation has been disconnected from environmental assessment and regulation (Dewick et al. 2004; von Gleich et al. 2007). This tradition has positioned environmental governance as retrospective and reactive (Davies 2009). However, there is a growing realization that environmental intervention at the nascent stages of technology development may be more effective. Therefore, there is a critical need to transcend retrospective models of environmental assessment and regulation by applying life-cycle assessment (LCA) to technologies at these early stages (Fleischer and Grunwald 2008; Meyer et al. 2011) such that life-cycle environmental trade-offs can be explored in modeling scenarios before significant investments in infrastructure create technological lock-in or result in stranded costs.

11.1.1 Toward Anticipatory Life-Cycle Assessment

LCA is increasingly recognized as the appropriate framework to understand the environmental impacts of processes, technologies, and industries (Curran 2004; Bauer et al. 2008; Eason et al. 2011) because it accounts for shifting of environmental burdens from one life-cycle phase to another. However, the majority of LCAs are *retrospective*—relying on detailed inventory data collected at scale and therefore insufficient for rapidly developing technologies (Wiek et al. 2008; Meyer et al. 2009). Efforts to incorporate scenario development and technology forecasting into LCA, referred to as *prospective* LCA, effectively orient environmental assessment toward the future impacts of decisions made today (Pesonen et al. 2000; Walser et al. 2011). Nonetheless, forward looking studies may still fail to identify alternative research agenda and communicate these to salient decision makers. Real-time assessment and governance of emerging technologies necessitates the development of novel *anticipatory* LCA methods that can be used to quantitatively explore environmental impact scenarios associated with distinct research agenda, policy interventions, or manufacturing pathways and relate findings in a decision-oriented manner (Canis et al. 2010; Linkov et al. 2011). (The distinction between retrospective, prospective, and anticipatory LCA is explored in more detail in Figure 11.1).

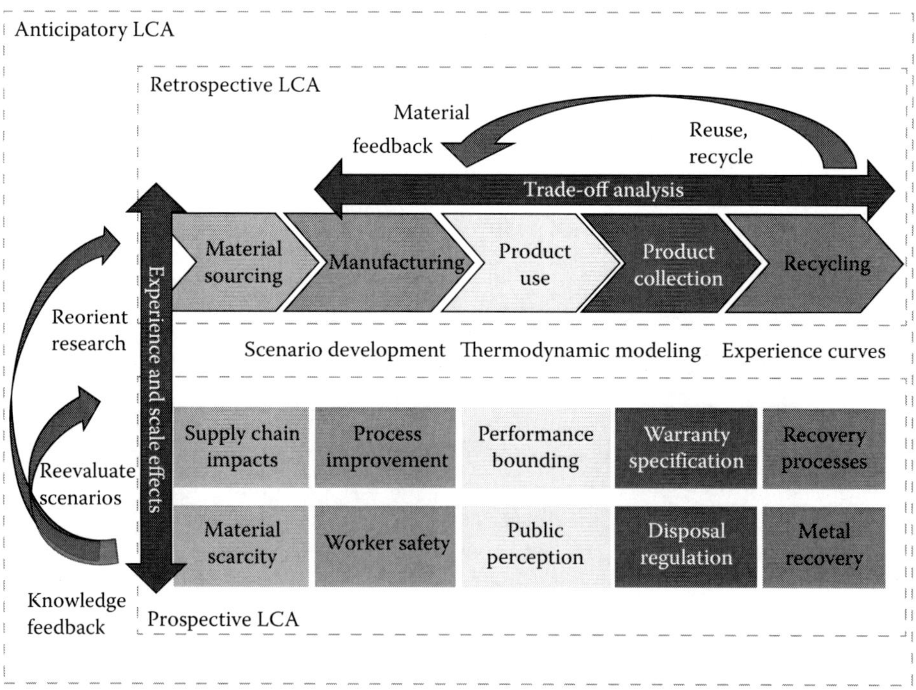

FIGURE 11.1 Cradle-to-grave components of anticipatory life-cycle assessment (LCA) for developing nanotechnologies.

11.1.2 Life-Cycle Assessment of Novel Nanotechnologies

A suite of examples illustrate the need for, and challenges that impede, the development of anticipatory LCA methods for nano-enabled energy technologies. A number of experts, including the U.S. Environmental Protection Agency (USEPA) and Woodrow Wilson International Center for Scholars, have called for the application of LCA to nanotechnology (Klopffer et al. 2007; Sengul et al. 2008; USEPA 2008; Theis et al. 2011). In practice, this is problematic for the following reasons:

- Uncertainty regarding the human and ecological health impacts of nanomaterials (Oberdorster et al. 2005, 2007; Wiesner et al. 2006; Stefani et al. 2011)
- High variability (e.g., purity and morphology) between engineered nanomaterials, even those with the same chemical composition (Landi et al. 2005; Powers et al. 2007)
- Uncertainty in extrapolating laboratory-scale inventory data to commercial scales (Seager and Linkov 2008, 2009; Gutowski et al. 2010)
- Selecting a functional unit relevant to the use phase of a nanomaterial that captures the potential benefits of engineered nanomaterials (Matheys et al. 2007; Wender and Seager 2011)

These challenges prevent cradle-to-cradle assessment of novel *nanoproducts*, and the majority of nano-LCAs focus on cradle-to-gate inventories of *nanomanufacturing* processes. Thus, few LCAs have overcome use-phase uncertainty in selecting a relevant functional unit. The exceptions are Walser et al. (2011), Reijnders (2010), Lloyd et al. (2005), and Lloyd and Lave (2003), but these analyses do not incorporate human health and toxicology research (Osterwalder et al. 2006; Singh et al. 2008; Plata et al. 2009). Likewise, early cradle-to-gate analyses of nanomaterials have called attention to the energy intensity of nanomanufacturing processes (Khanna et al. 2007; Krishnan et al. 2008b; Healy et al. 2008; Ganter et al. 2010; Grubb and Bakshi 2010; Anctil et al. 2011), but do not account for the potential benefits provided by nanomaterials in the use phase (Bell 2007; Helland et al. 2007). Finally, the environmental impacts of end-of-life recycling and processing of nanoproducts (Olapiriyakul and Caudill 2008; Ostertag and Husing 2008) are typically explored independent of research into exposure pathways (Kohler et al. 2008; Maynard 2009), which in turn is uninformed by research into social and market acceptance of nano-enabled technologies (Scheufele et al. 2007; Siegrist et al. 2007). Table 11.1 organizes the existing science and shows how the fragmented efforts that inform different aspects of nano-LCA have yet to be integrated in a comprehensive whole.

More importantly, Table 11.1 suggests that anticipatory LCA requires knowledge from multiple fields of study, as opposed to prospective LCA that is not inherently multidisciplinary. Different research questions and investigative methods are required at each life-cycle stage, and LCA of nanotechnology cannot proceed without parallel research in prerequisite specialty areas. That is, anticipatory LCA must incorporate social science, materials science, environmental science, and sustainability science perspectives to be applicable across *all* of Table 11.1.

TABLE 11.1
Relation of Nanostructured Material and Product Research Needs to LCA

	Life-Cycle Stage			
	Acquisition	**Purification and Manufacture**	**Use**	**End-of-life Disposition**
Material abundance and acquisition	Scarcity, criticality, and purity of materials (Krishnan et al. 2008b)	By-product and waste minimization	Risk assessment for emissions inventory and characterization, including source term characterization, fate and transport, exposure, and dose-response assessment (Bell 2007; Helland et al. 2007; Kohler et al. 2008)	
Bioavailability and toxicity				
Synthesis pathways	Energy and material intensity (Khanna et al. 2007; Healy et al. 2008; Krishnan et al. 2008a; Plata et al. 2009; Canis et al. 2010; Ganter et al. 2010; Grubb and Bakshi 2010; Gutowski et al. 2010; Anctil et al. 2011)			
Life-cycle characteristics		Technology comparison (Lloyd and Lave 2003; Lloyd et al. 2005)	Cost, functionality, and efficiency (Reijnders 2010; Walser et al. 2011)	Persistence, mobility, bioaccumulation (Oberdorster et al. 2005, 2007; Wiesner et al. 2006; Powers et al. 2007; Stefani et al. 2011)
Social context	Geopolitical sensitivities	Worker safety (Maynard 2009)	Market acceptance (Scheufele et al. 2007; Siegrist et al. 2007)	Disposal and take-back regulations (Olapiriyakul and Caudill 2008; Ostertag and Husing 2008)

11.1.3 Scale, Life-Cycle Assessment, and Thermodynamic Limits

Anticipatory LCA confronts the problem of data scarcity through a combination of scenario development and thermodynamic analysis of manufacturing processes and technology performance. Specifically, by coupling laboratory-scale inventory data with simplified technology performance modeling and projecting returns to scale, it is possible to provide upper and lower boundaries on environmental indicators of interest (e.g., embodied energy) for a specific technology. As shown in Figure 11.1, anticipatory LCA adapts existing LCA frameworks (termed *retrospective LCA*) and differs from *prospective LCA* in its inherent iteration and knowledge feedback.

Retrospective LCA is best suited for trade-off analysis—that is, identifying how decisions in one life-cycle phase affect the overall environmental profile (horizontal direction in Figure 11.1). Prospective and anticipatory LCA incorporates an additional dimension (vertical direction in Figure 11.1) probing how changes in scale and experience may shape environmental impact of future technologies. Environmentally-relevant questions associated with each lifecycle phase are listed as examples of how dynamic scenarios can capture broader impacts (e.g., supply chain and public perception) and thermodynamic modeling can focus in on process-level tradeoffs (e.g., manufacturing and recycling process improvements). Anticipatory LCA is unique in that it explicitly includes the process of knowledge feedback, where findings are communicated back to technology developers, policy makers, and laboratory researchers, with the goal of guiding nanotechnology development toward pathways of decreased environmental burden.

Three critical components of anticipatory LCA are as follows:

1. Combining laboratory-scale material and energy inventories with scenario development to explore potential changes in laboratory or pilot-scale thermodynamic degree of perfection. Those processes that are far from thermodynamic perfection might be expected to improve more quickly than those that are already approaching practical thermodynamic limitations (Gutowski et al. 2009, 2010).
2. Calculating upper and lower boundaries to use-phase performance based on theoretical limits and existing laboratory measurements coupled with thermodynamic modeling of use and manufacturing phases. Together these may identify life-cycle phases with the most potential for environmental improvement (Wender and Seager 2011).
3. Analogous experience curve modeling. It is well understood that high technology industries improve cost, material, and energetic efficiencies as total production knowledge accumulates. Analysis of experience curve patterns from more mature industries (e.g., aluminum and silicon) may result in estimates of the efficiency gains that accrue as emerging technologies are scaled-up (McDonald and Schrattenholzer 2001).

In situations of high uncertainty (e.g., nano-enabled energy technologies), this analysis can be used to develop scenarios of environmental burden and can call attention to environmentally problematic processes and technologies. Furthermore, by providing estimates of manufacturing and use-phase efficiency, these analyses can lead to prioritization of research needs that will result in the most meaningful environmental improvements. For example, an environmental agenda might call attention to research needs in manufacturing, rather than in product use-phase performance. Model results are ultimately incorporated into existing LCA tools (e.g., Simapro and EIO databases) to broaden system boundaries and account for supply chain impacts.

In the following case study, we apply these components of anticipatory LCA to single-wall carbon nanotube (SWCNT) manufacturing, compare the rapid improvements in SWCNT manufacturing to analogous material processing industries, and discuss the use of SWCNTs as an active anode material for advanced lithium ion batteries.

11.2 CASE STUDY: SINGLE-WALL CARBON NANOTUBES FOR LITHIUM ION BATTERIES

A major thrust of battery research is to increase the energy storage density of rechargeable batteries. This is motivated in part by consumer preference for lightweight electronics, but is increasingly important as electric and hybrid electric vehicles are implemented on larger scales. Recently, the energy density of batteries has increased dramatically—from lead acid batteries with a mass-based energy density up to 50 Wh/kg to lithium polymer batteries approaching 250 Wh/kg. Lithium ion batteries have emerged as the preferred chemistry because of their comparatively high energy densities per unit mass (USGS 2007). Further improvements will depend on increasingly sophisticated materials and manufacturing techniques. Engineered nanomaterials are appealing because of their large surface area and unique electrical properties. Specifically, SWCNTs can store lithium ions in interstitial spaces, collect charge carriers, and conduct charge to external circuits (Landi et al. 2008, 2009). SWCNT battery anodes could eliminate the need for charge collecting metal foil, thus reducing battery weight and increasing energy storage density. The potential gains in use-phase performance in SWCNT-enabled lithium ion batteries could justify increased energy investments in SWCNT manufacturing. However, there is no data available describing commercial scale manufacturing of SWCNT anodes, and only preliminary laboratory-scale data describing their use-phase performance. Thus, the systemic environmental consequences of SWCNT-enabled lithium ion batteries are inherently unclear and necessitate anticipatory LCA methods to quantitatively explore energy trade-offs between the manufacturing and use phases and how these may change with increased scale. Specifically, the aforementioned analyses can provide insights into future developments in nanomanufacturing processes (e.g., potential sources of efficiency gains) coupled with comprehensive use-phase modeling (e.g., from present capabilities to thermodynamic limits) to evaluate the promise of future nanotechnologies from cradle-to-use. Ultimately, these results can be incorporated into existing LCA tools to broaden system boundaries and include potential supply chain impacts of future technologies.

11.2.1 SWCNT Manufacturing from an Environmental Perspective

SWCNTs can be synthesized through at least four different pathways: chemical vapor deposition (CVD), high-pressure carbon monoxide (HiPCO), arc discharge, and laser vaporization. Early environmental assessments have called attention to the massive electricity consumption, high-purity input materials requirements, and low synthesis yields common to these processes (Healy et al. 2008; Cannis et al. 2010; Ganter et al. 2010). The majority of environmental impact is attributable to electricity consumption during SWCNT synthesis and to a lesser extent purification processes, while the most significant impact categories are climate change, airborne inorganics, and acidification. HiPCO demonstrates the comparatively lower environmental burdens because it is a continuous flow process with recycled exhaust gasses and thus has potential for scale-up to produce kilogram quantities of SWCNT (Singh et al. 2008).

11.2.2 Mechanisms of the HiPCO Process

The HiPCO process is a specialized form of CVD through which SWCNTs are produced at a high rate (0.45 g/h) from a carbon monoxide (CO) feedstock (Nikolaev et al. 1999; Smalley et al. 2004). Catalytic iron nanoparticles, formed in situ by the thermal decomposition of $Fe(CO)_5$ and aggregation of gas-phase Fe atoms, provide preferential sites for CO disproportionation, shown in reaction 11.1. The formation of solid carbon from CO gas in disproportionation promotes formation of SWCNT on the surface of the catalyst through the *Yarmulke* mechanism (Hafner et al. 1998; Moisala et al. 2006). Briefly, a hemispherical carbon cap forms on appropriately sized particles, and the cap is pushed away from the catalytic particle by the addition of carbon atoms until the particle becomes too large and overcoats with amorphous carbon or becomes too small and evaporates (Bladh et al. 2000; Bronikowski et al. 2001).

$$\begin{array}{l} 2CO\ (g) \leftrightarrow CNT\ (s) + CO_2\ (g) \\ \Delta b\ \ 275.1\ [kJ/mol\text{-}CO] 469.62\ [kJ/mol\text{-}C] 19.87\ [kJ/mol\text{-}CO_2] \end{array} \tag{11.1}$$

The standard exergies of formation of the reactants and products are also given in reaction 11.1. Overall, the reaction releases 60.7 kJ/mol-C (or 5.06 kJ/g-SWCNT) at standard conditions (Szargut et al. 1988; Gutowski et al. 2010) and consequently is spontaneous. However, the reaction rate is significant only at temperatures above 550°C (Renshaw et al. 1970) and increases with pressure, thus the HiPCO process requires high temperature (900–1100°C) and pressure (30–50 atm) conditions. Reaching and maintaining these conditions requires significant exergy inputs, currently orders of magnitude greater than energy released in CO disproportionation.

11.2.3 Degree of Perfection of the HiPCO Process

The degree of perfection provides a measure of the second law efficiency of manufacturing processes and is defined as the ratio of the chemical exergy of the product(s) at standard conditions to the sum of all exergy input (Szargut et al. 1988). Assuming the kinetic and potential exergy of the CO gas stream is negligible, the degree of perfection can be estimated as follows:

$$\mathrm{DoP} = \frac{b_{\mathrm{ch,SWCNT}}}{b_{\mathrm{ph,in}} + b_{\mathrm{ch,in}}} \quad (11.2)$$

where the standard chemical exergy of SWCNT ($b_{\mathrm{ch,SWCNT}}$) is 469.62 kJ/mol-SWCNT (Gutowski et al. 2010). Assuming ideal gas behavior, the minimum physical exergy (b_{ph}) required to heat and pressurize CO from standard conditions (25°C, 1 atm) to those at which SWCNT synthesis occurs (~1100°C, ~30 atm) is given by Equation 11.3 (Szargut et al. 1988).

$$b_{\mathrm{ph}} = c_{\mathrm{p}}\left[(T - T_0) - \frac{T_0 \ln T}{T_0}\right] + \frac{RT_0 \ln P}{P_0} \quad \left[\frac{\mathrm{kJ}}{\mathrm{mol\text{-}CO}}\right] \quad (11.3)$$

The total input exergy is then given by the sum of physical inputs and the standard exergy of CO feedstock multiplied by the mole ratio of CO to SWCNT (given by the inverse of the reaction yield), which results in the total exergy input per mole of SWCNT produced. Based on three publications describing the inputs and yields of the HiPCO process (Nikolaev et al. 1999; Bronikowski et al. 2001; Smalley et al. 2004), we calculate the degree of perfection and show its improvement over that time period in Figure 11.2.

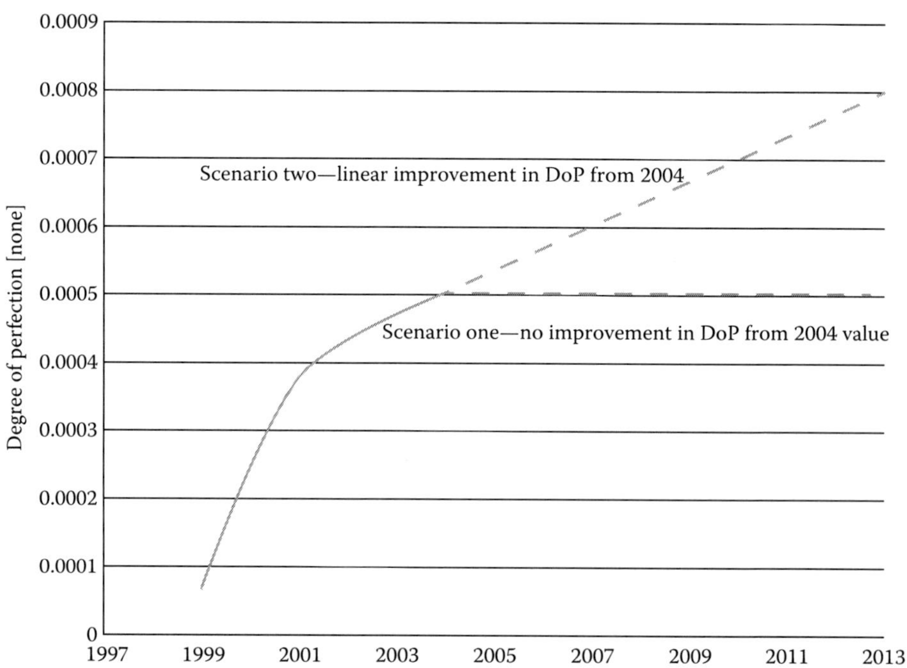

FIGURE 11.2 Improvements in the degree of perfection (DoP) of the high-pressure carbon monoxide process and two scenarios of future improvements.

When the HiPCO process was first reported in 1999, inputs were greater than 600,000 g CO/g SWCNT, and by patent application in 2004, CO inputs had fallen to tens of thousands of grams, which drives the observed improvements. The ideal (although never attainable) manufacturing process has a degree of perfection of one with lesser values indicating increased potential for efficiency gains. Presently, the degree of perfection for the HiPCO process is on the order of 10^{-3}–10^{-4}, which indicates significant room for improvement. By comparison, electric induction melting processes have a degree of perfection on the order of 10^{-1} (~.7) and are thereby approaching their second law limit.

11.2.4 Analogous Experience Curve Modeling

It is well understood that the thermodynamic and economic efficiency of material manufacturing processes improves with increased experience and scale (Haupin 2007; Smil 2008; Gutowski 2009). For example, the electricity demands of aluminum production through the Hall–Heroult process have asymptotically decreased toward the thermodynamic limit over 120 years. Likewise, the gross energy consumption of blast furnaces used for pig iron production decreased by orders of magnitude from early production values. The rapid gains in manufacturing efficiency early in process development, as shown in Figure 11.3, top, illustrate the challenge of environmental assessment of emerging technologies—early on LCA is trying to hit a moving target. Analogous to aluminum and pig iron production, SWCNT manufacturing may greatly improve in energetic efficiency with increases in scale and experience, scenarios for which are shown in Figure 11.3, bottom.

There are several historical examples of advances in material processing that subsequently enabled the development and growth of transformational industries. For example, improvements in aluminum processing enabled the aerospace industry and advances in pig iron production contributed significantly to the industrial revolution. Yet the improvements in aluminum and pig iron production accrued over centuries, whereas the HiPCO process was discovered less than 15 years ago. If carbon nanotubes are to have equally transformative effects as aluminum and steel industries, there is a critical need to identify sources of efficiency improvements *early* such that reductions occur rapidly. Section 11.2.5 will reveal that anticipatory LCA of SWCNT manufacturing and application in advanced batteries may call attention of research agenda that accelerate process improvement.

11.2.5 Use-Phase Performance Bounding of SWCNT Anode Lithium Ion Batteries

Half-cell testing of SWCNT anodes reveals a reversible capacity of 400 mAh/g SWCNT, compared to a theoretical limiting capacity of 1100 mAh/gSWCNT (Landi et al. 2008). Both values represent a significant improvement over traditional lithium ion battery anodes (made of mesoporus carbon beads) that provide a reusable capacity around 150 mAh/gC. The specific energy density of the battery is computed as the product of specific capacity and cell voltage, nominally 3.6 volts for $LiCoO_2$-carbon battery cells (Linden and Reddy 2002). Assuming complementary advances

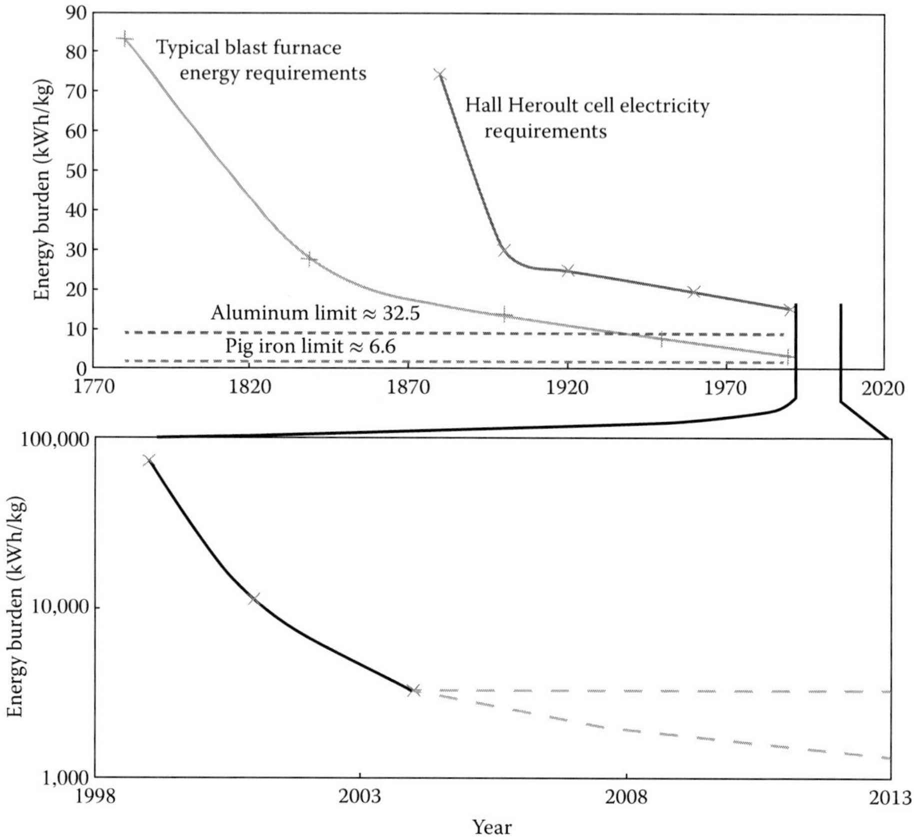

FIGURE 11.3 Historic reductions in aluminum and pig iron process energy and analogous improvements in the high-pressure carbon monoxide process.

in cathode technology and optimized battery geometry, SWCNT-enabled lithium ion batteries might store between 1.44 and 3.96 Wh/gSWCNT. Using these two limiting cases to provide upper and lower boundaries on battery performance, we convert existing cradle-to-gate inventory data describing SWCNT manufacturing (e.g., energy or material invested per gram of SWCNT produced) into a functional unit representative of battery performance, specifically kWh storage capacity. Dividing the manufacturing energy input per gram of SWCNT produced through the HiPCO scenarios presented in Figures 11.2 and 11.3 by the two limiting-case conversion factors provides a range of energy requirements per kWh storage capacity as shown in Figure 11.4.

Considering the best case scenario (e.g., the battery performs at its theoretical limit and the degree of perfection of SWCNT manufacturing improves linearly), SWCNT anodes will require roughly 250 MWh of electricity per kWh storage capacity in the battery (lowest dashed line in Figure 11.4). If the battery performs at its theoretical limit without improvements in manufacturing efficiency, the anode will require nearly 850 MWh/kWh storage capacity (middle line in Figure 11.4). Finally, if there are no improvements in manufacturing efficiency and battery performance, the anode

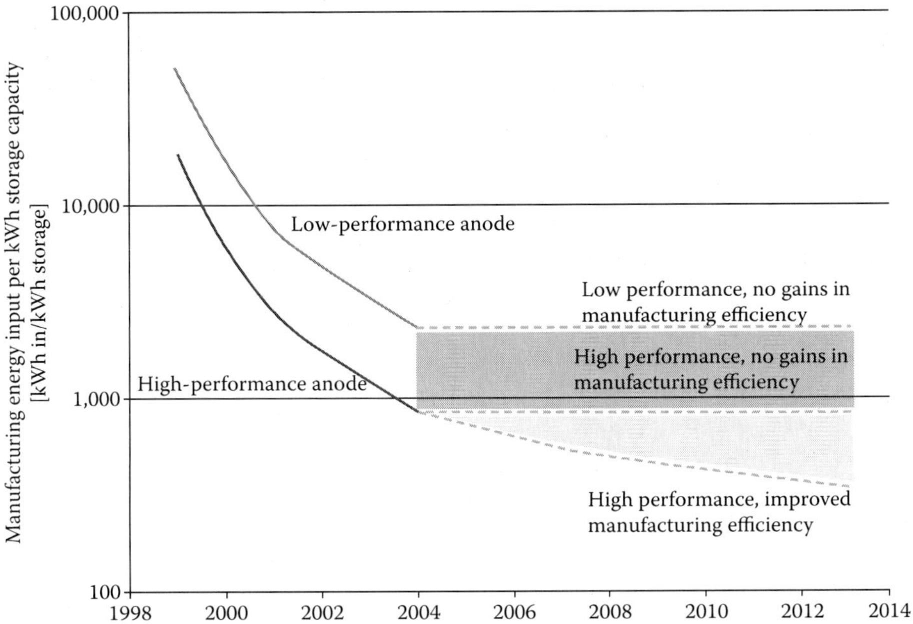

FIGURE 11.4 Energy requirements per kWh of storage capacity of single-wall carbon nanotube (SWCNT)-enabled lithium ion batteries, and two scenarios of future improvements in SWCNT synthesis via HiPCO.

will require over 2300 MWh/kWh storage capacity (uppermost line in Figure 11.4). These values represent the energy requirements for SWCNT-anode manufacturing alone and do not account for the remainder of battery manufacturing processes.

11.2.6 Discussion and New Directions

Figure 11.4 demonstrates that the SWCNT anodes, even if performing at their theoretical limit, are energetically infeasible unless significant improvements are made in SWCNT manufacturing. That is, despite nearly doubling the energy density of the battery, the life-cycle energy investments required for SWCNT anodes are environmentally problematic. A recent LCA of conventional carbon anode lithium ion batteries reports energy investments of .47 MWh/kWh storage capacity (Samaras and Meisterling 2008)—over two orders of magnitude less than SWCNT anodes alone. Thus, research improving the *functionality* of SWCNT anodes alone is unlikely to result in an environmentally viable technology. However, research efforts focused on decreasing the energy intensity of SWCNT manufacturing processes may result in technologies with practical potential to generate environmental benefits. Applying anticipatory LCA to (1) establish upper and lower boundaries to use-phase performance, (2) collect laboratory-scale manufacturing data, and (3) develop scenarios of future manufacturing improvements can be used to quantitatively explore trade-offs *before* the technology is implemented.

Ultimately, anticipatory LCA can be used to identify environmental hot spots and leverage points. For SWCNT manufacturing through the HiPCO process, the majority of exergetic losses are in the form of unconverted CO gases carried out as waste. Thus, a major pathway to reduce the environmental burden of SWCNT-enabled products is to improve the ratio of CO feedstock input to SWCNT produced, called the synthesis reaction yield (SRY). Pathways to improve the SRY of SWCNT manufacturing include purification and recycling of exhaust gases and tailoring catalyst size and structure for efficient feedstock use (Schauerman et al. 2009).

11.3 CONCLUSION

Research and development of nano-enabled energy technologies is inherently uncertain, and the tools necessary to conduct environmental assessment, specifically LCA, under such uncertainty have lagged behind nanotechnology development. Paradoxically, current approaches to LCA are least able to inform environmental understanding in the early stages of technology development, when LCA could most reduce the eventual systemic environmental burdens of the technology. This necessitates the development of *anticipatory* LCA methods, which employ thermodynamic analysis as a guidepost for understanding both the limits of manufacturing improvements and use-phases performance, thereby replacing a complete lack of data with potential scenarios. Ultimately, an anticipatory analysis may contribute to reorientation of laboratory research agenda toward pathways with decreased environmental burden. This chapter presented an example demonstrating the limits of a research agenda that focuses on improving use-phase performance of SWCNT-enabled lithium ion batteries alone, which is less valuable than research into lowering energy requirements of SWCNT manufacturing processes.

ACKNOWLEDGMENTS

Funding for this work was provided in part by the National Science Foundation (NSF) and Department of Energy (DOE) Quantum Energy and Sustainable Solar Technology Engineering Research Center at Arizona State University (ASU) (NSF CA EEC-1041895) and in part by the Center for Nanotechnology in Society (CNS) at Arizona State University (NSF NSEC 0531194 & NSF 0937591). Any opinions, findings, conclusions, and recommendations expressed in this material are those of the authors and do not necessarily reflect those of NSF or DOE. The authors would like to thank the students and faculty of the Sustainable Energy and Environmental Decision Science (SEEDS) collective at ASU as well as the anonymous reviewer for insightful comments.

REFERENCES

Anctil, A., Babbitt, C.W., Raffaelle, R.P., and Landi, B.J. 2011. "Material and energy intensity of fullerene production," *Environmental Science & Technology*, 45(6): 2353–2359.

Bauer, C. et al. 2008. "Towards a framework for life cycle thinking in the assessment of nanotechnology," *Journal of Cleaner Production*, 16(8–9): 910–926.

Bell, T.E. 2007. "Understanding risk assessment of nanotechnology," whitepaper, National Nanotechnology Coordination Office. Available at: http://whitepapers.hackerjournals.com/wp-content/uploads/2010/11/Understanding-risk-Assensment-of-nanotechnology.pdf.

Bladh, K., Falk, L.K.L., and Rohmud, F. 2000. "On the iron-encapsulated growth of single-walled carbon nanotubes and encapsulated metal particles in the gas phase," *Applied Physics A*, 70: 317–322.

Bronikowski, M.J., Willis, P.A., Colbert, D.T., Smith, K.A., and Smalley, R.E. 2001. "Gas-phase production of single-walled carbon nanotubes from carbon monoxide via the HiPCO process: A parametric study," *Journal of Vacuum Science and Technology A*, 19(4): 1800–1805.

Canis, L., Linkov, I., and Seager, T.P. 2010. "Application of stochastic multiattribute analysis to assessment of single wall carbon nanotube synthesis processes," *Environmental Science & Technology*, 44(22): 8704–8711.

Curran, M.A. 2004. "The status of life cycle assessment as an environmental management tool," *Environmental Progress*, 23(4): 277–283.

Davies, J.C. 2009. *Oversight of next generation nanotechnology*, Woodrow Wilson International Center for Scholars, Project on Emerging Nanotechnologies, Washington, D.C., 1–48.

Dewick, P., Green, K., and Miozzo, M. 2004. "Technological change, industry structure and the environment," *Futures*, 36: 267–293.

Eason, T., Meyer, D.E., Curran, M.A., and Upadhyayula, V.K.K. 2011. "Decision support framework for sustainable nanotechnology," whitepaper, USEPA National Risk Management Laboratory. Available at: http://nepis.epa.gov/Adobe/PDF/P100CH93.pdf. Accessed January 2013.

Evans, J.W. 2007. "The evolution of technology for light metals over the last 50 years: Al, Mg, and Li," *Journal of Minerals, Metals, and Materials Society*, 59(2): 30–38.

Fleischer, T., and Grunwald, A. 2008. "Making nanotechnology developments sustainable. A role for technology assessment?" *Journal of Cleaner Production*, 16(8–9): 889–898.

Ganter, M.J., Seager, T.P., Schauerman, C.M., Landi, B.J., and Raffaelle, R.P. 2010. "A life cycle energy analysis of single wall carbon nanotubes produced by laser vaporization." *Proceedings of the International Symposium on Sustainable Systems and Technology* (ISSST), IEEE, New York.

Grubb, G.F. and Bakshi, B. 2010. "Life cycle of titanium dioxide nanoparticle production," *Journal of Industrial Ecology*, 15(1): 81–95.

Gutowski, T.G. et al. 2009. "Thermodynamic analysis of resources used in manufacturing processes." *Environmental Science & Technology*, 43(5): 1584–1590.

Gutowski, T.G., Liow, J.Y.H., and Sekulic, D.P. 2010. "Minimum exergy requirements for the manufacturing of carbon nanotubes," *Proceedings of the International Symposium Sustainable Systems and Technology (ISSST),* IEEE Publishing, New York.

Hafner, J.H. et al. 1998. "Catalytic growth of single-wall carbon nanotubes from metal nanoparticles," *Chemical Physics Letters*, 292: 195–202.

Haupin, W. 2007. "History of electrical energy consumption by Hall–Héroult cells," In Peterson, W.S. and Miller, R.E., eds, *Hall–Héroult Centennial: First Century of Aluminum Process Technology*, John Wiley & Sons, Inc., PA, pp. 106–113.

Healy, M.L., Dahlben, L.J., and Isaacs, J.A. 2008. "Environmental assessment of single-walled carbon nanotube processes." *Journal of Industrial Ecology*, 12(3): 376–393.

Helland, A. et al. 2007. "Risk assessment of engineered nanomaterials: A survey of industrial approaches," *Environmental Science & Technology*, 42(2): 640–646.

Khanna, V., Bakshi, B.R., and Lee, L.J. 2007. "Life cycle energy analysis and environmental life cycle assessment of carbon nanofibers production," *Proceedings of the International Symposium on Sustainable Systems and Technology, (ISSST),* IEEE, New York.

Klopffer, W. et al. 2007. "Nanotechnology and life cycle assessment: A systems approach to nanotechnology and the environment." Woodrow Wilson International Center for Scholars, Washington, D.C. Available at: http://www.nanotechproject.org/file_download/files/NanoLCA_3.07.pdf. Accessed January 2013.

Kohler, A.R., Som, C., Helland, A., and Gottschalk, F. 2008. "Studying the potential release of carbon nanotubes throughout the application life cycle," *Journal of Cleaner Production*, 16: 927–937.

Krishnan, N. et al. 2008a. "A hybrid life cycle inventory of nano-scale semiconductor manufacturing," *Environmental Science & Technology*, 42(8): 3069–3075.

Krishnan, N., Williams, E.D., and Boyd, S.B. 2008b. "Case studies in energy use to realize ultra-high purities in semiconductor manufacturing," *Proceedings of the International Symposium on Electronics and the Environment (ISEE)*, IEEE, New York.

Landi, B.J., Ganter, M.J., Cress, C.D., DiLeo, R.A., and Raffaelle, R.P. 2009. "Carbon nanotubes for lithium-ion batteries," *Energy and Environmental Science*, 2: 638–654.

Landi, B.J., Ganter, M.J., Schauerman, C.M., Cress, C.D., and Raffaelle, R.P. 2008. "Lithium ion capacity of single wall carbon nanotube paper electrodes," *Journal of Physical Chemistry C*, 112: 7509–7515.

Landi, B.J., Ruf, H.J., Evans, C.M., Cress, C.D., and Raffaelle, R.P. 2005. "Purity assessment of single wall carbon nanotubes using optical absorption spectroscopy," *Journal of Physical Chemistry B*, 109(20): 9952–9965.

Linden, D. and Reddy, T. 2002. *Handbook of Batteries*, 3rd ed., McGraw-Hill, New York: 4.8.

Linkov, I., Bates, M.E., Canis, L.J., Seager, T.P., and Keisler, J.M. 2011. "A decision-directed approach for prioritizing research into the impact of nanomaterials on the environment and human health," *Nature Nanotechnology*, 6(12): 784–787. doi:10.1038/nnano.2011.163.

Lloyd, S.M. and Lave, L.B. 2003. "Life cycle economic and environmental implications of using nanocomposites in automobiles," *Environmental Science & Technology*, 37(15): 3458–3466.

Lloyd, S.M., Lave, L.B., and Scott Matthews, H. 2005. "Life cycle benefits of using nanotechnology to stabilize platinum-group metal particles in automotive catalysts," *Environmental Science & Technology*, 39(5): 1384–1392.

Matheys, J. et al. 2007. "Influence of functional unit on the life cycle assessment of traction batteries," *The International Journal of Life Cycle Assessment*, 12(3): 191–196.

Maynard, A.D. 2009. "Oversight of engineered nanomaterials in the workplace," *The Journal of Law, Medicine & Ethics*, 37: 651–658.

McDonald, A. and Schrattenholzer, L. 2001. "Learning rates for energy technologies," *Energy Policy*, 29: 255–261.

Meyer, D.E., Curran, M.A., and Gonzalez, M.A. 2009. "An examination of existing data for the industrial manufacture and use of nanocomponents and their role in the life cycle impact of nanoproducts," *Environmental Science & Technology*, 43(5): 1256–1263.

Meyer, D.E., Curran, M.A., and Gonzalez, M.A. 2011. "An examination of silver nanoparticles in socks using screening-level life cycle assessment," *Journal of Nanoparticle Research*, 13: 147–156.

Moisala, A. et al. 2006. "Single-walled carbon nanotube synthesis using ferrocene and iron pentacarbonyl in a laminar flow reactor," *Chemical Engineering Science*, 61: 4393–4402.

Nikolaev, P. et al. 1999. "Gas-phase catalytic growth of single-walled carbon nanotubes from carbon monoxide," *Chemical Physics Letters*, 313: 91–97.

Oberdorster, G., Oberdorster, E., and Oberdorster, J. 2005. "Nanotoxicology: An emerging discipline evolving from studies of ultrafine particles," *Environmental Health Perspectives*, 113(7): 823–839.

Oberdorster, G., Stone, V., and Donaldson, K. 2007. "Toxicology of nanoparticles: A historical perspective," *Nanotoxicology*, 1(1): 2–25.

Olapiriyakul, S. and Caudill, R.J.A. 2008. "A framework for risk management and end-of-life (EOL) analysis for nanotechnology products: A case study of lithium ion batteries," *Proceedings of the International Symposium on Sustainable Systems and Technology, (ISSST)*, IEEE Publishing, New York.

Ostertag, K. and Husing, B. 2008. "Identification of starting points for exposure assessment in the post-use phase of nanomaterial-containing products," *Journal of Cleaner Production*, 16(8–9): 938–948.

Osterwalder, N., Capello, C., Hungerbuhler, K., and Stark, W.J. 2006. "Energy consumption during nanoparticle production: How economic is dry synthesis?" *Journal of Nanoparticle Research*, 8: 1–9.

Pesonen, H.L. et al. 2000. "Framework for scenario development in LCA," *The International Journal of Life Cycle Assessment*, 5(1): 21–30.

Plata, D.L., Hart, A.J., Reddy, C.M., and Gschwend, P.M. 2009. "Early evaluation of potential environmental impacts of carbon nanotube synthesis by chemical vapor deposition," *Environmental Science & Technology*, 43(21): 8367–8373.

Powers, K.W., Palazuelos, M., Moudgil, B.M., and Roberts, S.M. 2007. "Characterization of the size, shape, and state of dispersion nanoparticles for toxicological studies," *Nanotoxicology*, 1(1): 42–51.

Reijnders, L. 2010. "Design issues for improved environmental performance of dye-sensitized and organic nanoparticulate solar cells," *Journal of Cleaner Production*, 18: 307–312.

Renshaw, G.D., Roscoe, C., and Walker Jr., P.L. 1970. "Disproportionation of CO over iron and silicon-iron single crystals," *Journal of Catalysis*, 18: 164–183.

Samaras, C. and Meisterling, K. 2008. "Life cycle assessment of greenhouse gas emissions from plug-in hybrid electric vehicles: implications for policy," *Environmental Science & Technology*, 43: 3170–3176.

Schauerman, C.M., Alvarenga, J., Landi, B.J., Cress, C.D., and Raffaelle, R.P. 2009. "Impact of nanometal catalysts on the laser vaporization synthesis of single wall carbon nanotubes," *Carbon*, 47(10): 2431–2435.

Scheufele, D.A. et al. 2007. "Scientists worry about some risks more than the public," *Nature Nanotechnology*, 2: 732–734.

Seager, T.P. and Linkov, I. 2008. "Coupling multicriteria decision analysis and lifecycle assessment for nanomaterials." *Journal of Industrial Ecology*, 12(3): 282–285.

Seager, T.P. and Linkov, I. 2009. "Uncertainty in life cycle assessment of nanomaterials," In Linkov, I. and Steevens, J., eds, *Nanomaterials: Risks and Benefits*, Springer, Amsterdam, pp. 423–436.

Sengul, H., Theis, T.L., and Ghosh, S. 2008. "Toward sustainable nanoproducts," *Journal of Industrial Ecology*, 12(3): 329–359.

Siegrist, M., Wiek, A., Helland, A., and Kastenholz, H. 2007. "Risk and nanotechnology: The public is more concerned than experts and industry," *Nature Nanotechnology*, 2: 67.

Singh, A. et al. 2008. "Environmental impact assessment for potential continuous processes for the production of carbon nanotubes," *American Journal of Environmental Sciences*, 4(5): 522–534.

Smalley, R.E. et al. 2004. "Single-wall carbon nanotubes from high pressure CO," *US Patent No. 7,204,970B2*.

Smil, V. 2008. *Energy in Nature and Society: General Energetics of Complex Systems*, MIT Press, MA.

Stefani, D. et al. 2011. "Structural and proactive safety aspects of oxidation debris from multi-walled carbon nanotubes," *Journal of Hazard Materials*, 189: 391–396.

Szargut, J., Morris, D.R., and Steward, F.R. 1988. *Exergy Analysis of Thermal, Chemical, and Metallurgical Processes*, Hemisphere Publishing Corp, New York.

Theis, T.L. et al. 2011. "A life cycle framework for the investigation of environmentally benign nanoparticles and products," *Physica Status Solidi RRL*, 5(9): 312–317.

U.S. Geological Survey Scientific Investigations and Reports 2008–5141. 2007. "Material use in the United States—Selected case studies for cadmium, cobalt, lithium, and nickel in rechargeable batteries," Available at: http://pubs.usgs.gov/sir/2008/5141/sir-2008-5141.pdf. Accessed January 2013.

USEPA. 2008. *Draft Nanomaterial Research Strategy*, United States Environmental Protection Agency, Office of Research and Development, Washington, D.C., EPA/600/S-08/002.

Von Gleich, A., Steinfeldt, M., and Petschow, U. 2007. "A suggested three-tier approach to assessing the implications of nanotechnology and influencing its development," *Journal of Cleaner Production*, 16: 899–909.

Walser, T., Demou, E., Lang, D., and Hellweg, S. 2011. "Prospective environmental life cycle assessment of nanosilver T-shirts," *Environmental Science & Technology*, 45(10): 4570–4578.

Wender, B.A. and Seager, T.P. 2011. "Towards prospective life cycle assessment: Single wall carbon nanotubes for lithium-ion batteries," *Proceedings of the International Symposium on Sustainable Systems and Technology (ISSST)*, IEEE, New York.

Wiek, A., Lang, D.J., and Siegrist, M. 2008. "Qualitative system analysis as a means for sustainable governance of emerging technologies: The case of nanotechnology," *Journal of Cleaner Production*, 16: 988–999.

Wiesner, M.R., Lowry, G.V., Alvarez, P., Dionysiou, D., and Biswas, P. 2006. "Assessing the risk of manufactured nanomaterials," *Environmental Science & Technology*, 40(14): 4336–4345.

12 Life-Cycle Assessment of Nanotechnology-Based Applications

Michael Steinfeldt

CONTENTS

Nanotechnology is frequently described as an enabling technology and a fundamental innovation, that is, it is expected to lead to numerous innovative developments in the most diverse fields of technology and areas of application in society and the marketplace. Nanotechnology is regarded as a substantial element for environmental relief. As a result, the following questions arise: How large are the possible relief effects on the environment by nanotechnological techniques? This contribution gives a current overview of existing studies of published life-cycle assessments (LCAs) of the manufacture of nanoparticles and nanocomponents and describe a new method of prospective technological assessment of nanotechnological processes in early innovation phases. The focus is placed on the potential environmental relief provided by nanotechnology-based products and processes.

12.1 LIFE-CYCLE ASSESSMENT METHODOLOGY

LCA is the most extensively developed and standardized methodology for assessing the environmental aspects and potential impacts throughout a product's life from raw material acquisition through production, use and recycling, and/or disposal (i.e., cradle to grave).

The LCA includes four stages:

1. Definition of the goal and scope of the investigation
2. Life-cycle inventory analysis (collection, compilation, and calculation of data on the input side; the consumption of raw and ancillary materials, including energy in-flows; and product data on the output side, including air and water emissions and waste data)
3. Life-cycle impact assessment (the analysis is organized [classification and characterization of the life-cycle inventory data according to its environmental relevance and environmental impact])
4. Life-cycle interpretation

Figure 12.1 makes clear the relationship between these steps. The directional arrows between the individual LCA steps should make the iterative nature of the process clear; that is, the results of further steps are always fed back into the process, possibly resulting in further changes and iterations.

A workshop comprising international experts from the fields of both LCA and nanotechnology concluded that the LCA ISO framework (DIN ISO 2006) is fully suitable to all stages of the life cycle of nanotechnology-based applications (Klöpffer et al. 2007). It has the advantage that by means of comparative assessments, an extrapolative analysis of eco-efficiency potentials in comparison to existing applications is possible. Typical impact categories include global warming/climate change, stratospheric ozone depletion, human toxicity, ecotoxicity, photo-oxidant formation, acidification, eutrophication, land use, and resource depletion (Rebitzer et al. 2004).

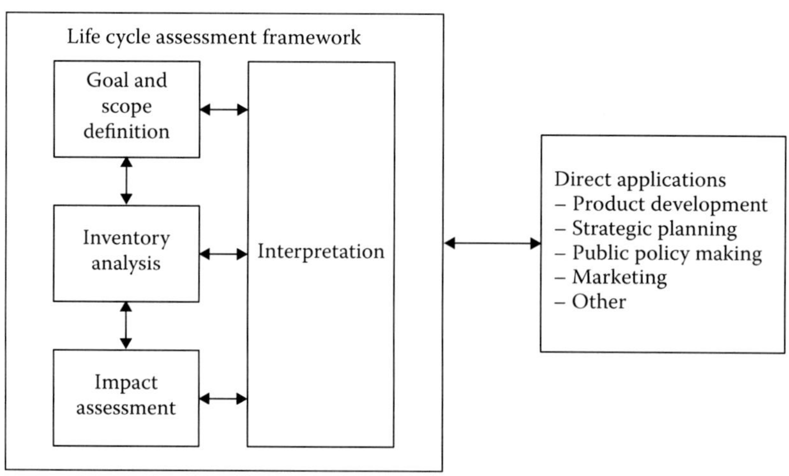

FIGURE 12.1 Steps in the preparation of a life-cycle assessment. (DIN EN ISO 14040 2006.)

12.2 EVALUATION OF NANOTECHNOLOGY-BASED PRODUCTS AND APPLICATIONS IN LCA STUDIES

In recent years, nanotechnology-based products and applications in LCA studies have been increasingly examined. A summary of studies of identified life-cycle aspects and environmental benefits is provided (Table 12.1).

The accomplished studies have mainly focused on cradle-to-gate assessments. Cradle to gate is an assessment of a partial product life cycle from manufacture to the factory gate. The use phase and the after-use phase (recycling, disposal) of the

TABLE 12.1
Overview of Studies about Life-Cycle Aspects of Nanotechnology-Based Applications

Nanoproduct	Approach	Tech Benefits	Environmental Benefits	References
Antireflex glass for solar applications as compared with traditional glass	No assessment, only indication of the environmental benefit	Increased solar transmission	6% higher energy efficiency	Bine 2002
Clay–polypropylene nanocomposite in light-duty vehicle body panels as compared with steel and aluminum	EIO-LCA	Reduced weight	Overall reduced environmental impact; large energy savings	Lloyd and Lave 2003
Nanoscale PGM particles in automotive catalysts	Eco-profile following LCA methodology	Reduced PGM loading levels by 50%	Overall reduced environmental impact (10%–40%)	Steinfeldt et al. 2003
Photovoltaic, dye photovoltaic cells as compared with multicrystalline silicon solar cells	Eco-profile following LCA methodology	Dye photovoltaic cells with better energy payback time, but smaller efficiency		Steinfeldt et al. 2003
Nanoscale PGM particles in automotive catalysts	EIO-LCA	Reduced PGM loading levels by 95%	Overall reduced environmental impact	Lloyd et al. 2005
Ultradur® High Speed plastic as compared with conventional Ultradur	BASF Eco-Efficiency Analysis	Significantly improved flowability, reduction of working time and energy consumption of injection molding process	Reduced environmental impact (1%, 5%–9%), only ozone depletion higher	BASF AG 2005, Steinfeldt et al. 2010a

(Continued)

TABLE 12.1 (*Continued*)
Overview of Studies about Life-Cycle Aspects of Nanotechnology-Based Applications

Nanoproduct	Approach	Tech Benefits	Environmental Benefits	References
Car tire with nanoscaled SiO_2 and carbon black	No Assessment, only indication of the environmental benefit	Increased road resistance	Up to 10% lower fuel consumption	UBA 2006
Nanocoatings as compared with conventional coatings	Eco-profile following LCA methodology	Necessary coating thickness smaller while maintaining functionality	5%–8% higher resource efficiency, 65% lower emissions of volatile organic compound	Steinfeldt et al. 2007
Styrene synthesis, CNT catalyst as compared with iron oxide-based catalysts	Eco-profile following LCA methodology	Change of type of reaction, reduction of the reaction temperature, change of reaction medium	≈50% reduced energy consumption of the synthesis process	Steinfeldt et al. 2007
White LED and quantum dots as compared with incandescent lamps and compact fluorescents	Eco-profile following LCA methodology	Higher lifetime	Higher energy efficiency compared to lamp, Higher energy efficiency compared to fluorescent lamp only with luminous efficacy higher 65 lm/W	Steinfeldt et al. 2007
OLED displays and nanotube field emitter displays as compared with cathode ray tube, liquid-crystal displays, and plasma screens	Eco-profile following LCA methodology	Increased energy efficiency, higher display resolution, reduced display thickness	Higher energy and resource efficiency; reduced material input by OLEDs, twice as high energy efficiency in the use phase	Steinfeldt et al. 2007
Ferrite adhesives as compared with conventional adhesives	Eco-profile following LCA methodology	Better curing of the adhesive by use of magnetic characteristics	12% (–40%) higher energy efficiency, dependent of size of adherent	Wigger 2007

TABLE 12.1
Overview of Studies about Life-Cycle Aspects of Nanotechnology-Based Applications

Nanoproduct	Approach	Tech Benefits	Environmental Benefits	References
Polypropylene nanocomposite in packaging film, agricultural film, and automotive panel as compared with conventional films	Environmental and cost assessments	Reduced weight, increased elasticity and strength of polypropylene	For agricultural film lower impact for five out of seven environmental categories (35%); for packaging film and automotive panel very small or no benefit	Roes et al. 2007
Nanodelivery system as compared with conventional micro-delivery system (vitamin E)	Screening LCA	Higher penetration and conversion rates	Potential for efficiency enhancement ≈34%	Novartis International AG et al. 2007
Comparison of different PVD coating TiN, TiAlN, Ti + TiAlN	LCA		TiN has the smallest environmental impact	Bauer et al. 2008
Field emission display screen with CNT compared with conventional screen technologies	Simplified LCA	Increased energy efficiency	Overall reduced environmental impact	Bauer et al. 2008
CNF polymer composites compared with steel for equal stiffness design (body panel of automobile)	LCA, SimaPro	Material substitution of steel	Energy savings from substitution dependent on polymer type, weight reduction, and CNF content rate	Khanna and Bakshi 2009
Nanotechnology-based disposable packaging (nano-polyethylene terephthalate bottle) as compared with conventional packaging	Environmental assessment, in particular CO_2-emissions	Improved barrier characteristics in particular against oxygen	Nano-PET bottle opposite to aluminum 1/3 and to glass of 60% fewer greenhouse gases	Möller et al. 2009

(*Continued*)

TABLE 12.1 (*Continued*)
Overview of Studies about Life-Cycle Aspects of Nanotechnology-Based Applications

Nanoproduct	Approach	Tech Benefits	Environmental Benefits	References
Glass coating with easy-to-clean effects compared with conventional glass surface	LCA, SimaPro	Easy-to-clean surface	Reduced environmental impact only at lower cleaning agent consumption	Klade et al. 2009
Nano-coating for wooden surfaces compared with conventional coatings	LCA, SimaPro		Overall reduced environmental impact except global warming impact is worse	Klade et al. 2009
Nickel nanoparticle deposition compared with conventional nickel phosphorus electroplating for facilitating diffusion brazing of arrayed microfluidics-assisted diffusion brazing	LCA, SimaPro software	Reduced usage of nickel and lower diffusion bonding energy requirements	Ca. 10%–15% reduced environmental impact	Haapala et al. 2009
Nanofluid solar hot water technology compared with conventional solar technology	Environmental and economic analysis	Improved efficiency	≈6% reduced embodied energy content and carbon dioxide emissions	Otanicar and Golden 2009
Manufacture of nanotechnology-based solderable surface finishes on printed circuit boards as compared with conventional surface finishes	Eco-profile following LCA methodology, Umberto software, qualitative preliminary risk assessment	Necessary layer thickness smaller with same functionality	In relation to qualitatively comparable procedures depending on environmental impact category around factors from 4 to 390 better	Steinfeldt et al. 2010a
Nanotechnology-based (MWCNT) conductive foils as compared with conventional foils	Eco-profile following LCA methodology, Umberto software, qualitative preliminary risk assessment	Necessary foils thickness smaller with same functionality	12%, 5%–20% reduced environmental impact	Steinfeldt et al. 2010a

TABLE 12.1
Overview of Studies about Life-Cycle Aspects of Nanotechnology-Based Applications

Nanoproduct	Approach	Tech Benefits	Environmental Benefits	References
Nanotechnology-based hybrid system city bus (Li-ion batteries) as compared with conventional diesel city bus	Prospective Eco-profile following LCA methodology, Umberto software	Reduction of fuel consumption by the hybrid system	≈20% reduced environmental impact by the future scenario	Steinfeldt et al. 2010a
Car air filter with nanofiber coating as compared with conventional car air filter	Environmental assessment, Umberto software	Reduction of air resistance and the associated fan power	8% reduced energy consumption of fan, environmental benefit of the entire system very small	Martens et al. 2010
Next-generation CNT composite materials as carrier tray for electronics components compared with conventional polycarbonates	Prospective Eco-profile following LCA methodology, Umberto software, qualitative preliminary risk assessment	New material, smooth and nondusty	Increase of possible production efficiency of electronics components (e.g., 1%) produces an enhancement of environmental impacts (also 1%)	Steinfeldt et al. 2010b
Electrodeposited Ni–MWCNT composite films for metal substrates for wind power plant compared with conventional material	Prospective Eco-profile following LCA methodology, Umberto software, qualitative preliminary risk assessment	Lower friction coefficients under dry conditions, i.e., superior solid lubrication	Increase of possible energy production efficiency of wind power plant of 0.25% produces an enhancement of environmental impacts between 3.7% and 11%	Steinfeldt et al. 2010b
QDPV compared with conventional PV	LCA, SimaPro software	Higher energy efficiency	Reduced environmental impact (CO_2, SO_x, NO_x emissions), but higher heavy metal emissions	Sengül and Theis 2011

(Continued)

TABLE 12.1 (*Continued*)
Overview of Studies about Life-Cycle Aspects of Nanotechnology-Based Applications

Nanoproduct	Approach	Tech Benefits	Environmental Benefits	References
Future nanocrystalline silicon-based multijunction PV compared with conventional PV	Life-cycle EPBT analysis	Great potential for increasing the efficiency of photon-to-electricity conversion	Assumptions to the plasma-enhanced chemical vapor deposition process have large influence of EPBT	Kim and Fthenakis 2011
Nanocrystalline materials in thin-film silicon solar cells compared with conventional solar cells	LCA, SimaPro, in particular cumulative energy demand and greenhouse gas emissions		Greenhouse gas emissions of nanocrystalline silicon PV higher than amorphous modules	Meulen and Alsema 2011
Silver nanoparticles in socks compared with conventional socks	Screening-level LCA, SimaPro	Enhanced biocidal treatment	Silver socks have overall higher environmental impacts, nanoscale silver content and production process have largest influence on the impacts	Meyer et al. 2011
Nanosilver T-shirts compared with conventional T-shirts	Prospective LCA	Enhanced biocidal treatment	Lower washing frequencies can compensate the increased climate footprint of Nanosilver T-shirt production	Walser et al. 2011

EIO-LCA, Economic input–output life-cycle assessment; PGM, platinum-group metal; LED, light-emitting diode; OLED, Organic LED; PVD, physical vapor deposition; CNF, carbon nanofiber; MWCNT, multiwalled carbon nanotube; QDPV, quantum dot photovoltaics; PV, photovoltaics; EPBT, energy pay-back time.

product are usually omitted (Meyer et al. 2009). For both phases, almost no data regarding environmental impact exist.

The results of the LCA comparisons make clear that nanotech applications neither intrinsically nor exclusively can be associated with the potential for a large degree of environmental relief. Nevertheless, for selected application contexts potentials for significant environmental relief can be ascertained using the chosen methods based on a comparative functionality of the different solutions. In addition, some of the studies have a strong prospective focusing. Often, only possible capabilities by nanotechnological applications are examined in scenarios with appropriate assumptions. The represented environmental benefits of nanotechnology-based applications are against the background various data gaps and methodical problems naturally with caution have to be judged.

LCA, like all methodologies, also has its weaknesses, blind spots, and deficiencies especially for the assessment of nanotechnologies. Some of these shortcomings need to be mentioned.

- There are impact categories for which generally accepted impact models do not yet exist. This is particularly true in the relevant categories of human and environmental toxicity. For example, with regard to scale, a consideration of the impact of fine dust particles (the particulate matter-10 risk dealing with the potential toxicity of particles smaller than 10 μm) in life-cycle impact assessments is therefore already doomed to failure because of its reference to weight and not particle size in nanotechnology applications. Other *qualitative* impacts that cannot be directly correlated with the levels of material and energy flows, such as structural impacts on ecosystems, also cannot adequately be captured.
- In comparative LCA analyses, in which a comparison of the new nanotechnological product with another conventional product is intended, the latter is mostly in the developed technology stage, for example, already implemented as large-scale plant. This means, that products at different developed technology stages are involved and compared to each other.
- Furthermore, in LCA, neither the technical risks nor the potency of applications are considered.

Current review publications on the state-of-the-art of LCA of engineered nanomaterials also clarify impressively the existing methodical gaps with the use of LCA at the nanoscale (Hischier and Walser 2012; Gavankar et al. 2012; Theis et al. 2011).

12.3 PROSPECTIVE LIFE-CYCLE ASSESSMENT AS AN APPROACH FOR PROSPECTIVE TECHNOLOGICAL ASSESSMENT OF NANOTECHNOLOGICAL TECHNIQUES

When applied to processes in the early innovation phases, as in the present case of mostly nanotechnology-based processes, the conditions for the application of these methods change radically toward a more prospective environmental (or technological) assessment and a more preliminary assessment of nanotechnology. Because of the interdisciplinary nature of nanotechnology, an enormous wealth of methods for the production of nanoscale products can be found in the literature. Products can, for example, be differentiated according to their nanoscale basic structure: particle-like structures (e.g., nanocrystals, nanoparticles, and molecules), linear structures (e.g., nanotubes, nanowires, and nanotrenches), layer structures (nanolayers), and other structures such as nanopores. Materials can also be produced from the gas phase, the liquid phase, or from solids in such a way that they are nanoscalar in at least one dimension.

These techniques can be classified based on the type of approach in top-down or bottom-up. Top-down processes achieve nanoscale dimensions through carving or grinding methods (e.g., lithography, etching, and milling). Bottom-up methods assemble matter at the atomic scale through nucleation and/or growth from liquid, solid, or gas precursors by chemical reactions or physical processes (e.g., gas-phase deposition, flame-assisted deposition, sol–gel process, precipitation, and self-organization techniques).

With the enormous variety of manufacturing processes of nanomaterials, it would be naturally very desirable to know as early as possible in the innovation process with which environmental impacts the specific nanotechnological techniques are connected.

In the new approach for prospective technological assessment, manufacturing processes of nanomaterials are modeled on the laboratory experiment and mini-plant level. On this basis, extrapolating estimations in different scenarios for the scale-up and optimization of the manufacturing process are investigated (Figure 12.2).

The different influences of the inputs and outputs of the process are considered by appropriate conversion factors (e.g., change of yield, change of energy efficiency, and change of efficiency of operating supplies). So it is possible to perform cradle-to-gate assessment of nanotechnological techniques for the production of nanoparticles and nanocomponents. Some first results of this new approach are also to be found in Figure 12.5.

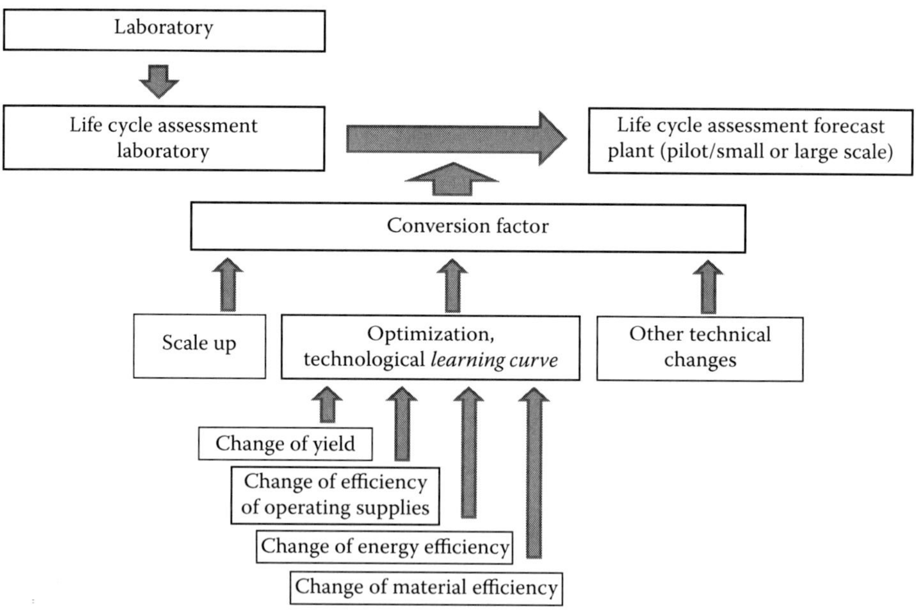

FIGURE 12.2 Principle method of life-cycle assessment forecast of plant based on laboratory data. (From Shibasaki, M., and Albrecht, S., Proceedings of European Congress of Chemical Engineering [ECCE-6], Copenhagen, 16–20 September, 2007; Shibasaki, M., *Method of LCA Prognosis of a Large Scale Plant based on Pilot Plant,* Shaker Verlag, Aachen [in German], 2009.)

12.4 EVALUATION OF SPECIFIC MANUFACTURED NANOPARTICLES

The largest groups of manufactured nanoparticles for industrial applications are inorganic nanoparticles (e.g., TiO_2, ZnO, SiO_2, and Ag), carbon-based nanomaterials (carbon nanofiber [CNF], multiwalled carbon nanotubes [MWCNTs], single-walled carbon nanotubes [SWCNTs]), and quantum dots (semiconductor nanoparticles with a specific size [e.g., CdSe, CdS, and GaN]). Besides qualitative environmental assessments of the different manufacturing methods (Steinfeldt et al. 2007; Sengül et al. 2008), quantified material and energy flow data exist only for a very small number of manufacturing processes and/or for individual nanomaterials. A summary of published studies is shown in Table 12.2. Particularly remarkable is that majority of the studies investigated the production of carbon-based nanomaterials.

Lekas evaluated substance flow analysis of CNTs throughout the economy from cradle to grave. The goal of the study was to gather production and use information of CNTs (current production, raw material inputs and quantities, end-use applications, and destination of materials) from literature and nanotube companies (Lekas 2005).

Osterwalder et al. (2006) performed cradle-to-gate assessments of titanium dioxide (TiO_2) and zirconium dioxide (ZrO_2) nanoparticle production. The goal of the study was to compare energy requirements and greenhouse gas emissions for the classical milling process with that of a novel flame synthesis technique using organic precursors. The functional unit of the study was 1 kg of manufacturing materials.

Roes et al. (2007) evaluated the use of nanocomponents in packaging film, agricultural film, and automotive panels. The goal of the prospective assessment was to determine whether the use of nanoclay additives in polymers (polypropylene, polyethylene, and glass fiber-reinforced polypropylene) is more environmentally advantageous than conventional materials. Specific material and energy flows of the nanoclay production were collected. The manufacture of nanoclay includes several processes, for example, raw clay (Ca-bentonite) extraction, separation, spray drying, organic modification, filtering, and heating.

Eckelman et al. (2008) have performed an E-factor analysis of several nanomaterial syntheses, as the E-factor is a measure of environmental impact and sustainability that has been commonly used by chemists. The E-factor (or waste-to-product ratio) includes all chemicals involved in production. Energy and water inputs are generally not included in E-factor calculations, nor are products of combustion, such as water vapor or carbon dioxide. Unfortunately, the results are not comparable with other studies.

Kushnir and Sandén (2008) modeled the requirements of future production systems of carbon nanoparticle and also used a cradle-to-gate perspective, including all energy flow-ups to the production and purification of carbon nanoparticles. All calculations are made for a functional unit of 1 kg of nanoparticles. Several production systems (fluidized bed chemical vapor deposition [CVD], floating catalyst CVD, high-pressure carbon monoxide [HiPCO], pyrolysis, electric arc, laser ablation, and solar furnace) are investigated and possible efficiency improvements are discussed. Carbon nanoparticles are found to be highly energy-intensive materials, in the order

TABLE 12.2
Overview of Studies of Published Life-Cycle Assessments of the Manufacture of Nanoparticles and Nanocomponents

Nanoparticle and/or Nanocomponent	Assessed Impact(s)	References
Carbon nanotubes	Substance Flow Analysis (SFA)	Lekas 2005
Metal nanoparticle production (TiO_2, ZrO_2)	Cradle-to-gate energy assessment, global warming potential	Osterwalder et al. 2006
Nanoclay production	Cradle-to-gate assessment, energy use, global warming potential, ozone layer depletion, abiotic depletion, photochemical oxidant formation, acidification, eutrophication, cost	Roes et al. 2007
Several nanomaterial syntheses	E-factor analysis	Eckelman et al. 2008
Carbon nanoparticle production	Cradle-to-gate energy assessment	Kushnir and Sandén 2008
CNT production	Cradle-to-gate assessment with SimaPro software, energy use, global warming potential, etc.	Singh et al. 2008
SWCNT production	Cradle-to-gate assessment with SimaPro software, energy use, global warming potential, etc.	Healy et al. 2008
CNF production	Energy use, global warming potential, ozone layer depletion, radiation, ecotoxicity, acidification, eutrophication, and land use	Khanna et al. 2008
Nanoscale semiconductor	Cradle-to-gate assessment, energy use, global warming potential	Krishnan et al. 2008
SWCNT synthesis process	Stochastic multiattribute analysis	Canis et al. 2010
Nanoscaled polyaniline production	Cradle-to-gate assessment with Umberto software, energy use, global warming potential, etc.	Steinfeldt et al. 2010a
MWCNT production	Cradle-to-gate assessment with Umberto software, energy use, global warming potential, etc.	Steinfeldt et al. 2010a
Nanoscaled titanium dioxide	Cradle-to-gate assessment, Ecoindicator 99 methodology, energy use, exergy	Grubb and Bakshi 2010
Fullerene (C_{60}, C_{70})	Cradle-to-gate assessment with SimaPro, energy use, material intensity	Anctil et al. 2011

Sources: Meyer, D.E., Curran, M.A., Gonzalez, M.A., *Environ Sci Technol* 43(5), 1256–1263, 2009; Own data.

SFA, Substance Flow Analysis; CNT, carbon nanotube; SWCNT, single-walled carbon nanotube; CNF, carbon nanofiber; MWCNT, multiwalled carbon nanotube.

of 2–100 times more energy intensive than aluminum, given a thermal to electric conversion efficiency of 0.35.

Singh et al. (2008) performed environmental impact assessments for two potential continuous processes for the production of CNTs. The HiP_{CO} disproportionation in a plug-flow reactor (CNT-PFR) and the fluidized bed reactor (CNT-FBR) with cobalt-molybdenum catalysts were selected for the conceptual design. The CNT-PFR has catalytic particles formed in situ by thermal decomposition of iron carbonyl. The CNT-FBR process uses the synergistic effect between cobalt and molybdenum giving high selectivity to CNTs from CO disproportionation.

Healy et al. (2008) have performed LCA of the three more established SWNT manufacturing processes: arc ablation (Arc), CVD, and HiP_{CO}. Each method consists of process steps that include catalyst preparation, synthesis, purification, inspection, and packaging. In any case, the inspection and packaging steps contribute minimally to the overall environmental loads of the processes. Although the technical attributes of the SWNT products generated via each process may not always be fully comparable, the study provides a baseline for the environmental footprint of each process. All calculations are made with a functional unit of 1 g of SWNTs.

Khanna et al. (2008) have performed a cradle-to-gate assessment of CNF production. The goal of the assessment was to determine the nonrenewable energy requirements and environmental impacts associated with the production of 1 kg of CNFs. Life-cycle energy requirements for CNFs from a range of feedstock materials are found to be 13–50 times higher than primary aluminum on an equal mass basis.

Krishnan et al. (2008) have presented a cradle-to-gate assessment and a developed library of materials and energy requirements, and global warming potential of nanoscale semiconductor manufacturing. The goal of the study was to identify potential process improvements. The functional unit selected was one silicon wafer with a 300-mm diameter, which can be used to produce 442 processor chips. The total energy required for the process is 14,100 MJ/wafer including 2,500 MJ/wafer accounting for the manufacture of fabrication equipment. The greenhouse potential is 13 kg CO_2 equivalent/wafer.

Canis et al. (2010) have discussed in relation to a product development problem of selecting the most advantageous technology for manufacturing SWCNTs. Four different synthesis processes were investigated: HiP_{CO}, arc discharge (Arc), CVD, and laser vaporization (Laser). The case study is an example of a stochastic multicriteria decision analysis-based method for situations where knowledge of the weights is lacking and uncertainty about criteria scores is significant. Unfortunately, the results are not comparable with that of other studies.

Steinfeldt et al. (2010a) have performed several in-depth LCAs of processes and products, including cradle-to-gate assessments of the production of nanoscaled polyaniline and of the production of MWCNT. With the cooperation of producers, it was possible to produce detailed models of the manufacturing processes for nanoscaled polyaniline and MWCNT, and to generate specific LCA data.

Grubb and Bakshi (2010) have investigated the LCA of the hydrochloride nanomanufacturing process for producing TiO_2 nanoparticles in comparison with conventional titanium dioxide. The functional unit of the study was 1 kg of manufacturing materials. This work also includes an exergy analysis to account for material resource consumption in the process.

Anctil et al. (2011) have performed a cradle-to-gate energy assessment of manufacturing of fullerenes and modified derivatives. The inventory is based on the functional unit of 1 kg of product (either C_{60} or C_{70}). Four synthesis methods were investigated. The embodied energy of 1 kg C_{60} after synthesis and separation is very different (pyrolysis with tetralin 12.7 GJ/kg C_{60}; pyrolysis with toluene 17.0 GJ/kg C_{60}; plasma Arc 88.6 GJ/kg C_{60}, and plasma radio frequency 106.9 GJ/kg C_{60}).

The data above can provide some insight regarding the potential burdens that must be addressed if the large-scale use of these types of nanoparticles and nanocomponents is to continue. For this purpose, the data from the studies are expressed in a common mass-based unit. Accordingly, energy demand is presented in MJ-equivalents/kg material and global warming potential is expressed as kg CO_2-equivalent/kg product. Energy consumption during the product life cycle is very important because it relates to the consumption of fossil fuels and the generation of greenhouse gases. Therefore, it is desirable to design manufacturing processes that minimize the use of energy. The data for energy consumption of the materials discussed above are shown in Figures 12.3 and 12.4. In addition, comparison of the data of conventional materials is included.

The represented cumulative energy requirements for various carbon nanoparticle manufacturing processes differ very strongly from each other. The various processes for the production of SWCNTs (excluding equipment fabrication) are by far the most energy-intensive processes as compared with the production of other carbon nanoparticles. A cause for the very large differences between the examined studies lies in the different process conditions (temperature and pressure) of the manufacturing processes. Furthermore, large differences are found in the assumptions of reactions and purification yields. The relative small reference value appears remarkable for the mass production of carbon black by means of flame synthesis. The production of MWCNTs based on catalyst CVD also surprises here with a relative small cumulative energy requirement.

The comparison of the cumulative energy requirements for the production of other conventional and nanoscaled materials and components make clear that the production of nanosemiconductors is also a very energy-intensive process. Only the extraction of the precious metal platinum, as an example, is still more complex. The production of nanoscaled polyaniline is likewise very energy intensive.

Comparison of the global warming potential for the production of various conventional and nanoscaled materials is shown in Figure 12.5. The production of silver nanoparticles has the largest impact when compared to other materials. However, the production of CNF and nanoscaled polyaniline shows a high global warming potential. The reason for the larger global warming potentials for silver nanoparticle, CNF, and polyaniline manufacturing is the much larger energy requirements when compared to other nanoparticle production. Also, the cleaning agents in the manufacturing processes have a large influence on the global warming potentials. The production of MWCNTs based on fluidized bed catalyst CVD also surprises here with a small global warming potential.

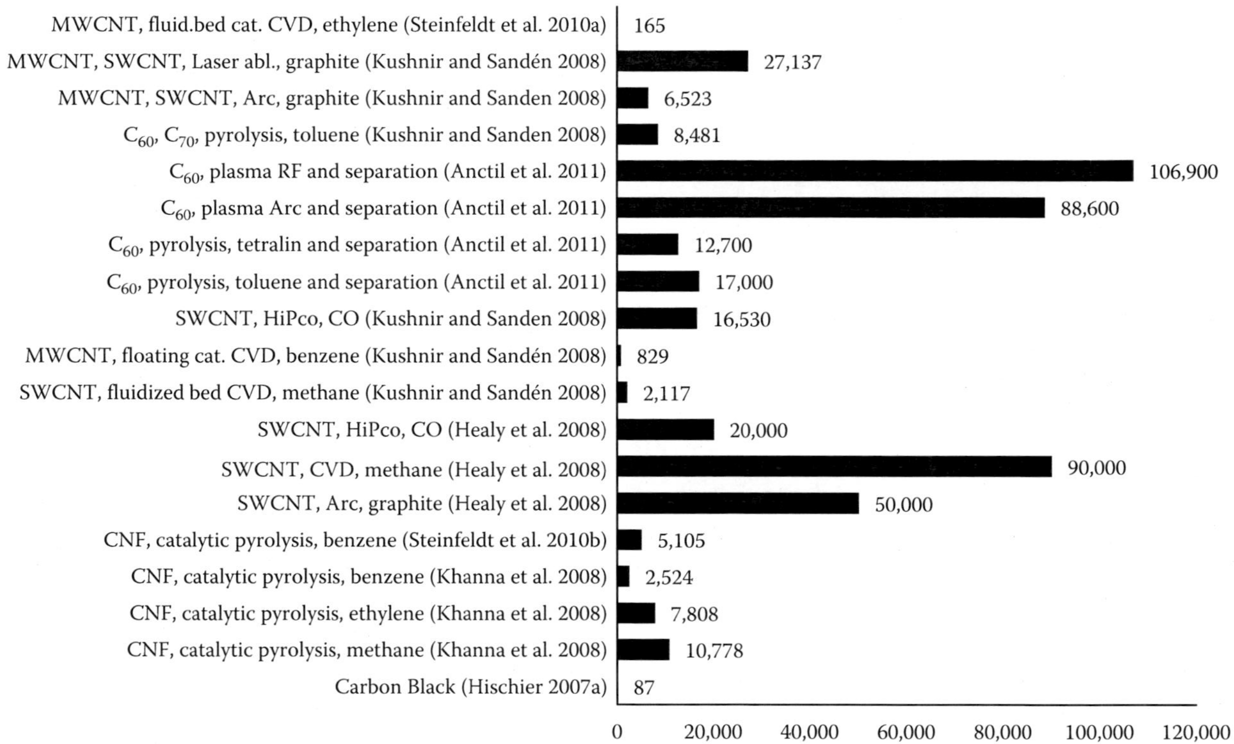

FIGURE 12.3 Comparison of the cumulative energy requirements for various carbon nanoparticle manufacturing processes (MJ-equivalent/kg material).

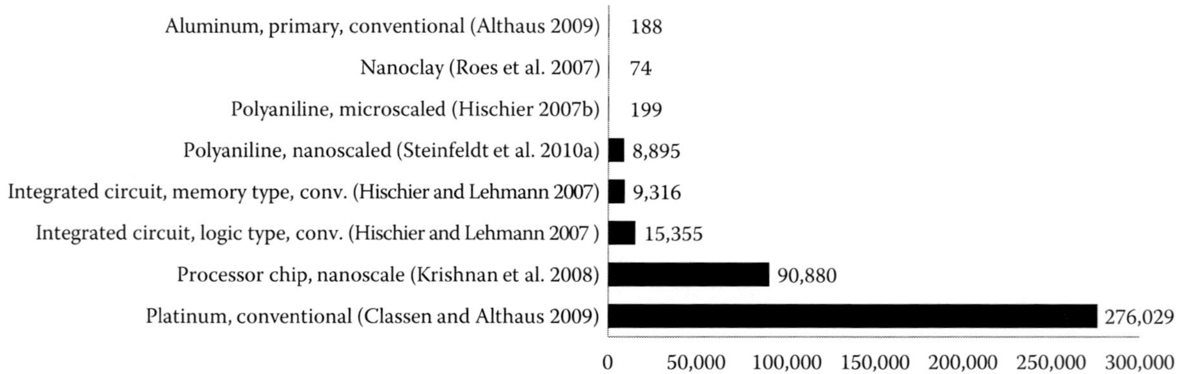

FIGURE 12.4 Comparison of the cumulative energy requirements for the production of various conventional and nanoscaled materials and components ([MJ-equivalent/kg product] in parts own calculation).

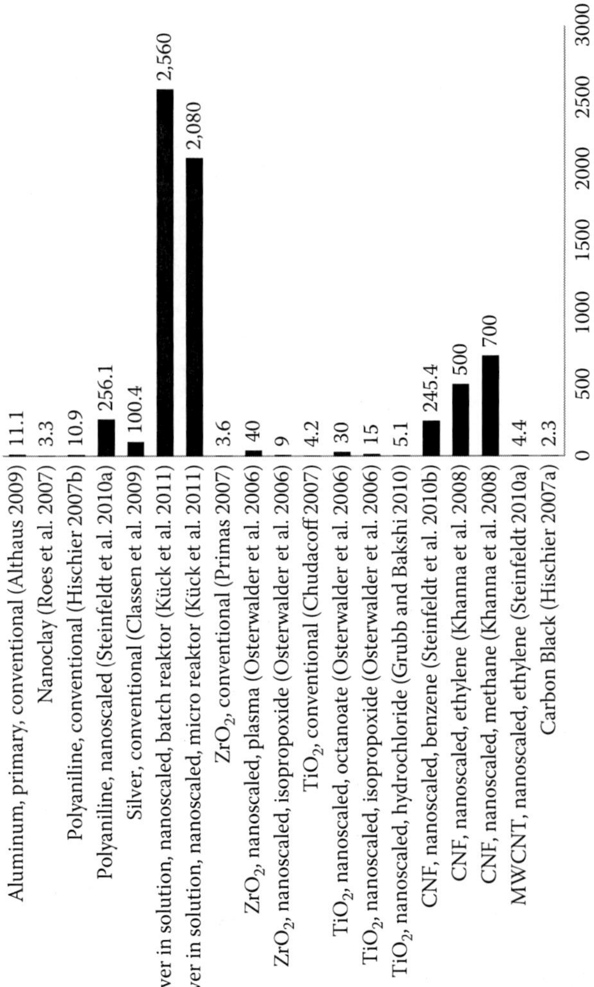

FIGURE 12.5 Comparison of global warming potential for the production of various conventional and nanoscaled materials ([CO_2-equivalent/kg product] in parts own calculation).

While these representative results present no comprehensive LCAs, the results do offer useful insight when considering the environmental impact of various nanomaterials and nanotechnology-based applications (see the overview of applications in Table 12.1). It is commonly pointed out that the nanocomponent is only a fraction of the total product (often only 2%, 3%, or 4%) implying that only a small fraction of the environmental impact of a nanoproduct can be attributed to the nanocomponent and its manufacture. The high specific energy demand for the production of nanoparticles is then related to the nanoproduct.

12.5 SUMMARY

The prevalence of diverse manufacturing routes for nanoproducts is a significant driver for nanotechnological innovations. All nanoproducts must proceed through various manufacturing stages to produce a material or device with nanoscale dimensions. Quantitative investigations of anticipated or still-to-be-realized environmental benefits arising from specific nanotechnological products and processes, as well as further-ranging environmental innovations such as product- and production-integrated environmental protection or energy-related solutions have so far been the exceptions.

Currently, a large number of data gaps exist when considering the application of LCA to nanoproducts. Specifically, only minimal data detailing the material and energy inputs and environmental releases related to the manufacture, release, transport, and ultimate fate of nanocomponents and nanoproducts exist. The presented new approach of prospective technological assessment of nanotechnological processes throughout their life cycle based on prospective scenarios and scaling-up models can help to close the existing data gaps.

Additional thought must also be given to address risk assessment and socioeconomic impacts/benefits that should be integrated with the LCA framework to provide a more comprehensive assessment tool for decision making when considering the use of nanoparticles for the manufacturing of nanoproducts (Gleich et al. 2008).

REFERENCES

Althaus HJ (2009). Aluminium. In: Althaus HJ, Blaser S, Classen M, Emmenegger MF, Jungbluth N, Scharnhorst W, Tuchschmid M: *Life Cycle Inventories of Metals*. Ecoinvent Report No 10, part I, Swiss Centre for Life Cycle Inventories, Dübendorf, Switzerland.

Anctil A, Babbitt CW, Raffaelle RP, Landi BJ (2011). Material and energy intensity of fullerene production. *Environ Sci Technol* 45(6):2353–2359.

BASF AG (2005). *Label Eco-Efficiency Analysis—Ultradur® High Speed*. Ludwigshafen, Germany.

Bauer C, Buchgeister J, Hischier R, Poganietz W, Schebek L, Warsen J (2008). Towards a framework for life cycle thinking in the assessment of nanotechnology. *J Clean Prod* 16(8–9):910–926.

BINE Informationsdienst (2002). *Antireflex Glass for Solar Applications*, Bonn, Germany (in German).

Canis L, Linkov I, Seager TP (2010). Application of stochastic multiattribute analysis to assessment of single walled carbon nanotube synthesis processes. *Environ Sci Technol* 44(22):8704–8711.

Chudacoff M (2007). Titanium dioxide. In: Althaus HJ, Chudacoff M, Hellweg S, Hischier R, Jungbluth N, Osses M, Primas A (eds.), *Life Cycle Inventories of Chemicals*. Ecoinvent Report No 8, Swiss Centre for Life Cycle Inventories, Dübendorf, Switzerland, pp. 761–768.

Classen M, Althaus HJ (2009). Platinum group metals (PGM). In: Althaus HJ, Blaser S, Classen M, Emmenegger MF, Jungbluth N, Scharnhorst W, Tuchschmid M: *Life Cycle Inventories of Metals*. Ecoinvent report No 10, part V, Swiss Centre for Life Cycle Inventories, Dübendorf.

Classen M, Tuchschmid M, Emmenegger MF, Scharnhorst W (2009). Gold and silver. In: Althaus HJ, Blaser S, Classen M, Emmenegger MF, Jungbluth N, Scharnhorst W, Tuchschmid M: *Life Cycle Inventories of Metals*. Ecoinvent Report No 10, part IX, Swiss Centre for Life Cycle Inventories, Dübendorf, Switzerland.

Eckelman, MJ, Zimmerman, JB, Paul T, Anastas, PT (2008). Toward green nano: E-factor analysis of several nanomaterial syntheses, *J Ind Ecol* 12(3):316–328.

Gavankar S, Suh S, Keller AF (2012). Life cycle assessment at nanoscale: Review and recommendations. *Int J Life Cycle Assess* 17:295–303.

Gleich A von, Steinfeldt M, Petschow U (2008). A suggested three-tiered approach to assessing the implications of nanotechnology and influencing its development. *J Clean Prod Sustain Nanotechnol Develop (special issue)* 16(8–9):899–909.

Grubb GF, Bakshi BR (2010). Life cycle of titanium dioxide nanoparticle production impact of emissions and use of resources. *J Ind Ecol* 15(1):81–95.

Haapala KR, Tiwari SK, Paul BK (2009). An environmental analysis of nanoparticle-assisted diffusion brazing. Proceedings of the 2009 ASME International Manufacturing Science and Engineering Conference, October 4–7, 2009, West Lafayette, IN.

Healy ML, Dahlben LJ, Isaacs JA (2008). Environmental assessment of single-walled carbon nanotube processes. *J Ind Ecol* 12(3):376–393.

Hischier R (2007a). Carbon black. In: Althaus HJ, Chudacoff M, Hellweg S, Hischier R, Jungbluth N, Osses M, Primas A: *Life Cycle Inventories of Chemicals*. Ecoinvent Report No 8, Swiss Centre for Life Cycle Inventories, Dübendorf, Switzerland, pp. 173–178.

Hischier R (2007b). *Life Cycle Inventories of Packaging and Graphical Papers*. Ecoinvent Report No 11, part II, Swiss Centre for Life Cycle Inventories, Dübendorf, Switzerland, pp. 194–197.

Hischier R, Lehmann M (2007). Electronic components. In: Hischier R, Classen M, Lehmann M, Scharnhorst W: *Life Cycle Inventories of Electric and Electronic Equipment: Production, Use and Disposal*. Ecoinvent Report No 18, part I, Swiss Centre for Life Cycle Inventories, Dübendorf, Switzerland, pp. 50–62.

Hischier R, Walser T (2012). Life cycle assessment of engineered nanomaterials: State of the art and strategies to overcome existing gaps. *Sci Total Environ* 425:271–282.

Khanna V, Bakshi BR (2009). Carbon nanofiber polymer composites: Evaluation of life cycle energy use. *Environ Sci Technol* 43(6):2078–2084.

Khanna V, Bakshi BR, Lee LJ (2008). Carbon nanofiber production: Life cycle energy consumption and environmental impact. *J Ind Ecol* 12(3):394–410.

Kim HC, Fthenakis VM (2011). Comparative life-cycle energy payback analysis of multi-junction a-SiGe and nanocrystalline/a-Si modules. *Prog Photovolt: Res Appl* 19(2):228–239.

Klade M, Meissner M, Stark S, Wallner A, Wenisch A, Veres E (2009). Matrix to the description of use and risks of nano-products (Matrix zur Darstellung von Nutzen und Risiken von Nano-Produkten) (NanoRate). Projektendbericht. Graz (in German), http://www.lebensministerium.at. Accessed 20 February 2012.

Klöpffer W, Curran MA, Frankl P, Heijungs R, Koehler A, Olsen SI (2007). *Nanotechnology and life cycle assessment: Synthesis of results obtained at a workshop in Washington. DC, 2–3 October 2006*. Woodrow Wilson International Center for Scholars, Washington.

Krishnan N, Boyd S, Somani A, Raoux S, Clark D, Dornfeld DA (2008). Hybrid life cycle inventory of nano-scale semiconductor manufacturing. *Environ Sci Technol* 42(8):3069–3075.

Kück A, Steinfeldt M, Prenzel K, Swiderek P, Gleich A von, Thöming J (2011). Green nanoparticle production using micro reactor. *J Phys: Conf Ser* 304 012074.

Kushnir D, Sandén BA (2008). Energy requirements of carbon nanoparticle production. *J Ind Ecol* 12(3):360–375.

Lekas D (2005). Analysis of nanotechnology from an industrial ecology perspective. Part II: Substance flow analysis study of carbon nanotubes. http://www.nanotechproject.org/process/assets/files/2720/36_nanotube_sfa_report_revised_part2.pdf. Accessed 20 February 2012.

Lloyd SM, Lave LB (2003). Life cycle economic and environmental implications of using nanocomposites in automobiles. *Environ Sci Technol* 37(15):3458–3466.

Lloyd SM, Lave LB, Matthews HS (2005). Life cycle benefits of using nanotechnology to stabilize platinum-group metal particles in automotive catalysts. *Environ Sci Technol* 39(5):1384–1392.

Martens S, Eggers B, Evertz T (2010). *Investigation of the Use of Nano-Materials in Environmental Protection*. UBA-Texte 34/2010, Dessau, Germany (in German).

Meulen R van der, Alsema E (2011). Life-cycle greenhouse gas effects of introducing nanocrystalline materials in thin-film silicon solar cells. *Prog Photovolt: Res Appl* 19(4):453–463.

Meyer DE, Curran MA, Gonzalez MA (2009). An examination of existing data for the industrial manufacture and use of nanocomponents and their role in the life cycle impact of nanoproducts. *Environ Sci Technol* 43(5):1256–1263.

Meyer DE, Curran MA, Gonzalez MA (2011). An examination of silver nanoparticles in socks using screening-level life cycle assessment. *J Nanopart Res* 13:147–156.

Möller M, Eberle U, Hermann A, Moch K, Stratmann B (2009). *Nanotechnology in the Food Sector*, vdf Hochschulverlag AG, Zürich, Switzerland (in German).

Novartis International AG, Ciba Spezialitätenchemie AG, Öko-Institute.V., Österreichisches Ökologie Institut, Stiftung Risiko-Dialog (2007). CONANO—COmparative challenge of NANOmaterials. A stakeholder dialogue project. Projektbericht, Wien (in German).

Osterwalder N, Capello C, Hungerbühler K, Stark WJ (2006). Energy consumption during nanoparticle production: How economic is dry synthesis? *J Nanopart Res* 8(1):1–9.

Otanicar TP, Golden JS (2009). Comparative environmental and economic analysis of conventional and nanofluid solar hot water technologies. *Environ Sci Technol* 43(15):6082–6087.

Primas A (2007). Zirconium oxide production from mineral sands. In: Althaus HJ, Chudacoff M, Hellweg S, Hischier R, Jungbluth N, Osses M, Primas A: *Life Cycle Inventories of Chemicals*. Ecoinvent Report No 8, Swiss Centre for Life Cycle Inventories, Dübendorf, Switzerland, pp. 861–875.

Rebitzer G, Ekvall T, Frischknecht R, Hunkeler D, Norris G, Rydberg T, Schmidt WP, Suh S, Weidema BP, Pennington DW (2004). Life cycle assessment: Part 1: Framework, goal and scope definition, inventory analysis, and applications. *Environ Int* 30(5):701–720.

Roes A, Marsili E, Nieuwlaar E, Patel MK (2007). Environmental and cost assessment of a polypropylene nanocomposite. *J Polym Environ* 15(3):212–226.

Şengül H, Theis TL, Ghosh S (2008). Toward sustainable nanoproducts: An overview of nanomanufacturing methods. *J Ind Ecol* 12 (3):329–359.

Şengül H, Theis TL (2011). An environmental impact assessment of quantum dot photovoltaics (QDPV) from raw material acquisition through use. *J Clean Prod* 19(1):21–31.

Shibasaki M (2009). *Method of LCA Prognosis of a Large Scale Plant based on Pilot Plant (Methode zur Prognose der Ökobilanz einer Großanlage auf Basis einer Pilotanlage in der Verfahrenstechnik—Ein Beitrag zur Ganzheitlichen Bilanzierung)*. Shaker Verlag, Aachen, Germany (in German).

Shibasaki M, Albrecht S (2007). Life cycle assessment prognosis for sustainability. Proceedings of European Congress of Chemical Engineering (ECCE-6), Copenhagen, 16–20 September.

Singh A, Lou HH, Pike RW, Agboola A, Li X, Hopper JR, Yaws CL (2008). Environmental impact assessment for potential continuous processes for the production of carbon nanotubes. *Am J Environ Sci* 4(5):522–534.

Steinfeldt M, Gleich A von, Henkle J, Endo M, Morimoto S, Momosaki E, (2010b). Environmental relief effects of nanotechnologies—factor 10 or only incremental increase of efficiency. *Proceeding of the 9th International Conference on EcoBalance Towards & Beyond 2020*, Tokyo, Japan.

Steinfeldt M, Gleich A von, Petschow U, Haum R (2007). *Nanotechnologies, Hazards and Resource Efficiency*. Springer, Heidelberg.

Steinfeldt M, Gleich A von, Petschow U, Pade C, Sprenger RU (2010a). *Environmental Relief Effects through Nanotechnological Processes and Products* (*Entlastungseffekte für die Umwelt durch nanotechnische Verfahren und Produkte*). UBA-Texte 33/2010, Dessau, Germany (in German).

Steinfeldt M, Petschow U, Hirschl B (2003). *Potential Applications of Nanotechnology based materials. Part 2: Analysis of Ecological, Social and Legal Aspects*. Schriftenreihe des IÖW 169/03, Berlin, Germany (in German).

Theis TL, Bakshi BR, Durham D, Fthenakis VM, Gutowski TG, Isaacs JA, Seager T, Wiesner MR (2011). A life cycle framework for the investigation of environmentally benign nanoparticles and products. *Phys Status Solidi RRL* 5(9):312–317.

Umweltbundesamt (UBA) (2006). *Nanotechnology: Prospects and Risks for Humans and Environment*, Berlin, Germany (in German).

Walser T, Demou E, Lang DJ, Hellweg S (2011). Prospective environmental life cycle assessment of nanosilver T-shirts. *Environ Sci Technol* 45(10):4570–4578.

Wigger H (2007). *Nanotechnological and Bionic Approaches in the Application Field Adhesive Bonding and their Potential Environmental Relief Effects*. Studienarbeit an der Universität Bremen, Bremen, Germany (in German).

Index

D

E

K

L

M

N

O

P